TP159
.C4
L653
1998

D1679668

TECHNICAL
INFORMATION CENTER

MAR 26 1999

A. E. STALEY MFG. CO.
DECATUR, ILL.

Industrial Centrifugation Technology

Industrial Centrifugation Technology

Wallace Woon-Fong Leung

McGraw-Hill
New York San Francisco Washington, D.C. Auckland Bogotá
Caracas Lisbon London Madrid Mexico City Milan
Montreal New Delhi San Juan Singapore
Sydney Tokyo Toronto

Library of Congress Cataloging-in-Publication Data

Leung, Wallace Woon-Fong.
   Industrial centrifugation technology / Wallace W.-F. Leung.
      p.   cm.
   Includes index.
   ISBN 0-07-037191-1 (alk. paper)
   1. Centrifuges. 2. Centrifugation. I. Title.
TP159.C4L48 1998
660'.2842—dc21   97-49814
                                                                                            CIP

McGraw-Hill

A Division of The McGraw-Hill Companies

Copyright © 1998 by The McGraw-Hill Companies, Inc. All rights reserved. Printed in the United States of America. Except as permitted under the United States Copyright Act of 1976, no part of this publication may be reproduced or distributed in any form or by any means, or stored in a data base or retrieval system, without the prior written permission of the publisher.

1 2 3 4 5 6 7 8 9 0   DOC/DOC   9 0 3 2 1 0 9 8

ISBN 0-07-037191-1

The sponsoring editor for this book was Robert Esposito, the editing supervisor was Paul Sobel, and the production supervisor was Tina Cameron. It was set in Century Schoolbook by Ron Painter of McGraw-Hill's Professional Book Group composition unit.

Printed and bound by R. R. Donnelley & Sons Company.

McGraw-Hill books are available at special quantity discounts to use as premiums and sales promotions, or for use in corporate training programs. For more information, please write to the Director of Special Sales, McGraw-Hill, 11 West 19th Street, New York, NY 10011. Or contact your local bookstore.

Information contained in this work has been obtained by The McGraw-Hill Companies, Inc. ("McGraw-Hill") from sources believed to be reliable. However, neither McGraw-Hill nor its authors guarantees the accuracy or completeness of any information published herein and neither McGraw-Hill nor its authors shall be responsible for any errors, omissions, or damages arising out of use of this information. This work is published with the understanding that McGraw-Hill and its authors are supplying information but are not attempting to render engineering or other professional services. If such services are required, the assistance of an appropriate professional should be sought.

 This book is printed on recycled, acid-free paper containing a minimum of 50% recycled, de-inked fiber.

In God, I Trust

Contents

Preface xi
Acknowledgments xv

Chapter 1. Introduction 1

Process Industries 1
Challenges of the Twenty-First Century 2
Stages of Separation 3
Functions of Solid-Liquid Separation 4
Types 5
Rotating Flow 7

Chapter 2. General Principles of Sedimenting Centrifuges 11

Centripetal and Centrifugal Acceleration 11
Solid-Body Rotation 13
Reaction Forces 15
Effect of Fluid Viscosity and Inertia 19
Gravitational Sedimentation 20
Centrifugal Sedimentation 21
Performance Criteria 28
Stress in Centrifuge Rotor 45
G Force versus Throughout 47
Materials of Construction 47
Critical Speeds 49
References 49

Chapter 3. Batch Sedimenting Centrifuges 51

Test-Tube and Clinical-Bottle Centrifuges 51
Preparatory Centrifuges and Ultracentrifuges 54
Separation Techniques in Preparative Ultracentrifuges 55
Zonal Centrifuge 56
Tubular-Bowl Centrifuge 56

Multibowl Clarifier Centrifuge	62
Solid-Bowl Batch Centrifuge	63
Fully Automatic High-G Basket Centrifuge	66
Transient Centrifugation Theory	67
References	70

Chapter 4. Continuous Sedimenting Centrifuges — 71

Disk Centrifuges	71
Continuous Solid-Bowl Decanter Centrifuges	99
References	121

Chapter 5. Applications of Sedimenting Centrifuges — 123

Three-Phase Separation	124
Two-Phase Separation	131
References	143

Chapter 6. Continuous-Feed Sedimentation Theory — 147

Plug-Flow Model	147
Boundary-Layer Model	151
Clarification Consideration	153
Recovery Prediction	154
Classification	157
Degritting	159
References	163

Chapter 7. General Principles of Filtering Centrifuges — 165

Bulk Filtration	165
Desaturation	172
Cake Washing	187
Refeences	191

Chapter 8. Batch Filtering Centrifuges — 193

Basket Cycle	193
Solid-Bottom Basket	196
Open-Bottom Basket	199
Semiautomatic and Fully Automatic Baskets	201
References	210

Chapter 9. Continuous Filtering Centrifuges — 211

Conical-Screen Centrifuges	211
Pusher Centrifuges	216
Screen-Bowl Centrifuges	222
Dewatering Models	224
References	227

Chapter 10. Applications of Filtering Centrifuges — 229

- Types of Filtering Centrifuges — 229
- Flue-Gas Desulfurization — 231
- Fully Automatic Basket Applications — 231
- Conical-Screen Centrifuges in Coal Prep Applications — 233
- Screen Scrolls for Chemical, Mineral, and Industrial Applications — 238
- Pushes for Salt, Soda Ash, Potash, and Polymers — 240
- Screen-Bowl Applications — 241
- References — 245

Chapter 11. Feed Acceleration — 247

- Ideal Feed Accelerator — 247
- Conventional Accelerators — 249
- Improved Feed Accelerators — 254
- Practical Considerations — 261
- Gravitational Droop — 261
- Side-by-Side Testing — 262
- Experimental Tests — 264
- Installation Experience — 265
- Benefits — 271
- References — 272

Chapter 12. Lab, Pilot, and Production Tests — 273

- Preliminary Screening — 273
- Spin-Tube Tests — 274
- Batch Solid-Bowl Tests — 281
- Basket Tests — 282
- Desaturation — 285
- Bucket Tests — 286
- Cake Void Fraction — 292
- Pilot Tests — 295
- Production Testing — 296
- Saturation Measurement for Continuous Centrifuges — 298
- Cake Washing — 298
- Conclusions — 299

Chapter 13. Centrifuge Selection and Sizing — 301

- Process Definition — 301
- Selection Based on Process Function — 304
- Selection Based on Size — 304
- Selection by Feed Solids Concentration — 307
- Sizing — 308
- Costs — 311
- References — 316

Chapter 14. Optimization and Troubleshooting — 317

- Optimization of Solid-Bowl Decanter — 317
- General Process Functions — 327
- Troubleshooting — 337
- References — 337

Chapter 15. Kaolin Processing — 339

- Flow Sheet — 339
- Product Types — 341
- Classification — 342
- Degritting — 349
- Quantity versus Quality — 352
- Cake Discharge — 352
- Dewatering — 356
- Desliming — 357
- References — 357

Chapter 16. Dewatering of Compactible Solids — 359

- Characteristics of High-Solids Decanter — 359
- Economics and Optimal Operation — 380
- Combined Centrifugation and Drying — 385
- Economics of Combined Centrifugation and Drying — 385
- References — 387

Chapter 17. Cake Compaction Theory — 389

- Model — 389
- Transient Consolidation — 392
- Steady-State Consolidation — 393
- References — 399

Appendix A. Troubleshooting Industrial Centrifuges — 401

- Troubleshooting Pusher Centrifuges — 401
- Decanter Process Troubleshooting — 402
- Screen-Bowl Centrifuge Process Troubleshooting — 406
- Disk Centrifuge Process Troubleshooting — 407
- Basket Centrifuge Process Troubleshooting — 408

Appendix B. Symbols — 411

Appendix C. Conversion Factors — 417

Appendix D. Sieve and Particle Sizes — 421

Name Index   423
Subject Index   425

Preface

By the use of rotation which generates high centrifugal force with magnitude amounting to several-hundreds times to hundred-thousands times of earth's gravity, centrifuges have been used for fluid/particle separation. It has gained wide acceptance in industrial applications as well as in analytical clinical use.

Despite its popularity, the knowledge on centrifugation is limited. Part of this is due to the fact that the subject is rather complex. Separation of the solids from the liquid phase takes place in the clarification or separation zone where the flow field is strongly affected by rotation. Unlike gravitational separation, the centrifugal field is not constant; it varies linearly with radius from the pool surface to the bowl wall. This gives some rather unexpected yet interesting results. Also flow occurs in very thin viscous boundary layers which has no comparison to the normal flow field as found in other processes. On the other hand, when the solids get settled from a slurry, the cake continuously gets concentrated in solids while its rheological behavior changes over time for the case of batch centrifuges, or over the distance with which the cake is transported for continuous-feed centrifuges. Linear Newtonian flow behavior is an exception rather than the rule even for dilute feed slurry. Mechanical components can vary from a simple design to a complicated and sophisticated design. The dynamics of a rotor is complicated enough, it is further compound by the ever-changing process occurring inside the centrifuge, which can throw a sensitive mechanical balanced rotor into dynamic unbalance under certain conditions. Other than the complexity encountered in centrifugation, because of its diversity, the required engineering skills in this field encompass that of chemical, mechanical, aerospace, electronics and control, material, and manufacturing engineering; and furthermore science skills from chemistry, physics and biology.

In the past decades, significant effort has been made in advancing the understanding of centrifugation and the application knowledge.

Centrifuges is no longer being treated as a "black box" with inflows and outflows. Also, improvement on existing designs as well as innovative new designs and technologies addressing specific needs have been developed. These results have been disseminated in limited forms in technical presentations in conferences, trade shows and marketing presentations. Despite this, the process know-how is largely kept confidential within the users as well as the centrifuge manufacturers.

Several years ago, I have put together a three-day industrial centrifuge course for the Center for Professional Advancement. Since then the course has been taught extensively in Europe and the United States. Concurrently, while I was revising the centrifugal separation sections for the Perry's and Green's *Chemical Engineers Handbook* and Schweitzer's *Handbook of Separation Techniques for Chemical Engineers*, both of which are published by McGraw Hill, it occurred to me that it would be useful to assemble all these materials in an even more comprehensive reference textbook.

Centrifuges are divided into sedimenting and filtering types each of which can be divided into two modes of operation - batch feed and continuous feed. Chapters 2, 3, 4, 5 and 6 of *Industrial Centrifugation Technology* are devoted to, respectively, general principles, batch feed, continuous feed, applications and theory of continuous-feed sedimenting centrifuge; whereas Chaps. 7, 8, 9, and 10 are devoted to the parallel for filtering centrifuge. Feed accelerator technology is discussed in Chap. 11. In Chap. 12, laboratory, pilot and production tests are discussed. Chapter 13 is devoted to selection and sizing of the equipment. Chapter 14 is taken up for optimization of centrifuges with special emphasis on decanters with some useful troubleshooting guidelines included in App. A. Kaolin separation using centrifuges is discussed in Chap. 15 while Chap. 16 is dedicated to dewatering of wastewater sludges and biomass to high cake dryness and Chap. 17 to the theory of cake compaction.

I have to admit that the selection process on the materials to be included in *Industrial Centrifugation Technology* is not an easy task. Low-speed, high-rate as well as high-speed, low-rate centrifuges are considered in details. Over 250 illustrations in figures and tables are included in the text. Various generic designs with their typical applications are also included in the text. Even zonal centrifuges and ultracentrifuges, both of which are operating at very high centrifugal gravity treating a small sample and which could have been considered outside the scope of industrial centrifuges, are also discussed. I have deliberately left out certain topics and materials in order to streamline the contents providing a coherent presentation of the subject matter. Various details on specific centrifuge designs, which are considered proprietary to centrifuge manufacturers, are also omitted. Yet, the text still embodies to a large part the most important aspects as well as the fundamental principles in which all centrifuge designs are based on.

Wallace Woon-Fong Leung

Acknowledgments

I want to thank Dr. Ascher Shapiro, Institute Professor Emeritus of MIT, my colleague at Bird and also my friend, whom I have the honor to be associated with for the past twelve years. I knew Dr. Shapiro since the 1970's at MIT but never had the opportunity of working together. At Bird, we have worked closely in developing an understanding of the centrifuge mechanics. We shared frustrating moments as well as rewarding experiences. I also like to give my sincere gratitude to Dr. Frank Tiller, Professor Emeritus at University of Houston and Dr. Werner Stahl, Head Professor of the Institute of Mechanics at University of Karlsruhe, Germany for their constant support. I also like to thank all my colleagues at Bird Machine Company, who have supported my work throughout these years. Also special thanks go to my German friends from University of Karlsruhe and Bird Humboldt.

I thank Dr. Donald Dahlstrom, Professor Emeritus at University of Utah and Dr. Albert Rushton, Professor Emeritus at UMIST, United Kingdom for carefully reviewing my manuscript and providing invaluable suggestions and comments.

In addition, I'd like to thank my parents from Hong Kong for their constant encouragement and invaluable support. Finally, I thank Stella, Jessica and Jeffrey for their patience and understanding throughout the process of completing the book.

Chapter

1

Introduction

Industrial centrifuges have been used widely in process industries for separating solids from liquids, especially in solids processing. This includes, in a broad sense, liquid–liquid separation, such as the extraction of dissolved solids from one liquid to another solvent liquid, liquid–liquid–solid separation, and solid–solid–liquid separation. While gas–solid separation is outside the scope of our discussion, many phenomena dealt with in this book also pertain to separation using a gas centrifuge, such as isotopes.

In this chapter the solid–liquid separation challenges that the industry is facing today as well as the directions that the industry is taking will be discussed. In general centrifuges have been used in four different stages of separation—pretreatment, thickening, separation, and posttreatment. The various functional aspects involving centrifuges in these four stages as well as their specific objectives are summarized, and related process applications are discussed throughout the book.

Industrial centrifuges can be classified into two main types—sedimenting and filtering centrifuges. Both have similarities with respect to sedimentation under the earth's gravity and pressure filtration. However, centrifugal separation takes place in a rotating flow where additional complications arise from the motion of the liquid, with solid particles sedimenting or compacting in the liquid. The chapter finishes with a common example demonstrating the complicated yet intriguing phenomena associated with the movement of solids in a rotating flow.

Process Industries

The process industries can be subdivided into several categories:

- *Chemical processing* produces important raw ingredients such as polyvinyl chloride, polystyrene, polypropylene, polyethylene, polyester, terephthalic acid, sodium carbonate, salts, acids, oil refinery by-products, and so on.

- *Industrial processing* deals with both valuable and waste streams concerning oil–water–solids, synthetic materials, textiles, pulp and paper, green liquor, lime mud, iron ore and steel, rubber, gypsum, magnesium hydroxide, and so on.
- *Mining and mineral processing* mines (and precipitates) calcium carbonate, kaolin, alumina, potassium chloride, talc, phosphoric acid, copper, cobalt, lead, manganese, and so on.
- *Food processing* deals with poultry, starch and protein, palm oil, edible and inedible rendering, juice, fruits, salads, wine, brewery, fish, snack and pet food, canning, and so on.
- *Pharmaceutical and biotechnology industries* manufacture penicillin, mycelia, cells, skins, tissues, medicines, and so on.
- *Solid-fuel industry* processes coal, tar sands, synthetic fuels, flue gas desulfurization, and so on.

Frequently water or other liquid is required for "wet" solids processing, and more importantly the wastewater or spent liquid generated needs to be further cleaned or processed through multistage (primary, secondary, or tertiary) treatment before discharge or recycling.

Most importantly, whenever liquids (mostly water) or solvents are involved in solids processing, ultimately one or more stages of *solid–liquid separation* will be required. Centrifuges using enhanced centrifugal gravity G of between 1 and 500,000 g [$1g$ (the earth's gravity) = 9.8 m/s^2 or 32.2 ft/s^2] have been widely used for separation in these process industries. This is especially true for applications that are considered difficult to separate, for processing high volumes of slurries, where space for the equipment is a limitation, and for applications demanding high-quality separation.

Challenges of the Twenty-First Century

Among the many challenges we are facing today in solid–liquid separation, these are some of the key ones:

- Finer particle processing
- Better separation quality
- Poorer grades of raw materials
- Tighter environmental restrictions
- Escalating energy costs
- Reduction in space

- Same centrifuge for processing several grades of similar, or perhaps different, materials, each with different characteristics
- Higher labor cost
- Tighter equipment budget

Centrifuge technology is geared to meet these challenges through improvements on existing designs as well as by developing new and innovative centrifuge designs and addressing areas such as:

- Fine-particle separation and filtration
- Higher centrifugal forces for separation
- Energy-efficient and effective separation
- Insensitivity to variations in feedstock quality, with greater flexibility, especially for batch equipment
- Full automation, with sensing and control capabilities for batch and continuous-feed equipment to reduce operator attendance
- More flexible and versatile electric [ac variable-frequency drive and dc] and hydraulic drive options, in order to accommodate custom cycle requirements for batch centrifuges as well as various needs for continuous units
- More compact equipment
- Batch and continuous centrifuges with clean-in-place (CIP) and sanitary-in-place (SIP) provisions for food and pharmaceutical applications
- Higher reliability, with less frequent mechanical breakdowns
- Longer life of mechanical components and parts
- Higher solid and volumetric throughput equipment, with comparable or slightly lower operating cost per unit mass of dry solids processed, as compared to smaller centrifuges
- Systems approach, with centrifuges working compatibly with various other noncentrifuge separation equipment in the same circuit, to accomplish the required solid–liquid separation task so that the process becomes more "process flow sheet" oriented, as opposed to equipment oriented

Stages of Separation

Solid–liquid separation can be divided into four stages each having various steps, with one or more forming the basis of a solid–liquid separation operation:

1. *Pretreatment:* Chemicals are added to form an aggregation of

particles through the use of bivalent and trivalent metallic ions, polyelectrolytes, and pH control. This step can also be considered a physical process by controlling the size and length of the polymer through crystallization or reaction, or by wet milling process.

2. *Thickening:* Excess liquid is removed, or slurry is concentrated to a suitable consistency for downstream dewatering. Centrifuges are often used for this process, especially when large volumes of slurry need to be treated and total power consumption becomes an issue.

3. *Separation and dewatering:* Centrifugation is often used for this stage. Two-phase or three-phase separation is very common. Dewatering of semisolids or concentrated materials depends on the compactible behavior of the material.

4. *Posttreatment:* For example, polishing the centrate or separating liquids by a two-phase or three-phase separation process requires a high-G centrifuge for the ultimate polishing. Posttreatment also utilizes a dewatering centrifuge to remove, for example by drainage, as much liquid as possible before downstream processing, such as thermal drying, incineration, hauling, landfill, or recrystallization.

Functions of Solid–Liquid Separation

Centrifuges have key roles in solid–liquid separation. The separation process can be classified into the following eight general functions:

1. *Separation:* Separation of slurry into liquid or concentrated solid stream by either two-phase separation (solid–liquid separation) or three-phase separation (solid–liquid–liquid or solid–solid–liquid separation)

2. *Clarification:* Purification or polishing of effluent liquid to remove residual solids from the liquid stream

3. *Classification or fractionation:* Separation of particles in accordance with their respective sizes and densities

4. *Degritting:* Removal of oversized particles, heavier media, or foreign particles from a feed slurry

5. *Thickening or concentration:* Concentrating feed slurry and removing extra liquid for downstream processing

6. *Dewatering or deliquoring:* Removing moisture from cake solids to produce a dry cake

7. *Washing:* Removal of impurities by washing in the centrifuge

8. *Separation and repulping:* Removal of impurities by stages of separation followed by reslurrying

Types

Centrifuges for the separation of solids from liquids are classified into two general types—sedimenting and filtering centrifuges. As shown in Fig. 1.1a, under centrifugal force the solid phase, assumed to be denser than the liquid phase, settles out of the bowl wall. This is referred to as sedimentation. Concurrently the lighter more buoyant liquid phase is displaced toward the smaller diameter. Most centrifuges run with an air core and the free surface of the liquid pool at atmospheric pressure, whereas some centrifuges run with slurry filled to the structural hub or even to the axis, in which pressure above atmospheric pressure can be sustained.

In a filtering centrifuge, solid–liquid separation does not require a difference in densities between the two phases. However, when there

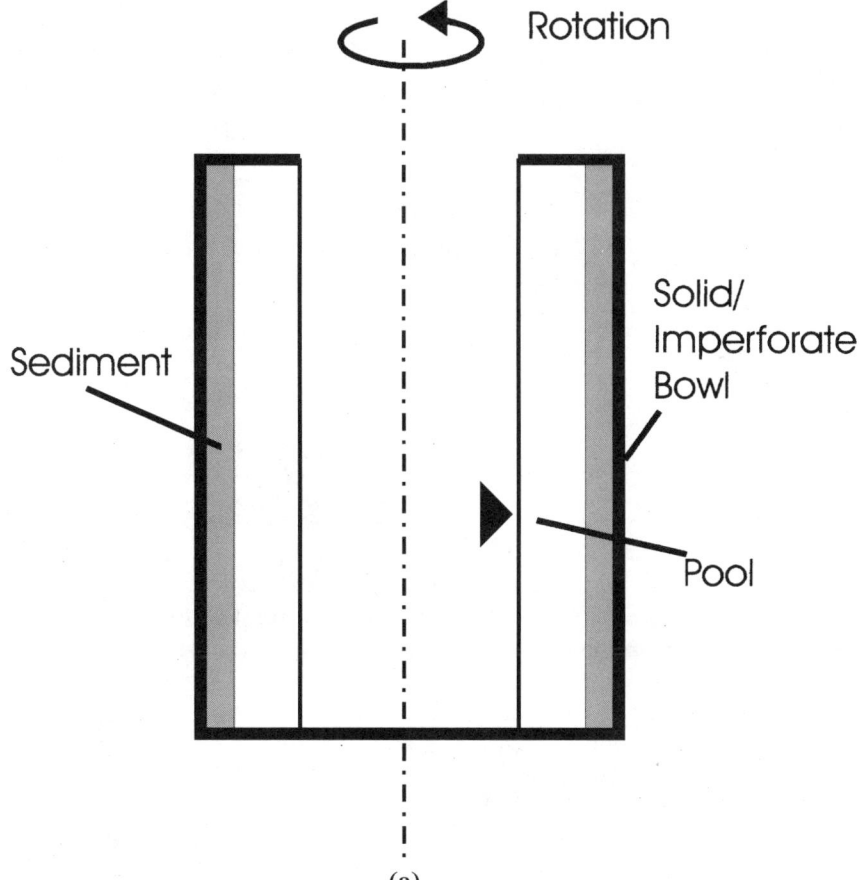

Figure 1.1 (a) Sedimenting solid-bowl centrifuge.

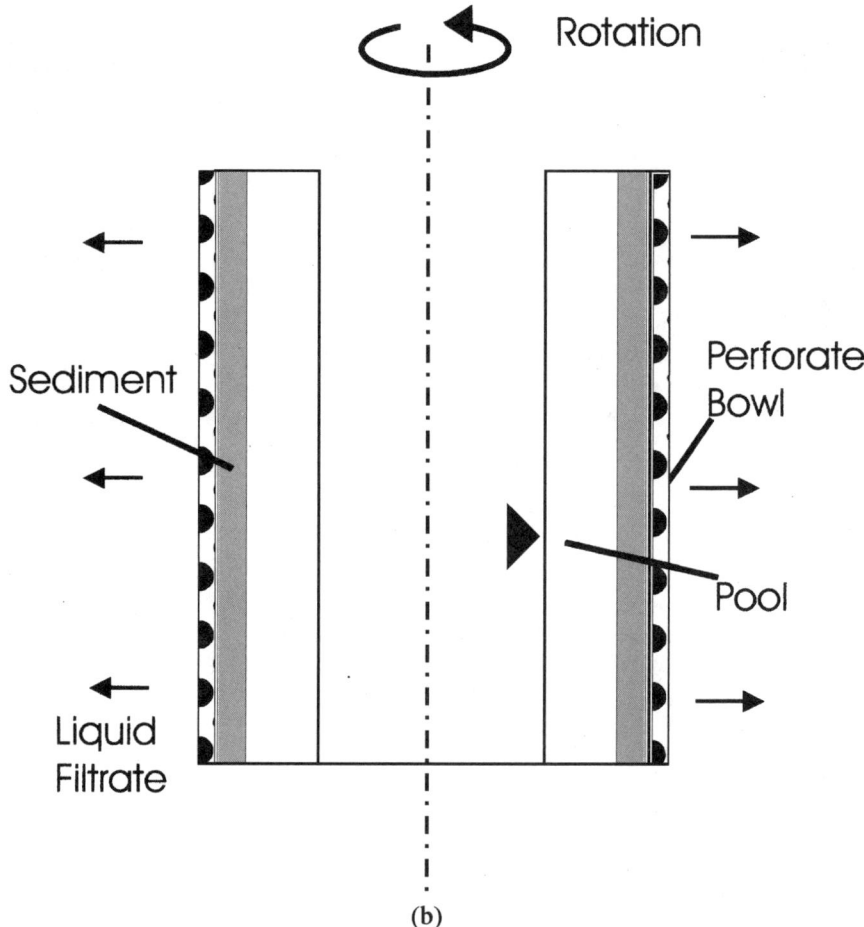

Figure 1.1 (*b*) Filtering perforate-bowl centrifuge.

is a density difference between solid and liquid phases, sedimentation and filtration take place simultaneously, with the former usually occurring at a more rapid rate than the latter. For the case when the solid is heavier than the liquid, as shown in Fig. 1.1*b*, this results in phase segregation with the concentrated heavier sediment gravitating toward a larger radius at the perforate bowl wall; simultaneously the lighter liquid displaces toward a smaller radius. Vice versa, when the solid is lighter, the concentrated solid floats above the liquid pool toward a smaller radius. In both cases, the solids are retained by the filter medium while the liquid flows through the cake and subsequently the filter. Therefore it is common practice to separate floatable solids by filtering centrifuges as the density difference does not play a role in the separation. However, it is also feasible to use a specially designed sedimenting centrifuge to separate floatable solids, as will be discussed in Chap. 4.

The built-up cake layer acts as an effective filter, preventing further loss of finer solids into the filtrate. This also allows the use of a filter medium with larger pore openings compared to the smallest particles in the slurry to be filtered. Hence, prior to cake formation finer solids may pass through the filter, with the production of temporary solid-laden filtrate. Subsequently the filtrate gets clarified, with the finer solids captured by the cake formed from deposition of the solids on the filter medium surface. Consequently, in certain applications the filtrate is recycled back to the feed during startup until clear filtrate appears.

Rotating Flow

Separation in centrifuges is by means of centrifugal gravity. The rotating flow in centrifuges can be quite complicated and nonintuitive, unlike those encountered under earth gravity. In the past two decades, a basic understanding of the processes incurred in centrifugal separation has been established and this has led to better operation of centrifuges and, in many instances, to highly innovative centrifugal technologies.

To demonstrate the complexity of rotating flow, consider as an example a fluid introduced into a rotating bowl mounted vertically, as shown in Fig. 1.2a. The fluid forms an annular pool as it begins to accelerate to speed. Concurrently a complicated flow pattern also starts. The fluid adjacent to the top and bottom of the bowl gains angular momentum as it contacts these rotating surfaces, which act as momentum "source." The fluid gets flung out to the large diameter by centrifugal force and runs axially along the inner bowl wall at the large diameter toward the center. Concurrently the fluid at the surface of the liquid pool immediately replenishes the void created at the top and bottom surfaces of the bowl. All these flow streams are confined in extremely thin "boundary" layers, shown by the dashed lines in the Fig. 1.2a. To satisfy conservation of mass or continuity, a "secondary flow" is generated in the interior of the liquid pool, whereby fluid flows radially inward from the bowl wall toward the pool surface. It is this secondary flow that helps to expedite the transfer of momentum and energy from the rotating surfaces of the bowl to the body of fluid.

When the bowl comes to an abrupt halt, the rotating fluid decelerates again through a similar yet reversed flow mechanism, as depicted in Fig. 1.2b. Here the fluid adjacent to the top and bottom surfaces of the bowl flows radially inward. To replenish the fluid, the boundary layer at the bowl wall flows axially toward the corners at the two ends of the bowl while that at the pool surface flows axially toward the center. The secondary flow in the fluid interior flows radially out-

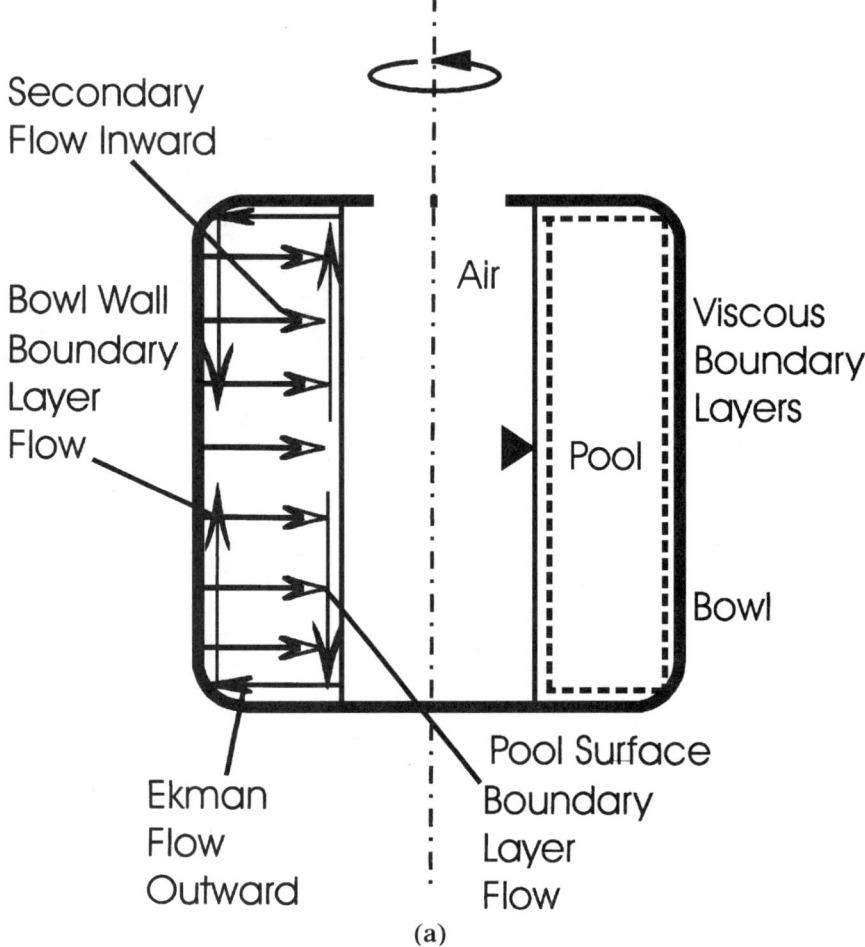

Figure 1.2 (a) Fluid accelerated from rest in contact with rotating bowl.

ward, carrying momentum of the rotating fluid to the boundary layers at the bowl wall, which in turn reduces the momentum acting as a momentum "sink" and returns the lower momentum fluid back to the interior via the boundary layers at the two end surfaces of the bowl and the pool surface. This is exactly the reverse of the acceleration flow pattern shown in Fig. 1.2a.

Furthermore, consider the case of a tea cup with tea leaves. When the cup is rotated by hand, the tea leaves are found settled out at the large diameter of the tea cup when the content is brought to speed. Likewise, after stirring the tea cup with a spoon and the content eventually has come to rest, the tea leaves are found deposited at the cen-

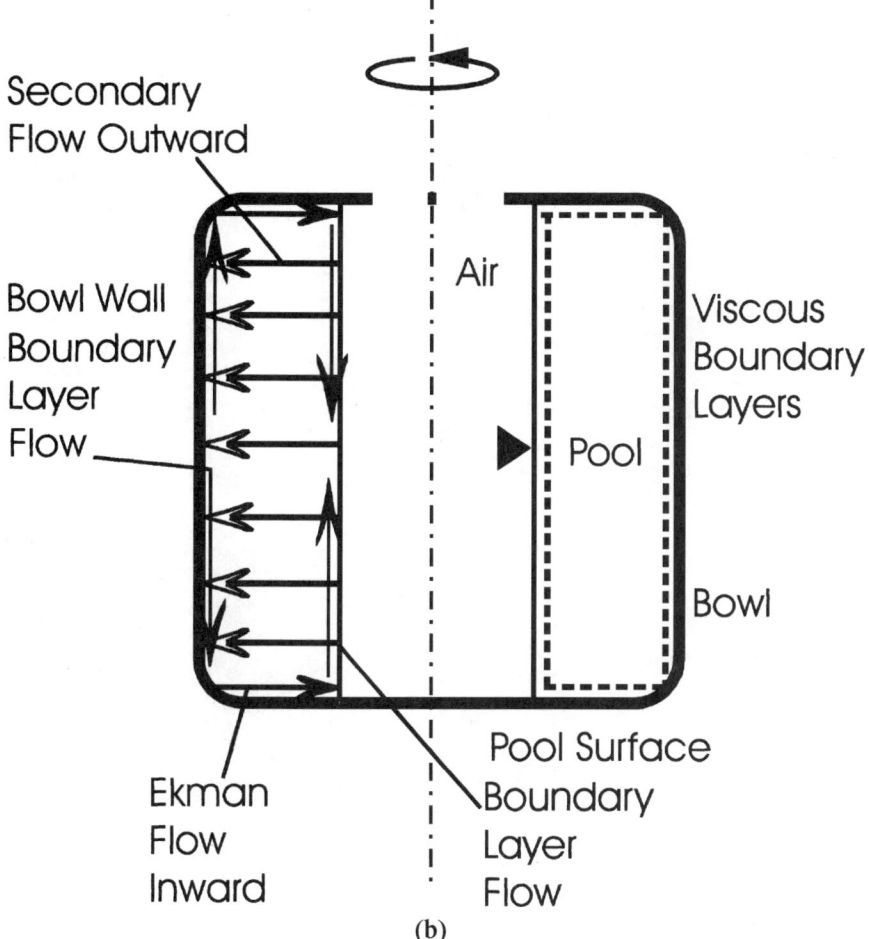

Figure 1.2 (b) Rotating fluid decelerated in contact with stationary bowl.

ter of the cup. This again can be explained from the acceleration and deceleration flow patterns illustrated in Fig. 1.2a and b. While the complex mechanism is involved, it effectively provides the rapid transfer of momentum characteristics of a rotating environment. Otherwise, much longer process times would be needed for normal mechanisms in a nonrotating environment, such as by viscous diffusion, to accomplish the same task.

Chapter 2

General Principles of Sedimenting Centrifuges

In this chapter, the fundamentals of a sedimenting centrifuge are discussed. It is important to lay out the general framework before discussing the different types and designs of centrifuges in use today. Their common thread are the basic principles governing their behavior. In this chapter, as well as in the next four chapters, the focus will be on sedimenting centrifuges with solid bowl walls.

Centripetal and Centrifugal Acceleration

In a stationary laboratory reference frame, a body moving along a circular trajectory with a constant tangential speed V_θ experiences an instantaneous change in the velocity vector equal to $V_\theta\, d\theta$, where $d\theta$ is an incremental change in the angular position. The rate of change of the tangential velocity, or the centripetal acceleration, is $V_\theta\, d\theta/dt$. This further simplifies to V_θ^2/R based on the kinematics condition that $V_\theta\, dt = R\, d\theta$. From the incremental change in the vector it can be shown that the centripetal acceleration is directed radially inward, that is, toward the axis. The centripetal force acting on the mass m is thus mV_θ^2/R.

A car negotiating a curve with a 12.2-m (40-ft) radius at 40 km/h (24.9 mi/h) acquires a centripetal acceleration of 10.1 m/s² (33.2 ft/s²). This is about $1g$, where g is the earth's gravity, which is 9.8 m/s² (32.2 ft/s²). The centripetal force is contributed from friction between the tire and the road. For a 1360-kg (3000-lb) car, this amounts to 13,706 N (3096 lbf). Usually the curve is embanked at an angle toward the center of the turn so that in addition to road friction, the component of the body force directing along the incline also contributes to the centripetal force. This relieves the prime dependence on road friction, which can decrease in wet weather. In process separation, the feed is introduced into a hydrocyclone 0.3 m (12 in) in radius at a tangential speed of, say, 30.5 m/s (100 ft/s). The centripetal acceleration is therefore 3050 m/s² (10,000 ft/s²), or 310g.

The preceding analysis holds for the motion of a body in an inertial reference frame, for example, the laboratory. On the other hand, it is most desirable to consider the process in a centrifuge, and the dynamics associated with such, in a noninertial reference frame, rotating at the same angular speed as the centrifuge. In this case all phenomena, including fluid–solid transport and separation, appear to be under steady state. This provides a simple scenario for analysis as the transient or time-dependence effect drops out. Hereafter, unless specified otherwise, the rotating reference frame is adopted, and the speed of the reference frame is taken to be the steady speed of rotation of the centrifuge.

In a rotating noninertial reference frame, however, additional forces and accelerations arise, some of which are absent in the inertial frame. Analogous to centripetal acceleration, an observer in the rotating frame experiences a centrifugal acceleration G directed radially outward from the axis of rotation with magnitude

$$G = \Omega^2 R \tag{2.1}$$

where Ω is the angular speed of the rotating frame and R the radius from the axis of rotation. The centrifugal acceleration G, as stated earlier, is conveniently measured in multiples of the earth's gravity g,

$$\frac{F_G}{F_g} = \frac{mG}{mg} = \frac{G}{g} = \frac{\Omega^2 R}{g} \tag{2.2}$$

As shown by Eq. (2.2), this ratio is also a measure of the relative centrifugal force with respect to the earth's gravitational force. When a centrifuge with a bowl diameter D ($= 2R$), measured in meters, rotates at angular speed Ω, measured in revolutions per minute (rpm), the centrifugal gravity at the bowl wall is $G/g = 0.000559 \Omega^2 D$. (In U.S. customary units $G/g = 0.0000142 \Omega^2 D$ where D is expressed in inches and Ω in rpm.) G can be as low as 100–300g for slow-speed large-diameter basket units, 500–5000g for most industrial decanter centrifuges, 10,000g for high-speed small decanter centrifuges and disk centrifuges, and as much as 20,000g for tubular centrifuges. Because $G/g \gg 1$, the effect due to the earth's gravity, 1g, is usually negligible. In analytical ultracentrifuges used to separate materials with very small density differences between solid and liquid phases, G can be as high as 500,000g to effectively separate the two phases.

Figure 2.1a provides a useful log–log chart, where the G/g ratio is plotted against speed for a given bowl diameter D. (Note a conversion

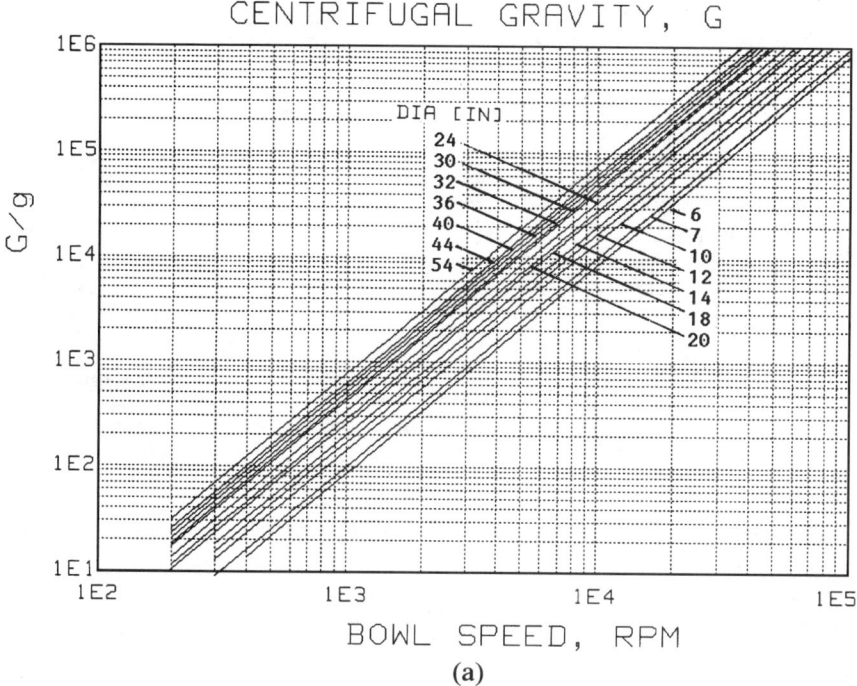

Figure 2.1 (a) Centrifugal gravity as a function of bowl diameter and speed (1 in = 25.4 mm).

chart between U.S. customary units and metric units on some common centrifuge bowl diameters is given in Table 2.1.) Given $G = \Omega^2 R$, the relationship provides a straight line with a slope of 2 on a log–log plot. The diameter covers from 0.08 m (3 in) to 1.37 m (54 in), and the rotational speed ranges from 100 to 100,000 rpm. The speed of rotation of the centrifuge is best determined by a stroboscope, or a tachometer sensing the reflection from a reflective tape mounted on the rotor. Short for such equipment, the speed can be estimated by the product of the motor speed and the ratio of the diameter of the driver-to-driven sheaves for belt-drive arrangements with insignificant belt slippage. Similarly, the motor speed multiplied by the ratio of the teeth of the driver-to-driven gears is also used to estimate the centrifuge revolution speed for the case with a gear-drive arrangement.

Solid-Body Rotation

When a fluid rotates about an axis in a solid-body mode, the tangential or circumferential velocity of the fluid V_θ is linearly proportional to the radius. Thus,

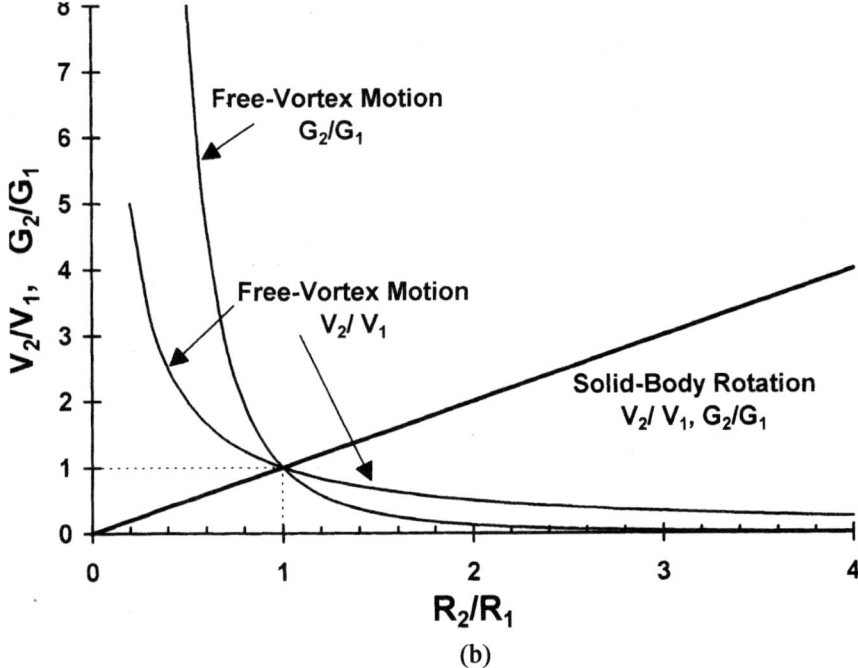

Figure 2.1 (b) Comparison of tangential velocity and G between solid-body rotation and free-vortex motion.

**TABLE 2.1
Centrifuge Bowl
Diameter, D**

Inch	MM
3	76
4	102
6	152
7	178
9	229
10	254
12	305
14	356
18	457
20	508
24	610
30	762
32	813
36	914
40	1016
44	1118
54	1372

$$V_\theta = \Omega R \tag{2.3}$$

This is similar to a system of particles in a rigid body under rotation. Under this condition, the magnitude of the centripetal acceleration, $V_\theta^2/R = (\Omega R)^2/R = \Omega^2 R$ is equal to the centrifugal acceleration, despite these two accelerations being considered in entirely different reference frames. However, when the fluid is not rotating in a solid-body mode with respect to its surrounding, such as during feed acceleration in a centrifuge, the laboratory reference frame should be used. V_θ is less than ΩR, and the difference $\Omega R - V_\theta$ is the slip velocity. In this underaccelerated condition, the effective centrifugal acceleration and the centrifugal force are V_θ^2/R and mV_θ^2/R, respectively.

In a cyclone separator, the motion of the fluid slurry to be separated follows that of a free vortex in absence of any external torque. Thus, when a slurry at an initial position at radius R_1 and tangential speed V_1 (note subscript θ is deleted from V for clarity) with $G_1 = V_1^2/R_1$ is brought to a smaller radius R_2, the tangential speed increases to $V_2 = V_1 R_1/R_2$ in lieu of the free vortex motion (or conservation of angular momentum), and the centrifugal gravity developed at the smaller radius R_2 becomes $G_2 = V_2^2/R_2 = (V_1 R_1/R_2)^2/R_2 = G_1(R_1/R_2)^3$ where $R_2 < R_1$. This demonstrates that the G-force for separation increases dramatically (inversely to the third power) with reducing radius in a cyclone.

In contrast, in a centrifuge the liquid slurry rotates in a solid-body rotation where $G = \Omega^2 R$. Hence, the centrifugal gravity G_1 and G_2 of the fluid slurry, respectively, at radius R_1 and R_2, with $R_2 < R_1$, are related by $G_2 = G_1(R_2/R_1)$. This indicates that in a centrifuge reducing radius results in a decrease in G-force for separation, and vice versa. The comparison of the tangential velocity and G-force between the solid-body rotation in a centrifuge and the free vortex in a cyclone is shown in Fig. 2.1b.

Free vortex motion (VR = constant) can be induced in the liquid pool of a centrifuge as the liquid originally at a state of solid-body rotation at radius R_1 is abruptly brought to a different radius R_2, where solid-body rotation cannot be maintained. This can occur at liquid effluent discharge where the liquid tangential speed increases above and beyond that of a solid-body rotation when it is brought to a smaller discharge radius. Likewise, when the feed is introduced to the separation pool at a larger radius from the discharge ports of the feed chamber which is at a much smaller radius, the feed tangential speed slows down. The associated G-force also changes accordingly with change in tangential speed but at a much significant rate with radius (that is, $G \propto 1/R^3$) as compared to that of the velocity (that is, $V \propto 1/R$), see Fig. 2.1b.

Reaction Forces

In a centrifuge all the phenomena are driven by the centrifugal force or the centrifugal gravity. This gives rise to other reaction forces, some of which act within the liquid phase and others through the solid structure, yet they all originate from $G(=\Omega^2 R)$, which is proportional to the second power of the angular speed.

Forces in liquid phase

Frictional force. In the liquid phase, dissipative viscous friction or stress arises as liquid slurry flows over a solid wall moving at a different velocity. This can be flow over the wall of a structure in a centrifuge, or simply denser particles settling in a lighter liquid phase in a centrifuge bowl.

Pressure force. Under the earth's constant gravitational field of 1g, the hydrostatic pressure p in a liquid increases linearly with increasing liquid depth h. Thus $dp/dh = \rho g$, in which $p = \rho g h$, with the ambient pressure taken to be zero and ρ the liquid density. On the other hand, under centrifugal gravity, $dp/dR = \rho \Omega^2 R$, from which $p = \frac{1}{2}\rho\Omega^2(R^2 - R_o^2)$, where $p = 0$ at the radius corresponding to the liquid surface $R = R_o$. Note that the fluid pressure increases as a quadratic function of radius R.

Coriolis force. The Coriolis acceleration arises in a rotating frame and has no parallel in an inertial frame. When a body moves at a linear velocity u relative to a reference frame rotating at angular speed Ω, it experiences a Coriolis acceleration

$$\vec{a} = -2\vec{\Omega} \times \vec{u} \qquad (2.4)$$

The Coriolis acceleration vector is perpendicular to the plane formed by the velocity vector and the rotation vector, which abides by the right-hand rule in accordance with the direction of rotation. If the rotation of the reference frame is anticlockwise, then the Coriolis acceleration is directed 90° clockwise from the velocity vector, and vice versa when the frame rotates clockwise. The Coriolis acceleration distorts the trajectory of the body as it moves rectilinearly in the rotating frame. Figure 2.2a shows a ball making a curve trajectory as it rolls over a turntable. It is only before entering or after exiting the turntable that the ball follows a rectilinear path. Figure 2.2b shows a denser particle settling in a centrifuge. Due to the influence of the Coriolis acceleration, the particle moves backward with respect to rotation (retrograde motion)[1] as it migrates toward the bowl wall at a larger radius. Likewise, the displaced liquid moves forward with respect to rotation (prograde motion) as it is displaced radially inward. The Coriolis effect as observed in a rotating reference frame is related to a more general experience in the stationary inertial frame, namely,

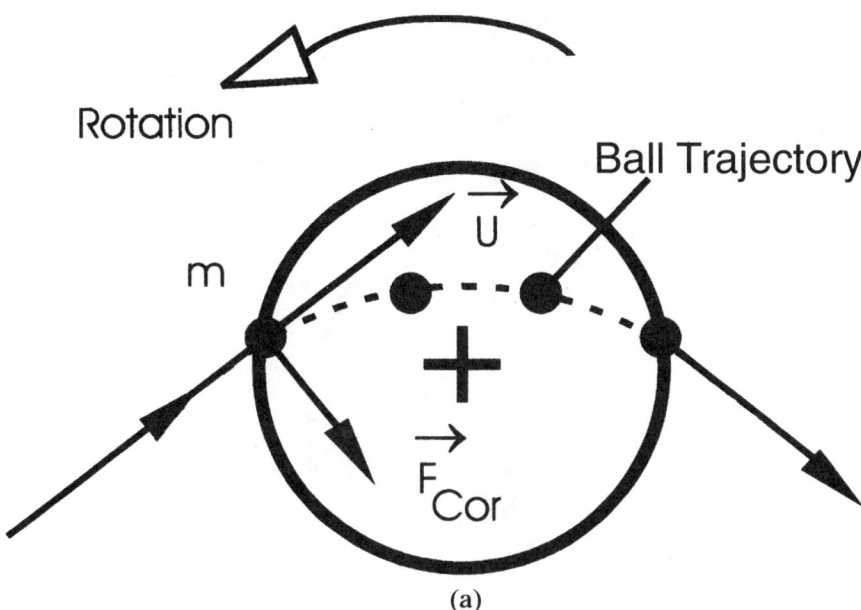

Figure 2.2 (*a*) Coriolis force on a ball traveling over a turntable.

the conservation of angular momentum. In the absence of external torque, the angular momentum of a body with mass m is conserved, that is, $mV_\theta R$ = constant or $V_\theta \sim 1/R$, given m is constant. An example of conservation of angular momentum is the free vortex motion in a fluid as discussed in the preceding section. It follows that as a particle settles to a larger radius R_1 from its original position R_o, V_θ is reduced by the factor R_o/R_1. Likewise, its tangential speed increases as it is displaced radially inward. The momentary underspeeding or overspeeding particle would eventually be brought back to the appropriate solid-body rotation speed, wherein $V_\theta = \Omega R$, by the viscous friction and any induced secondary flow.

One other interesting yet peculiar phenomenon is the well-known Taylor–Proudman column.[2] Here a body of fluid in a centrifuge is considered to be instantaneously removed from its position, thereby leaving a void, it is reasonable to expect that the neighboring fluid, from locations at either a larger or a smaller radius, will flow in and fill the void. Instead, the void is immediately filled by the liquid at the same radius, yet at a neighboring distance along the axis. This suggests that in rotating flow there is a certain amount of stiffness along the radial direction (that is, radial stratification) which resists changes, yet fluid at the same radius can communicate along the axial direction much more freely with minimal resistance. Another way of demonstrating the Taylor–Proudman column is that when the

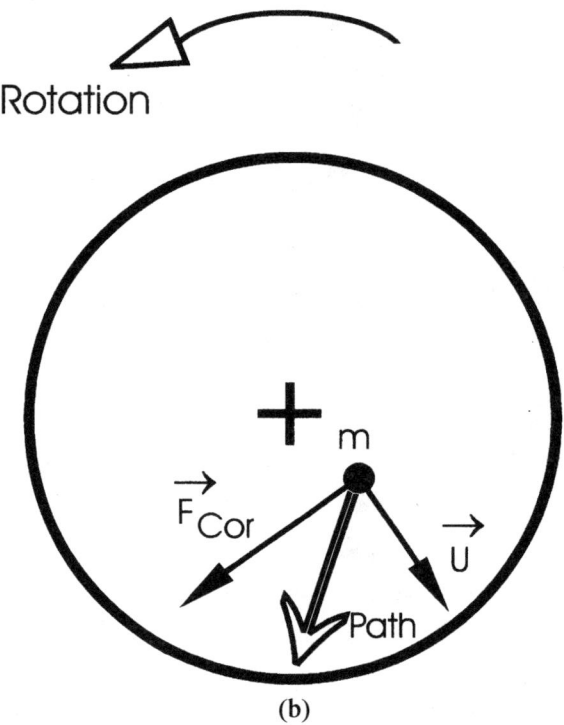

Figure 2.2 (*b*) Coriolis force acting on a sedimenting particle in a rotating pool.

base of a cylinder containing a rotating fluid has a physical protrusion into the fluid, a column of fluid forms along the axial direction from the protrusion and extends along the entire length of the cylinder. Dye introduced into the fluid demonstrates that the dye flows around the column as though the dye trace encountered a physical solid column.

Forces in solid structure

Compression and tensile stress are generated at the bowl wall and supporting structures of the centrifuge. This results from forces and moments (bending and torsional) from the centrifugal loading due to the mass of the centrifuge and its contents. Centrifugal force also acts to compact the solid sediment in the centrifuge bowl. This produces a compacted dewatered cake suitable for downstream processing. Given that all the forces within the supporting structure in the centrifuge, or along the particle-to-particle contact in the sediment, are due to centrifugal gravity, it is expected that the stress must be

General Principles of Sedimenting Centrifuges

at least proportional to G or, equivalently, the quadratic power of the speed of rotation. Thus the centrifugal loading is very sensitive to speed change.

Effect of Fluid Viscosity and Inertia

The dynamic effect of viscosity on a rotating liquid slurry as found in a sedimenting centrifuge is confined to very thin fluid layers, known as Ekman layers.[2] These fluid layers are adjacent to rotating surfaces that are perpendicular to the axis of rotation, as found in bowl heads (end covers), flanges, circular baffles and weirs, and conveyor blades. The thickness of the Ekman layer δ is on the order of

$$\delta = \sqrt{\frac{\nu}{\Omega}} \qquad (2.5)$$

where $\nu = \mu/\rho$ is the kinematic viscosity of the liquid. For example, with water at room temperature, $\nu = 1 \times 10^{-6}$ m²/s, and for a surface rotating at $\Omega = 3000$ rpm, δ is only 0.05 mm. Although this layer is very thin, it is nevertheless responsible for the transfer of angular momentum between the rotating surfaces and the fluid during both acceleration and deceleration. It works in conjunction with larger-scale inviscid bulk flow, transferring and exchanging momentum in a rather complicated way. This is demonstrated in Chap. 1 by the tea cup example, where the content of the cup is brought to rotation as it is being stirred, and the rotating content eventually comes to a halt after stirring has stopped. The viscous effect is characterized by the dimensionless Ekman number Ek,

$$\text{Ek} = \frac{\nu}{\Omega L^2} \qquad (2.6)$$

where L is a characteristic length. The Ekman number is equal to the second power of the ratio of the Ekman layer thickness to the characteristic length L. It relates the viscous diffusion ν to that of the bulk convection flow ΩL^2.

The effect of fluid inertia manifests itself during abrupt changes in the velocity of the fluid. It is characterized by the Rossby number Ro, which is defined as

$$\text{Ro} = \frac{u}{\Omega L} \qquad (2.7)$$

Typically $0 < \text{Ro} < 10$, where the high end of the range shows the possible significance of inertia. On the other hand, Ek is very small, usually less than 10^{-6}. Therefore it follows from both Eqs. (2.5) and (2.6) that the viscous effect becomes important only in thin boundary layers with a thickness of $\delta = Ek^{1/2}L$.

Gravitational Sedimentation

The most celebrated relationship governing the separation of solid particles from the liquid phase by sedimentation is Stokes' law.[3] For a spherical particle with diameter d, the viscous frictional drag on the particle in laminar flow is $3\pi\mu V_{sg}d$. This depends on the liquid viscosity μ, the settling velocity V_{sg}, and the particle diameter d. On the other hand, the body weight, discounting the buoyancy force, is $\pi(\rho_s - \rho)gd^3/6$, which depends on the density difference between the solid particle and the liquid $(\rho_s - \rho)$, the gravitational acceleration g, and the particle diameter d. When the particle reaches equilibrium or steady state, the body weight balances the buoyancy and drag forces from which the "terminal" settling velocity becomes

$$V_{sg} = \frac{(\rho_s - \rho)gd^2}{18\mu} \tag{2.8}$$

Note that the settling velocity is proportional to the density difference and gravity, to the particle diameter to the second power, and, inversely, to the liquid viscosity. Therefore increasing the density difference and the particle size and decreasing the liquid viscosity will enhance the settling rate of the particles in a given slurry. It may be shown that the above expression, strictly speaking, applies to the case where the Reynolds number, defined by $Re = \rho V_{sg} d/\mu$ is less than 0.2.

Example. Assume coal particles with a density of 1500 kg/m^3 in water having a density of 1000 kg/m^3 and a viscosity of 0.001 kg/m · s at room temperature. The earth's gravity is 9.8 m/s^2. Therefore the settling velocity of the 10-μm particle is

$V_{sg} = 1/18(1500 - 1000 \text{ kg/m}^3)(9.8 \text{ m/s}^2)(10^{-5} \text{ m})^2/10^{-3} \text{ kg/m} \cdot \text{s}$

$= 2.72 \times 10^{-5} \text{m/s} \ (= 1.10 \times 10^{-3} \text{ in/s})$

The foregoing calculation is repeated for particles of 100, 33, 3.3, and 1 μm. The results are tabulated in Table 2.2.

In the above example it is clear that as the particle size is reduced by $\frac{1}{3}$ the settling rate is reduced approximately by $\frac{1}{10}$. The bigger 100-μm and 33-μm particles can be separated with conventional gravity settling tanks or clarifiers. On the other hand, the smaller 10-μm particles require more effective separation technology, such as the

TABLE 2.2

Size, μm	V_{sg}, m/s	V_{sg}, in/s	Comment
100	2.72×10^{-3}	1.1×10^{-1}	Fast
33	2.72×10^{-4}	1.1×10^{-2}	Medium
10	2.72×10^{-5}	1.1×10^{-3}	Slow
3.3	2.72×10^{-6}	1.1×10^{-4}	Very slow
1	2.72×10^{-7}	1.1×10^{-5}	Extremely slow

lamella settler, with reduced settling distance or increased settling area. Sedimentation of particles less than 5μm in size requires centrifugation to increase the effective "gravity" for separation.

The Reynolds number for the 10-μm particle is determined to be 0.00027 whereas for the 100-μm particle, it is 0.27. Note the latter is outside the range of validity of Stokes' law where Re needs to be less than 0.2, nevertheless the exercise serves the point of highlighting the strong effect of particle diameter, within the 100-μm range, on settling rate.

Centrifugal Sedimentation

The Stokes' settling velocity V_{sg} at $1g$ is extended to an effective settling velocity V_{sG} in centrifugation by replacing gravity g with centrifugal gravity G. Thus $V_{sG} = V_{sg} (G/g)$. It is of interest to determine the required Gs to settle a given particle size. A particle typically resides in the clarification or separation zone of a centrifuge for only a few seconds. Let us assume that on average the residence time is 5 s. During this time, if the particle settles through a radial distance of 25 mm (1 in), then it is likely that the particle is separated out from the fast-moving liquid layer phase into a somewhat stagnant pool of liquid and eventually is collected as sediment or cake against the bowl wall. This suggests that the settling velocity of the particle should be at least 0.025 m in 5 s, that is, 0.005 m/s (0.2 in/s). Could V_{sG} reach this required velocity by operating at the appropriate G and, if so, what is this G? Table 2.3 provides the result.

In Table 2.2, as the particle size is reduced by a factor of 3.3, the required centrifugal gravity increases by a factor of 10 because the velocity varies as the second power of the particle size. This poses a stringent separation requirement on particles, especially those under 3 μm, with G values of up to 10,000 g. Obviously, in reality the G force needs not be that high as the settling area can be increased by using a longer clarifier or processing can be performed at a reduced feeding rate. Despite the simplicity of this exercise, it brings out the important fact that separation by sedimentation depends critically on

TABLE 2.3

Size, μm	V_{sG}, m/s	G/g	Equipment
100	0.005	1.7	Gravity clarifier
33	0.005	17.0	Gravity clarifier
10	0.005	170	Lamella/hydrocyclone/low-speed centrifuge
3.3	0.005	1,700	Industrial centrifuge
1	0.005	17,000	Industrial centrifuge with large settling area, or small high-speed centrifuge

the particle size, especially the finer spectrum of the particle-size distribution.

Furthermore, rearranging Stokes' law under centrifugal gravity,

$$V_{sG} = \frac{(\rho_s - \rho)R(\Omega d)^2}{18\,\mu} \qquad (2.9)$$

It is clear that the separation velocity is very sensitive to the speed of rotation and the particle size. It varies as the second power of both quantities. Therefore high-speed operation and flocculation of particles help to increase the separation efficiency. Reducing the viscosity of the liquid also enhances separation. The viscosity of water and of most organic solvents decreases sharply with increasing process temperature. Figure 2.3 charts the variation of the viscosity of water with temperature. The viscosity of water drops to ½ and ⅓ of its value at room temperature [20°C (68°F)] when the temperature rises to 55°C (131°F) and 83°C (181°F), respectively. Thus it is most effective to carry out separation of a highly viscous slurry, such as grease or fat, at elevated temperature.

One of the many driving factors for sedimentation is the density difference between liquid and solid phases. Tables 2.4 and 2.5 show the densities of some common liquid and solid materials used in separation. The most pertinent property for separation is the density difference, which can be expressed as the ratio $(\rho_s - \rho)/\rho$. With water as the suspending liquid phase, this ratio varies between 0.01 and 0.05 for sewage, and between 0.5 and 3 for most mineral slurries.

Equation (2.9) can be rewritten as

$$d = \sqrt{18\,\frac{V_{sG}\,R\mu}{V_{rim}\,(\rho_s-\rho)}} \qquad (2.9'')$$

given the rim speed $V_{rim} = \Omega R$. To facilitate the settling out of fine particles of size d, the machine should actually have a smaller radius R, a high G and a high rim speed V_{rim}, given that V_{sG} is determined by

General Principles of Sedimenting Centrifuges

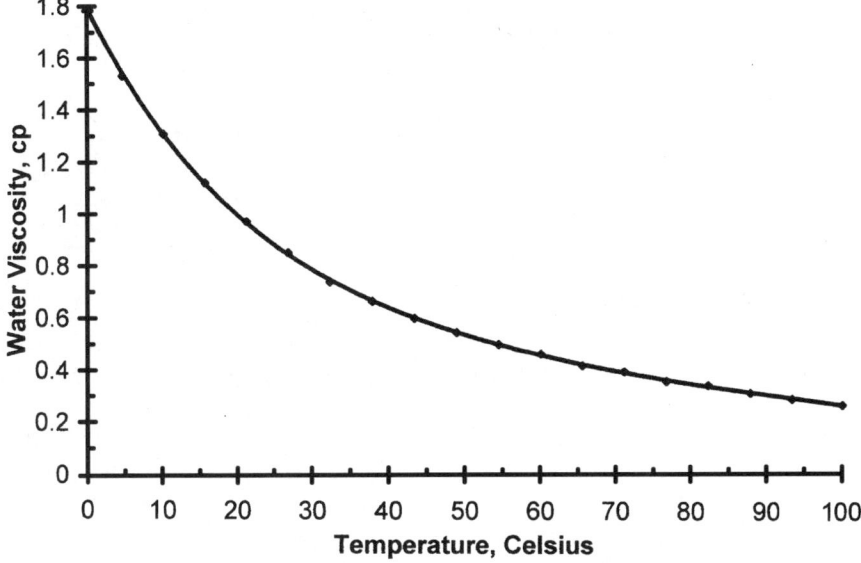

Figure 2.3 Water viscosity as a function of temperature.

TABLE 2.4

Liquid	ρ, kg/m³
Hexane	633
Oil	700–990
Water	1000
Brine	1000–1500
Molasses	1500

TABLE 2.5

Solid	ρ_s, kg/m³	$(\rho_s - \rho)/\rho$
HDPE	950–970	0.5–0.53*
Sewage	1050–1200	0.05–0.2
DMT	1200	0.2
PVC	1300–1400	0.3–0.4
Coal	1400–1500	0.4–0.5
Potash	1640	0.64
Silica	2170	1.17
Kaolin (clay)	2580	1.58
$CaCO_3$	2710	1.71
Fe_2O_3	2800	1.80
Steel waste	3300	2.3
Al_2O_3	3910	2.91
Barite	4400	3.4

*HDPE in hexane, liquid for all others is assumed to be water.

the separation requirement. As will be shown later, the maximum V_{rim} is limited by the strength of the construction materials. On the other hand, the smaller-diameter centrifuge unfortunately yields lower throughput. This conclusion is also applicable for the hydrocyclone, where V_{rim} should be modified to be the tangential speed of the feed stream, which depends on the pumping power. In addition, from Eq. (2.9″) it follows that reducing the liquid viscosity and increasing the density difference further enhances the separation of the finer particles.

Sedimentation of irregularly shaped particles

Stokes' law does not account for many complications that arise in practice. Strictly speaking, Stokes' law applies to spherical particles. For irregularly shaped particles, a simple modification can be made by using the "equivalent" diameter in Eqs. (2.8) and (2.9″). Despite this, Stokes' law does not account for form drag on particles, with streamlines separated downstream in their wake. An example is the platelet-shaped clay particle settling with the geometric diameter of the disk perpendicular or at an oblique angle to the G field. This is shown in Fig. 2.4a. Here the streamlines separate at the wake of the particle, as opposed to reattaching when the disk diameter is stream-

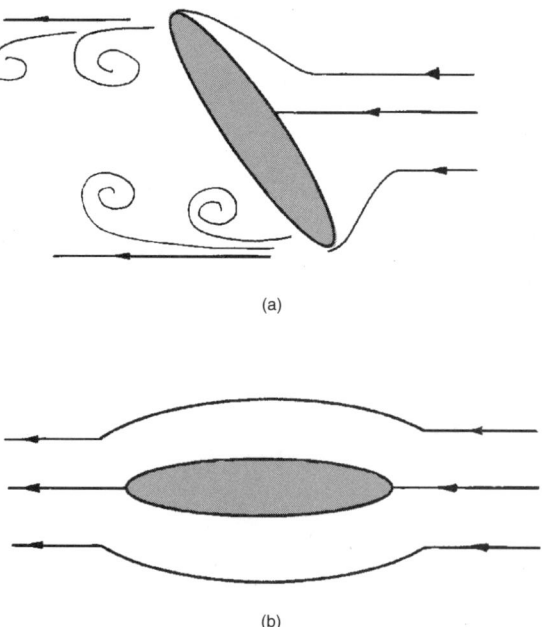

Figure 2.4 Streamlines of flow in the vicinity of a disk-shaped particle. (a) Longer axis approximately perpendicular to flow. (b) Longer axis parallel to flow.

lined with the body force vector, as illustrated in Fig. 2.4b. The former incurs additional form drag, which in many instances is much greater than the viscous frictional drag (from liquid viscosity) on which Stokes' law is based. This occurs even at small Reynolds numbers (Re = 1 to 10), where viscous drag is expected to dominate for smooth rounded particles. In fact, the resistance to flow is twice as high for the orientation shown in Fig. 2.4a as for the streamline orientation shown in Fig. 2.4b.

The net driving force (weight minus buoyancy) acting on an irregularly shaped particle settling in a dilute suspension is $(\rho_s - \rho)VG$, where $V = C_p L^3$ is the volume of the particle, with C_p the particle shape factor and L the characteristic dimension. On the other hand, the viscous drag force acting on the settling particle equals $C_v \mu V_{sG} L$, where C_v is the viscous drag coefficient depending on the particle shape. Balancing these two forces,

$$C_p(\rho_s - \rho)L^3 G = C_v \mu V_{sG} L$$

and the terminal velocity becomes

$$V_{sG} = \left(\frac{C_p}{C_v}\right)\frac{(\rho_s - \rho)L^2 G}{\mu} \qquad (2.10a)$$

For a spherical particle, the ratio of the two coefficients is $1/18$, and L is the particle diameter.

When the form drag dominates over the viscous drag (in which Stokes' law is based on) for nonstreamline, irregularly shaped particles, the drag force equals $C_f \rho V_{sG}^2 L^2$ where C_f depends on the particle Re. Specifically, $C_f = 7.26/Re^{0.6}$ for $0.2 < Re < 500$ (intermediate region), and $C_f = 0.173$ for $Re > 500$ (turbulent flow). L is the equivalent diameter of the particle. The force balance then yields

$$C_p(\rho_s - \rho)L^3 G = C_f \rho V_{sG}^2 L^2$$

and the terminal velocity becomes

$$V_{sG} = \sqrt{\left(\frac{C_p}{C_f}\right)\frac{(\rho_s - \rho)LG}{\rho}} \qquad (2.10b)$$

Note that when form drag prevails, with nonrounded particles the terminal settling velocity depends much less on the density difference between the solid and liquid phases, the particle dimension, and the G force. Nevertheless increasing the particle size in the slurry by flocculation using polymer or increasing G, or both, is still beneficial. When form drag prevails with irregular particles, it produces less than desirable size separation between product and reject streams. Interestingly, viscosity and, thus, temperature have no effect on separation when the flow around the particle is fully turbulent, with Re > 500.

Sedimentation with concentrated feed solids

There are three modes of sedimentation[4]—discrete, hindered, and compression—which are affected by both the solid concentration and the degree of flocculation. This is summarized in Fig. 2.5. When the volume concentration of particles in a nonflocculated slurry reaches the point at which the flow field of one particle affects the flow field of adjacent particles, the sedimentation velocity of the particles is reduced. The particles, however, are still settling as discrete particles, with the larger particles settling faster than the smaller ones. The most commonly used empirical relationship is the equation by Richardson and Zaki,[5] which states

$$\frac{V_s}{V_{sG}} = (1 - \varepsilon_s)^{4.6} \tag{2.11a}$$

where V_{sG} is Stokes' settling velocity under G and ε_s is the solids volume fraction. This is graphed in Fig. 2.6. This relationship generally holds for ε_s up to 0.3–0.4 under 1 g. It is also assumed to hold for centrifugal gravity G. Classification of particles by size is still possible, but at a much reduced rate, depending on the solids volume fraction in the

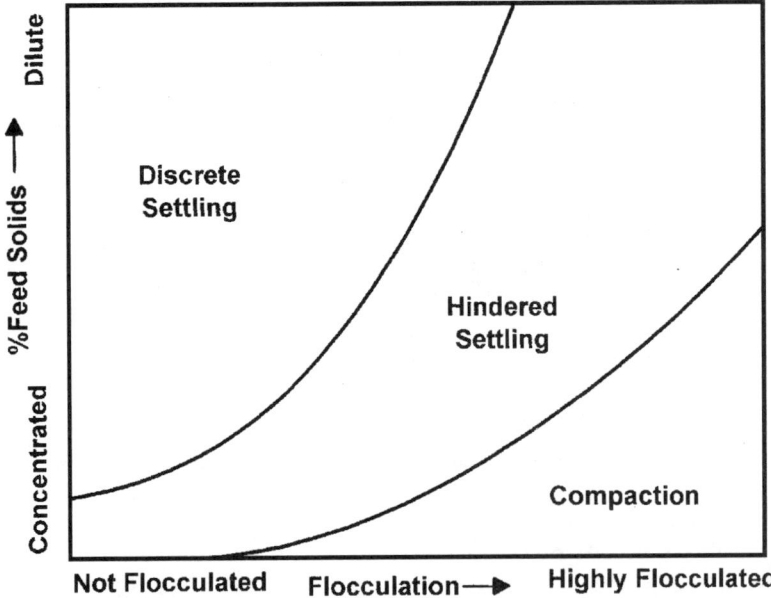

Figure 2.5 Classification of particle sedimentation.

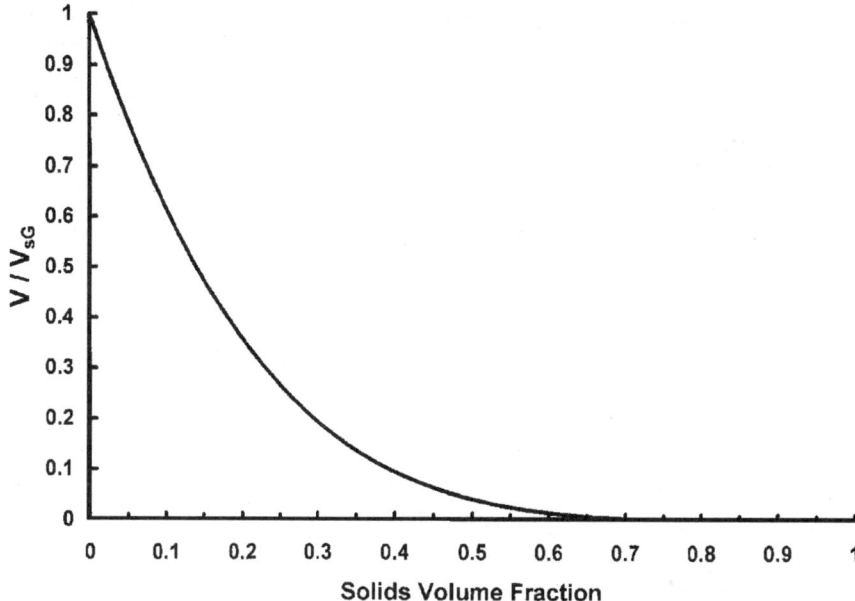

Figure 2.6 Richardson and Zaki's relation of sedimentation rate as affected by particle concentration.

slurry. For example, in high-density classification of kaolin solid-bowl centrifuges are employed, where the feed slurry ε_s approaches 0.5.

At higher solids concentration for unflocculated slurry, there is a concentration at which the entire slurry, regardless of particle size, settles as a "blanket." This is commonly referred to as hindered settling. Under the hydrodynamic influence the smaller particles are settling as fast as the larger ones. Classification by size is not possible as all the particles, regardless of size, are settling at the same rate. At even higher concentration, the particles are physically touching each other, resulting in a "matrix structure" slowly subsiding and compacting downward along the G field, with liquid expressing concurrently upward through the voids in the cake. This cake compaction and expression phenomena will be discussed in greater depth in Chap. 17.

When the slurry is flocculated, again depending on the solids concentration, it goes through domains of discrete settling, hindered settling, and compaction. Yet the solids concentration at which the slurry changes in behavior from discrete to hindered settling and from hindered settling to compaction takes place at a much lower solids concentration than with the unflocculated counterpart. This is due to flocculation which modifies the structure and size of the particles and which promotes earlier particle-particle interactions. These different domains of sedimentation are illustrated in Fig. 2.5.

From an engineering perspective, the viscosity term can be modified to include this particle-particle interaction due to solids concentration and flocculation as a measure of deviation from Stokes' law based on the dilute limit. For a given slurry, the measured slurry viscosity μ can be approximated by a power-law relationship,

$$\frac{\mu}{\mu_o} = \frac{1}{(1-\varepsilon_s)^n} \qquad (2.11b)$$

In the dilute limit as $\varepsilon_s \ll 1$, $n = 2.5$, and this reverts back to Einstein's result $\mu = \mu_o(1 + 2.5\varepsilon_s)$. Typically $n = 2.5$–5 for most slurry, and n depends on the slurry type and the degree of flocculation.

Performance Criteria

The separation of a solid–liquid slurry is typically measured by the purity of the separated liquid phase in the centrate or effluent, and the separated solids in the cake or sediment. In addition, there are other important considerations. In general the following criteria, depending on the objectives of the process, are used in various combinations to assess the centrifuge performance:

1. Cake solid and cake moisture
2. Centrate solids or total solids recovery
3. Size recovery or yield
4. Polymer dosage
5. Dissolved cake impurities
6. Volumetric and solids throughput
7. Torque load
8. Power consumption

Cake solid and cake moisture

There are two cake dewatering (or deliquoring) mechanisms—compaction and drainage. (Note that the liquid phase needs not be water, and the term "dewatering" is used generically. The term "deliquoring" is also being used in the literature to cover liquids other than water.)

In compaction, a rearrangement of the loosely packed solid structure from sedimentation takes place to form a much tighter packed arrangement under centrifugal force. Concurrently, liquid trapped in the sediment cake is expressed out of the tight structure. This is referred to as expression. In general, the solid particle by itself is generally not compressible (that is, no further reduction of the solid particle volume under compression) with the exception of flocculated solids. Compaction refers to repacking of the sediment to form a tighter structure for both incompressible or compressible solid particles. Dewatering by compaction is accomplished by high centrifugal compressive and shear stress and sufficiently long time for liquid expression.

When the cake is lifted out of a liquid pool, the liquid originally saturating the pores between particles in the cake drains out under the centrifugal force, leaving the pores subsequently filled by air. The drainage or desaturation of the liquid in the cake is usually not complete, with residual liquid remaining in the cake pores. The amount of residual liquid saturation depends on the centrifugal versus capillary force, wetting angle, dewatering time, cake height, cake permeability, particle surface roughness, particle porosity, and the mechanism for trapping inherent (or bound) moisture, such as particle porosity and particle surface charges.

Given these two dewatering mechanisms, all centrifuged cakes can be roughly divided into compactible and incompactible, and drainable and nondrainable cakes. There are a total of four different permutations.

Compactible and nondrainable cake. Sewage, fruit, animal waste, pet food, starch, food processing, cells, all biological materials, clay, hydroxides, and organic salts are typical examples of compactible and nondrainable cake. Solid-bowl centrifuges, batch or continuous feed, are used to dewater compactible and nondrainable cake. High compaction stress generated by centrifugation on a thick cake layer is transmitted and exerted on the bottom cake stratum adjacent to the solid bowl wall, with liquid expressing simultaneously away from the bowl wall toward the cake surface, which is less consolidated and therefore more open to flow. Subsequently the liquid is removed or decanted off the cake surface. Filtering centrifuges are typically not used for this application. This is because as the cake is filtered under high G, the majority of the cake layer experiences minimal pressure drop whereas a large portion of the pressure drop occurs at the highly compacted "skin" layer adjacent to the filter medium. This reduces filtrate flow, leading to slow drainage because of the skin layer and less effective compaction in the cake interior.

Incompactible and nondrainable cake. There is no cake which is genuinely incompactible and nondrainable; otherwise, it becomes impossible to dewater the cake. Fine calcium carbonate and titanium dioxide with particles of several micrometers and below form nondrainable and relatively incompactible cake. Solid-bowl centrifuges with moderate to high Gs are used with continuous rearrangement and shear on the cake structure by scrolling.

Incompactible and drainable cake. Most cakes are in this category, such as polyvinyl chloride (PVC), dimethyl terephthalate, high-density polyethylene (HDPE), coarse coal, oxide, carbonate, and other minerals having particle sizes greater than 45 μm, inorganic salts, sulfates, and chlorides (such as from sodium, potassium, magnesium, or calcium). When the minimum particle size in the slurry is greater than 75 μm, continuous filtering equipment is used with filtering screens such as pusher and screen scroll. For particle sizes less than 75 μm but greater than 20 μm, screen bowls can be used. For particle sizes between 10 and 40 μm, drainage takes a considerable time due to the low cake permeability. Batch basket centrifuges with ring dams holding off the pool are used and the dewatering time can take from a few minutes to an hour. For coarser solids above 50 μm, solid-bowl centrifuges are used to dewater the cake by drainage at the dry beach. The blade tip and the bowl wall form a filtering funnel through which the liquid from the cake drains off.

Compactible and drainable cake. Fibrous and some polymeric material are examples of drainable and compactible cakes. Filtering centrifuges at relatively low G and sedimenting centrifuges at higher G are used for these applications with a thin cake operation.

In the above processes cake dryness is commonly measured by the solids fraction by weight (W_s) or by volume (ε_s), for example, kg (lb) of dry solid per kg (lb) of wet cake (solid and liquid), or cm³ (in³) of dry solid per cm³ (in³) of wet cake. On the other hand, cake moisture is measured by the liquids fraction by weight (W_m) and is therefore equal to $1-W_s$. Another form of measurement is the liquids content M_c, which is kg (lb) of liquid per kg (lb) of dry cake solid. Also, cake porosity ε ($= 1 - \varepsilon_s$) and saturation S are conveniently employed. Cake porosity is the volume fraction of the pores or voids in the wet cake, that is, cm³ (in³) of void per bulk volume cm³ (in³) of wet cake. Liquid saturation S is defined as the fraction of total pore or void volume occupied by the liquid in a unit volume cake. A simple materials balance reveals the following relationships among these variables:

$$\varepsilon + \varepsilon_s = 1 \tag{2.12a}$$

$$W_m + W_s = 1 \tag{2.12b}$$

$$W_s = \frac{\rho_s \varepsilon_s}{\rho_s \varepsilon_s + \rho(1 - \varepsilon_s)S} \tag{2.12c}$$

$$M_c = \left(\frac{1 - \varepsilon_s}{\varepsilon_s}\right)\left(\frac{\rho}{\rho_s}\right)S \tag{2.12d}$$

When $S = 1$ (cake fully saturated with liquid), it follows from Eq. (2.12c) that

$$\varepsilon_s = \frac{1}{(1/W_s - 1)\rho_s/\rho + 1} \tag{2.12e}$$

In fact, Eq. (2.12e) holds also for both the feed and the centrate, which are in slurry form with $S = 1$. In such a case, the subscript s in Eq. (2.12e) needs to be replaced, by f for feed or e for centrate.

With nondrainable and compactible cake as well as with drainable and compactible cake, the cake moisture W_m decreases with increasing G for a fixed dewatering time, or with increasing dewatering time at fixed G (see Fig. 2.7a). At high G or t, the moisture curve reaches an asymptote (see dotted line in Fig. 2.7a). It would not be possible to differentiate these two types by examining W_m alone. However, a closer examination of the cake porosity ε and saturation S provides further clues as shown in Figs. 2.7b and c. With nondrainable and compactible cake, ε decreases with G or t, but S stays constant at approximately 100% (horizontal line in Fig. 2.7c). For drainable and incompactible

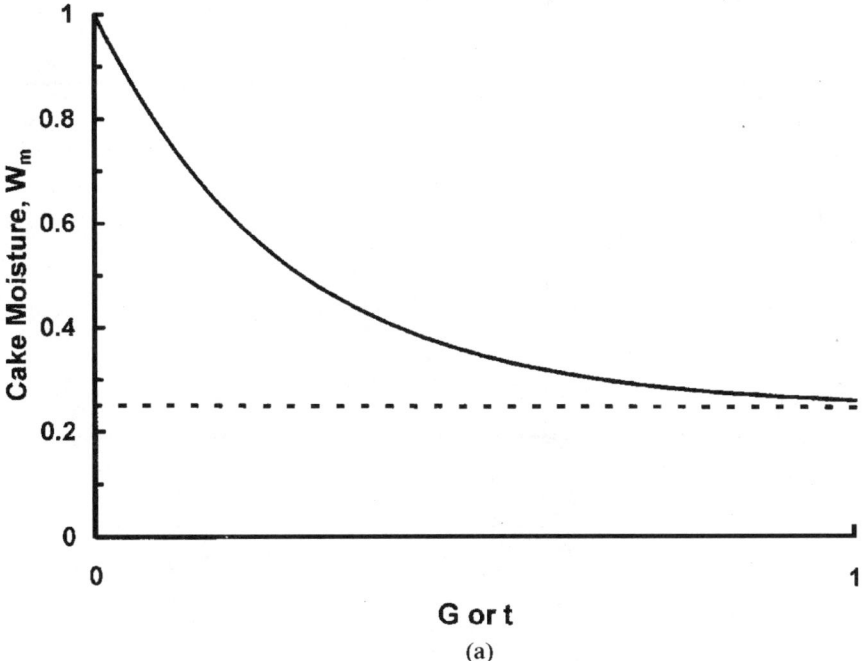

Figure 2.7 (a) Moisture weight fraction in cake during dewatering.

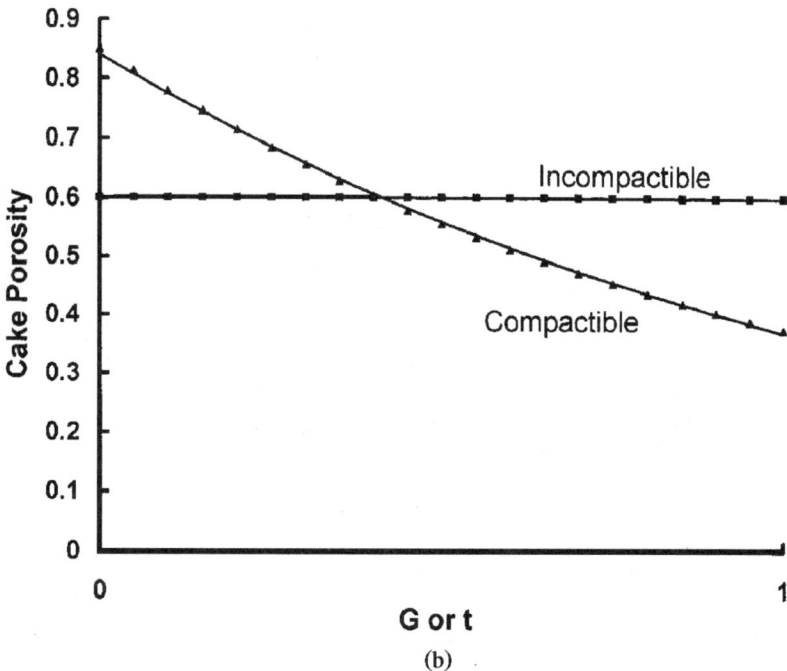

Figure 2.7 (b) Void fraction in cake during dewatering.

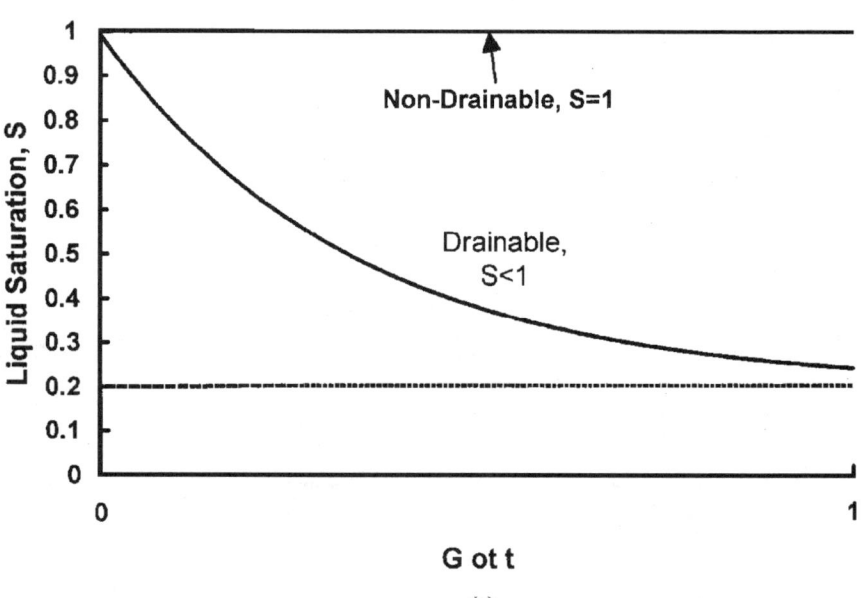

Figure 2.7 (c) Liquid saturation in cake during dewatering.

cake, S decreases with G or t, which is similar in behavior to W_m, but the cake porosity ε stays constant (horizontal line in Fig. 2.7b).

Centrate solids or total solids recovery

In clarification, the clarity of the centrate is measured by the centrate solids W_e. It is more appropriate to normalize the centrate solids by the feed solids because the centrate solids concentration increases with higher feed solids concentration. For example, it is useful to determine W_e/W_f, as opposed to W_e alone, as a function of the feed rate for a given centrifuge geometry and operating G. This can be seen in Fig. 2.8. Another often used index is the total solids recovery Rec_s in the cake, which is defined as

$$\%\text{Rec}_s = 100 \, \frac{m_s W_s}{m_f W_f} \qquad (2.13a)$$

where again the subscripts e, s, and f denote centrate, cake solid, and feed, respectively, m is the bulk mass throughput rate in kg/s (lb/h). In contrast, for classification and degritting the valuable product is the fine particles in the centrate stream, with the cake solid as the reject. In this case, the product recovery is measured by the solids recovered in the centrate, or

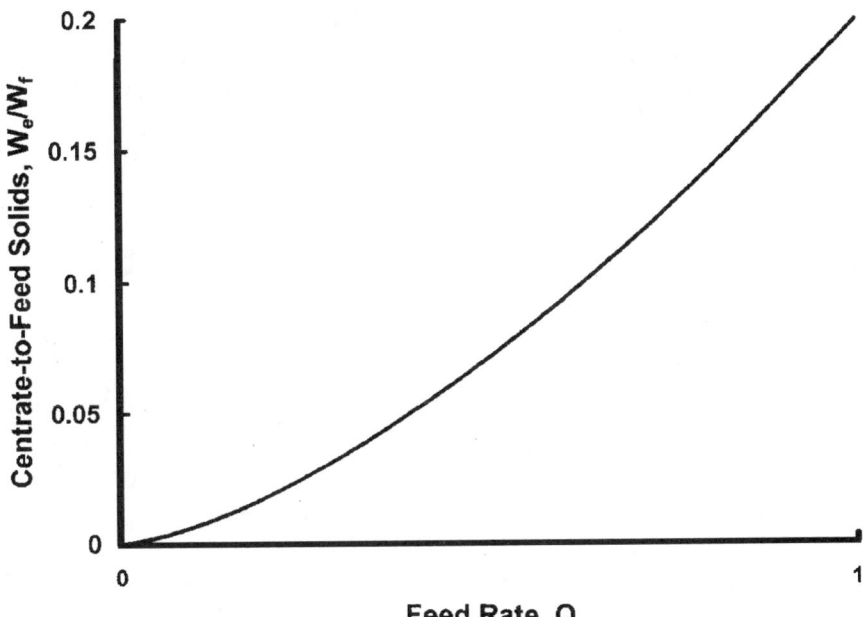

Figure 2.8 Effect of flow rate on centrate solids.

$$\%\text{Rec}_e = 100 \frac{m_e W_e}{m_f W_f} \qquad (2.13\text{b})$$

Under steady state, the mass balance of both solids and liquid is

$$m_f W_f = m_s W_s + m_e W_e \qquad (2.14\text{a})$$

$$m_f = m_s + m_e \qquad (2.14\text{b})$$

It follows that the ratio of throughput of the cake solid wet mass to that of the feed is

$$\frac{m_s}{m_f} = \frac{W_f - W_e}{W_s - W_e} \qquad (2.15\text{a})$$

When $W_e/W_f \ll 1$ (that is, a liquid centrate almost completely free from solids), we obtain the simple result that $m_s/m_f = W_f/W_s$. Otherwise Eq. (2.15a) needs to be used. Likewise, the ratio of the throughput of the centrate wet mass (or liquid effluent) to that of the feed is

$$\frac{m_e}{m_f} = \frac{W_s - W_f}{W_s - W_e} \qquad (2.15\text{b})$$

Equations (2.15a) and (2.15b) allow one to determine the wet mass throughput of the cake and the centrate from the measured solids concentration of all three streams together with the measured feed mass throughput. In some high-rate applications, there are mechanical and process limitations on the cake mass rate and the centrate volumetric rate that the centrifuge can handle. Under these conditions it is important to determine this quantity in order to avoid overloading the centrifuge. These limitations will be discussed later.

Using Eqs. (2.15a) and (2.15b), the recovery of the solid in the cake and the centrate as defined by Eqs. (2.13a) and (2.13b), respectively, is

$$\%\text{Rec}_s = 100 \left(\frac{W_s}{W_f}\right)\left(\frac{W_f - W_e}{W_s - W_e}\right) = 100 \left(\frac{1 - W_e/W_f}{1 - W_e/W_s}\right) \qquad (2.16\text{a})$$

$$\%\text{Rec}_e = 100 \left(\frac{W_e}{W_f}\right)\left(\frac{W_s - W_f}{W_s - W_e}\right) = 100 \left(\frac{W_s/W_f - 1}{W_s/W_e - 1}\right) \qquad (2.16\text{b})$$

For example, with $W_f = 20\%$, $W_e = 1\%$, and $W_s = 60\%$, $\%\text{Rec}_s = 96.6\%$ from Eq. (2.16a). Note that when $W_e \ll W_s$, which is valid for applications where the centrate solids content is small as compared to that of the cake solids, Eq. (2.16a) can be approximated by $\text{Rec}_s = 1 - W_e/W_f$. For the previous example, this approximation becomes $\text{Rec}_s = 95\%$. This approaches the exact value as the centrate solids concentration gets smaller. On the other hand, when one considers the recovery of valuable product in the centrate, such as with fine mineral (kaolin and calcium carbonate) classification and degritting, W_e is comparable in magnitude to W_f. In fact, $W_e > W_f$, and Eq. (2.16b) should be used.

Stringent requirements on centrate quality or the capture of valuable solid product often require the recovery to exceed 95% and in some cases 99+%. In the latter case the centrate solids are typically measured in parts per million (ppm). Dewatering of PVC slurry requires PVC centrate solids concentration to be less than 50 ppm.

Particle-size distribution (PSD) affects the level of the recovery. A frequently used PSD is the cumulative weight fraction $F(d)$ less than a given particle size d. It is measured in the laboratory by particle-size analysis, with equipment which operate on principles such as sedimentation or optical scattering.

Figure 2.9a shows the %cumulative-under-size distribution versus particle size for a monodispersed PSD slurry. It exhibits an approximate step-shape behavior. (The frequency plot, not shown, reveals a corresponding peak at the most popular size.) Interestingly, the %Rec$_s$ versus feed rate in Fig. 2.9a also demonstrates a dramatic change in recovery at a critical feed rate Q_c. Below this rate, %Rec$_s$ remains at

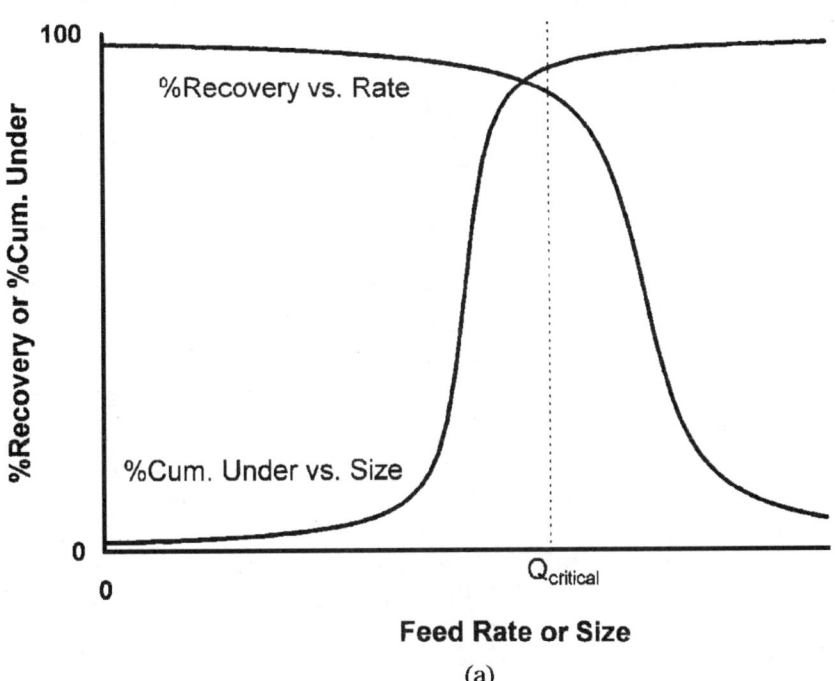

Figure 2.9 Effect of particle-size distribution on solids recovery. (*a*) Monodispersed.

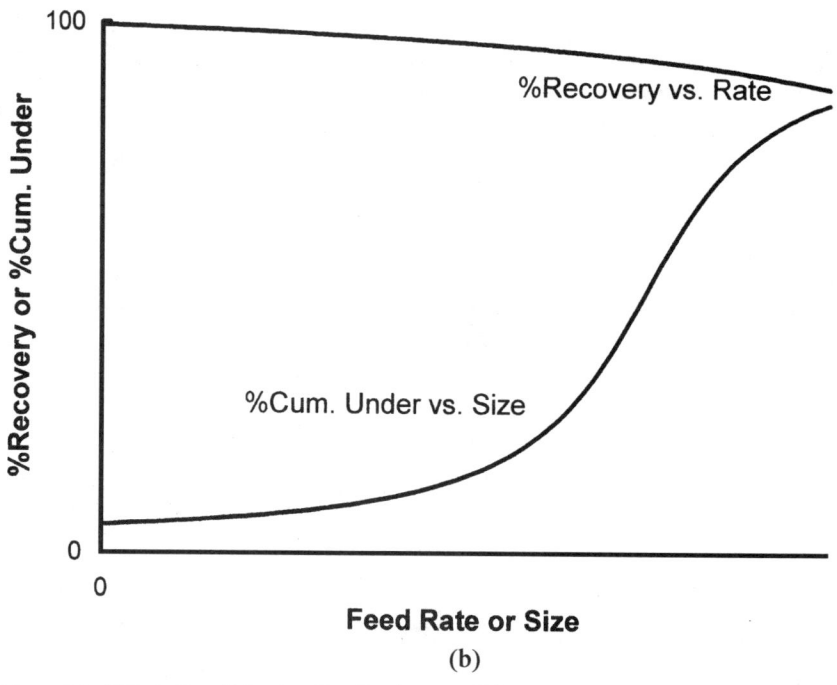

Figure 2.9 Effect of particle-size distribution on solids recovery. (b) Polydispersed.

nearly 100%, whereas as the feed rate exceeds Q_c, %Rec_s drops off precipitously to a low value. This step change in recovery is almost a mirror reflection of the %cumulative-under-size distribution of the monodispersed PSD slurry shown in Fig. 2.9a. On the other hand, with a polydispersed PSD the size distribution curve is more evenly distributed. This is shown in Fig. 2.9b. Correspondingly, the mirror image with polydispersed slurry demonstrates that %Rec_s decreases gradually with increasing feed rate. There is no critical feed rate as found in separating monodispersed PSD slurry.

Size recovery or yield

Centrifuges have been applied to classify polydispersed fine particles. In kaolin classification, the product is typically measured as a certain percentage less than a given size (for example, 95% less than 2 μm and 90% less than 1 μm). Each combination of percent and size cut represents a condition to which the centrifuge would have to be tuned to meet the product specification.

The process performance is evaluated based on the yield %Y, which is defined as the mass fraction of feed particles of a given size below which reports to the centrate product. For example, 1000 kg/h of less than 2 μm is fed to the centrifuge and the product centrate with particles less than 2 μm exits the centrifuge at a rate of 700 kg/h (1540 lb/h). Therefore the yield of less-than-2-μm product is 70%. In general, %Y is defined as

$$\%Y = 100 \, \frac{m_e W_e F_e}{m_f W_f F_f} \qquad (2.17a)$$

From material balance, the PSD of the feed and the centrate as well as the total solids recovery determine the yield,

$$\%Y(d) = \frac{F_e(d)}{F_f(d)} \, \%\text{Rec}_e \qquad (2.17b)$$

A material balance can be achieved within the centrifuge with the following three important quantities in each stream: (1) the mass rate m (solid and liquid), (2) the solids weight fraction W, and (3) the size distribution under $F(d)$. They have to be measured or inferred when determining the solid, liquid, and particle-size balance.

The complementary measure of process performance is the cumulative capture efficiency $\%Z(= 100\% - \%Y)$. This is the percent of feed particles of a given size and smaller captured in the cake, which in thickening and dewatering applications is the product stream.

Polymer dosage. Cationic and anionic polymers have been used commonly to coagulate and flocculate fine particles in the slurry. This is especially pertinent to biological materials such as in wastewater treatment. Here cationic polymers are often used to neutralize the negative charge ions left on the surface of the colloidal particles. In some applications both cationic and anionic polymers are used in a two-step treatment, with one being used as a conditioner and the other as a neutralizer. Polymer dosage is measured by kg of dry polymer per 1000 kg of dry solids cake (lb_m of dry polymer per ton of dry solids cake). Polymer exists in either solid powder, liquid, or liquid emulsion form. The solid polymer requires adequate dissolution of the powder in water to form an active solution. Mixing and the associated kinetics, that is, time and the level of energy input associated with mixing, are all important. The resultant polymer concentration should typically be no higher than 0.3% and should also be free from agglomerates in the form of "fish eyes," which are often associated with a poorly prepared solution. With liquid polymers, the equivalent (active) dry solid polymer is used to calculate the dosage. If the activity of the liquid polymer is not known, the dosage is calculated as though the full liquid polymer were active. The result is qualified as "neat."

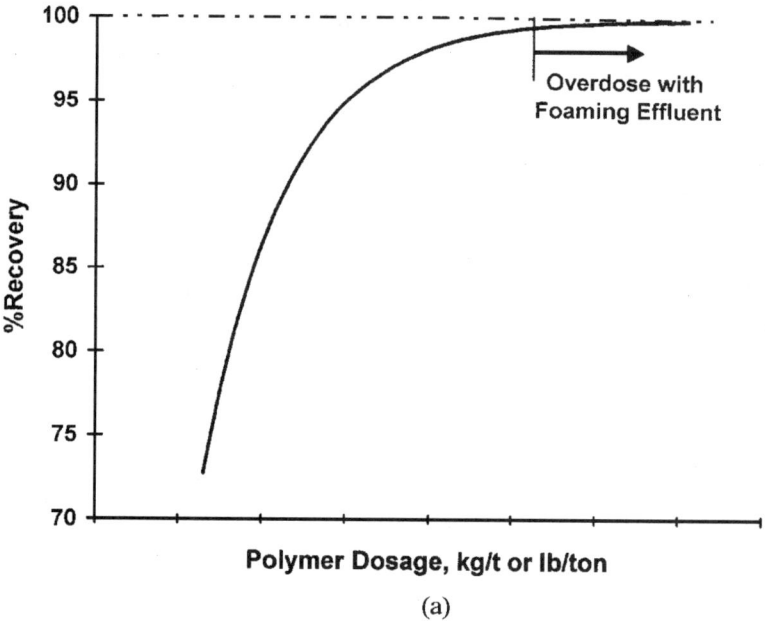

Figure 2.10 Polymer dosage. (a) %recovery.

Usually because the activity is typically 40–50% with liquid polymer, the neat dosage value is generally 2 to 2.5 times that, determined using a solid polymer, with 100% activity. In addition, polymer in liquid emulsion form is also used. There is a minimum polymer dosage which necessitates agglomeration and capture of the fines in the sediment. Figure 2.10a shows typical behavior with both organic and inorganic polymer. %Rec_s increases with the dosage until it reaches a maximum beyond which the centrate turns foamy, revealing that unused excess polymer solution leaves with the centrate.

With organic polymer, the cake solids increase with higher polymer dosage, as shown in Fig. 2.10b. It is possible that for certain polymer-feed combinations overdosage results in a wetter cake. With inorganic solids, such as alum sludge in water treatment, cake solids decrease monotonically with increasing polymer dosage. As such, this trade-off often limits the dosage of inorganic polymer, as it adversely affects cake dryness despite solids recovery benefits from the addition of polymer.

The range of optimum dosage is dictated by the polymer type, the solids in the slurry, the physical properties of the slurry such as pH and ionic strength, and the operating conditions and characteristics of the centrifuge. Flocculated particles (flocs) obtained from certain polymer-feed slurry may be more sensitive to mechanical shear than oth-

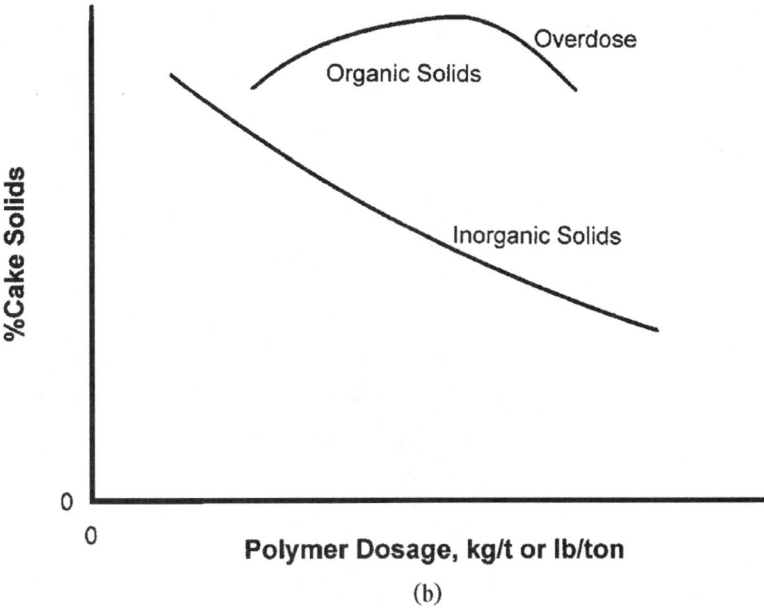

Figure 2.10 Polymer dosage. (*b*) %cake solids.

ers. This occurs especially during feed acceleration where the flocculated slurry is brought abruptly to high centrifugal gravity, typically 2500–4000g. A more "gentle" and "efficient" feed accelerator is beneficial for this type of polymer. Also, polymer can be introduced to the feed at various locations, either internal or external to the centrifuge. The latter provides additional time for the feed to react with the polymer solution to form a more stable flocculated particle structure before entering the centrifuge.

Dissolved cake impurities

Cake washing is used to remove cake impurities dissolved in mother liquor or in precipitated form adhered to the solid particles. Cake washing can be carried out most effectively with filtering centrifuges. As will be discussed in Chap. 4, in decanter centrifuges cake washing has to be conducted outside the annular pool in the dry beach. Given the limited dry beach, cake washing is not effective. Alternatively, dissolved impurities in the mother liquor can be reduced effectively in sedimenting centrifuges through stages of separating and repulping.

Let M_t be the total mass of the slurry containing suspended solids with weight fraction W_{ss} and dissolved solids with weight fraction W_{ds}, and let W_c be the cake solids by weight after centrifugal separation.

1. *Before centrifugation:*

Mass of solids	$W_{ss}M_t$
Mass of liquor containing dissolved solids	$(1 - W_{ss})M_t$
Mass of dissolved solids	$W_{ds}M_t$

2. *After centrifugation:*

Mass of solids (unchanged)	$W_{ss}M_t$
Mass of liquor containing dissolved solids	$M_t W_{ss}(1 - W_c)/W_c$
Mass of dissolved solids	$W_{ds}M_t/\text{RF}$

Where the repulp factor RF is the ratio of the mass of dissolved solids before centrifugation to that after centrifugation,

$$\text{RF} = \left(\frac{1 - W_{ss}}{W_{ss}}\right)\left(\frac{W_c}{1 - W_c}\right) \qquad (2.18)$$

Therefore the dissolved solid is reduced by a factor RF(>1) after each separation and repulping. Staging separation and repulping proves to be an effective means to remove dissolved contaminants. After the nth separation and repulping, the dissolved solids dropped exponentially down to $1/(\text{RF})^n$ of its original level. A countercurrent separation and repulping results when the centrate from each stage is used as a repulp liquid for the preceding stage. In this case, the removal of dissolved solids is less. However, it minimizes the use of fresh repulp liquid as well as the subsequent handling of the contaminated liquid.

Volumetric and solids throughput

As with any physical system, there is a trade-off between quality and quantity, and there is no exception with centrifugal separation. The maximum volumetric and solids throughput to a centrifuge is dictated by one or more governing process factors. The most common ones are centrate solids concentration and cake dryness. Figure 2.11 depicts solids recovery (top curve) and cake solids (middle curve) both decrease with increase in solids throughput [dry solid (DS)] for a continuous-feed centrifuge. The solids throughput in the figure increases from left to right. The abscissa can also be interpreted as cycle time, which increases from right to left. A long cycle time corresponds to low solids throughput and vice versa. In Fig. 2.11, upon feed dilution, in certain instances, the solids recovery curve shifts upward, showing better centrate clarity or higher solids capture; this is due to an increase in the sedimentation rate with less particle-particle interference during settling. On the other hand, solids recovery is getting worse at high solids

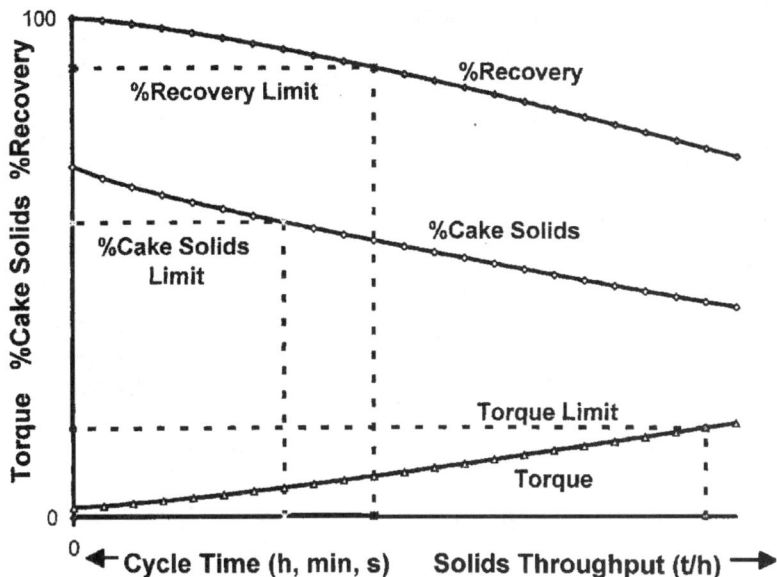

Figure 2.11 Solid recovery, cake solids, and torque as a function of solids rate.

throughput, resulting in classification with finer particles leaving in the centrate. The cake can dewater to higher dryness due to higher permeability and better drainage with the coarser particles remaining in the cake makeup, as compared to the case without classification, where both fine and coarse particles are present. From the process standpoint, either the recovery or the cake solids can set the limit on the solids throughput to the centrifuge, besides mechanical torque limitations.

Torque load

Torque is provided by the drive to accelerate the centrifuge and maintain the centrifuge at speed, overcoming any additional resistance. In a scroll-type centrifuge the scroll conveyor is rotating at a differential speed as compared to the bowl. This differential speed between scroll and bowl provides the conveyance mechanism by which the sediment is transported from the clarification end of the machine to the solids discharge end. In some designs braking torque is also provided during machine shutdown to bring the rotor to a prompt halt; otherwise, it will take much longer for windage (that is, drag on rotor from ambient air mass) and friction to brake the machine.

The lowest curve in Fig. 2.11 shows the trend of increasing torque with increasing solids throughput. In some applications, such as PVC dewatering, there is also a fluctuating torque component (chatter torque) in addition to the time-averaged torque (also referred to as the steady component). The chatter torque manifests when there is a stick-and-slip mechanism, such as with the scroll centrifuge, where the cake is conveyed by the differential speed between conveyor and bowl. This chatter mechanism is due primarily to the cake, which stores and releases energy periodically in a torsional mode during conveyance. This shows up as a fluctuating component in the torque measurement.[6]

Power consumption

In a centrifuge, power is consumed (1) to overcome windage and friction from bearing and seal, (2) to accelerate the feed stream from zero speed to full tangential speed as the feed is discharged to the pool so as to establish the required G force for separation, and (3) to convey cake solids in continuous-feed centrifuges. The power to overcome windage and bearing friction is usually established through tests for a given centrifuge geometry, at different speeds of rotation. It is proportional to the mass of the centrifuge, the first power of the speed for the bearing friction, and the second power of the speed for windage. It also depends on the bearing diameter. The friction due to the seals in use for chemical applications is usually negligible compared to the other contributions.

Feed acceleration. The torque and power required for feed acceleration can be determined through the fundamental principles of mechanics. Consider the feed accelerator as a "control volume" through which a mass $\rho_{sL} Q\, dt$, where ρ_{sL} is the density of the slurry and Q is the volumetric flow rate, is fed for a duration dt with an initial zero tangential speed $V_\theta = 0$ and, therefore, zero initial angular momentum. The exit mass has tangential speed V_θ and angular momentum $\rho_{sL} Q\, dt\, V_\theta R$, where R is the exit radius of the accelerator. Thus the required torque T_{in} to effect the rate of change of angular momentum is $(\rho_{sL} Q\, dt\, V_\theta R - 0)/dt = \rho_{sL} Q\, V_\theta R$. Also the required power is $P_{in} = T_{in}\Omega = \rho_{sL} Q V_\theta \Omega R$, where Ω is the rotational speed of the centrifuge. Note that ΩR is the tangential speed of the feed accelerator at the exit radius. For the feed stream to be properly accelerated to solid-body rotation, it should attain this speed as it leaves the accelerator. If the feed accelerator efficiency is 100%, $V_\theta = \Omega R$, $T_{in} = \rho_{sL} Q \Omega R^2$, and $P_{in} = \rho_{sL} Q (\Omega R)^2$. Otherwise these equations should be used with the term V_θ ($<\Omega R$) to be determined. Interestingly, the exit stream after properly accelerated has kinetic energy $\tfrac{1}{2} \rho_{sL} Q (\Omega R)^2$ at exit radius R. Consequently this output stream carries at best only half the input power, with the other 50% dissipated as heat in the process of feed acceleration. This is unfortunately very inefficient. Note that unlike a centrifugal pump, the feed stream has to be accelerated to a given tangential speed (the solid-body speed of rotation of the rotor), irrespective of the volumetric feed rate. In contrast, a pump delivers a certain pressure head, which depends on the flow rate.

For engineering calculation, the horsepower for feed acceleration is given by

$$P_{acc} = 5.984 \times 10^{10} \times \text{sg}(\Omega R_p)^2 Q \qquad (2.19)$$

where sg is the specific gravity of the feed slurry, Q the volumetric flow rate of feed in gal/min (L/s), and Ω the speed in rpm; R_p, in inches (meters), corresponds to the radius of the pool surface for a sedimenting centrifuge, or to the radius of the cake surface for a filtering centrifuge. (Note to convert horsepower to kilowatts multiply by 0.746.)

As can be seen in Eq. (2.19), the power consumption for feed acceleration varies as the second power of the pool radius when the pool is at solid-body rotation. When the effluent is brought from an initial radius R_1 to a final smaller radius R_2, the power should be reduced by the factor $(R_2/R_1)^2$. Theoretically, when the effluent is discharged at the axis of the machine with $R_2 = 0$, the power consumption should be zero. Therefore, in practice R_2 is made as small as possible. However, when the effluent liquid is brought to a smaller radius it has a tendency of undergoing a free-vortex motion. This is because in absence of external torque applied to the fluid, the angular momentum $mV_\theta R$ of the fluid is constant resulting in no further reduction in power P since $P \propto V_\theta R$. Therefore, for a free-vortex motion the power consumption on acceleration of the fluid is constant with $P_2/P_1 = 1$ regardless of the radius, or radii ratio R_2/R_1, where "1" corresponds to the initial position and "2" the final position. This gives a horizontal line for a free-vortex motion on a P_2/P_1 versus R_2/R_1 plot as illustrated in Fig. 2.12. In contrast, for a solid-body rotation, because $P \propto V_\theta R \propto R^2$, $P_2/P_1 = (R_2/R_1)^2$ and reducing discharge radius R_2 from an initial radius R_1 can result in significant reduction in power consumption; and vice versa when the discharge radius R_2 increases, see Fig. 2.12. In practice, radial and longitudinal vanes are used to enforce a solid-body rotation in the effluent liquid discharge for power savings purpose, especially when the effluent is brought abruptly to a smaller discharge radius. (Note Fig. 2.12 for the power ratio comparison between free-vortex motion and solid-body rotation of a fluid should complement Fig. 2.1b where the tangential velocity and G ratios are compared.)

The horsepower for cake conveyance for a decanter centrifuge with a conveyor screw is

$$P_{con} = 1.587 \times 10^{-5} \times T_{sp}\Delta \qquad (2.20)$$

where Δ is the differential speed in rpm between the screw conveyor and the bowl, and T_{sp} is the conveyance torque measured at the spline

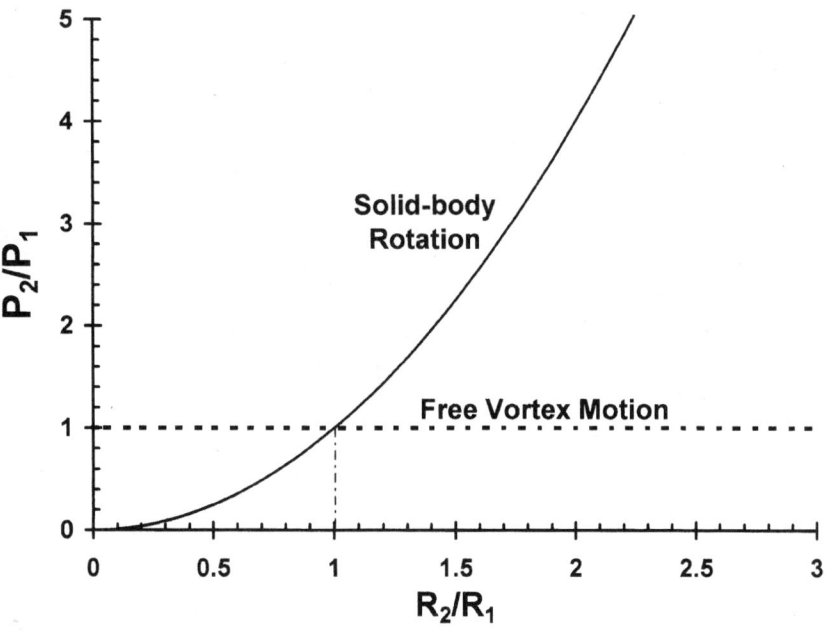

Figure 2.12 Power consumption for acceleration, a free-vortex motion versus a solid-body rotation.

in in · lb$_f$ (N · m). For centrifuges where the cake is discharged differently, the conveyance power is simply

$$P_{con} = MGC_f V \qquad (2.21a)$$

where M is the mass of the cake, G the centrifugal acceleration, C_f the coefficient of friction, and V the cake advance velocity. From Eqs. (2.20) and (2.21a) it can be inferred that

$$T_{sp} = \frac{m_s GC_f L}{\Delta} = m_s \Omega R \, gr \, C_f L \qquad (2.21b)$$

where m_s is the cake solids rate (dry basis), Ω the rpm, R the radius, gr the gear ratio of the gear box, and L the distance through which the cake has to be conveyed. The spline torque is inversely related to the differential speed and directly proportional to acceleration G, cake velocity, and cake mass. Furthermore, the torque at the pinion of the gear box can be evaluated,

$$T_p = \frac{T_{sp}}{gr} = m_s \Omega R C_f L \qquad (2.21c)$$

where T_p is the torque at the pinion. In deriving Eq. (2.21b) the relationship $\Delta = (\Omega_b - \Omega_p)/gr$ has been used with the pinion held fixed, that is, $\Omega_p = 0$. Note that the pinion torque in Eq. (2.21c) is proportional to the mass throughput rate of the solids (dry basis), rotation speed, bowl radius, coefficient of friction, and finally the length of the bowl, the distance through which the cake solids have to be conveyed.

Stress in Centrifuge Rotor

The stress in the centrifuge is quite complex. Analytical methods, such as the finite-element method, are used to analyze the mechanical integrity of a given rotor design. Without getting into an involved analysis, some useful insights can be gained from a simple analysis based on the hoop stress of a rotating bowl under load. At equilibrium the tensile hoop stress σ_h of the cylindrical bowl wall with thickness t_b is balanced by the centrifugal body force acting on the mass of the bowl wall with density ρ_m and its contents (cake, slurry, or liquid) with equivalent density ρ. Considering the circular segment of radius R_b containing a load at radius R_s, the unit subtended angle, and the unit axial length, in Fig. 2.13, a force balance leads to the relationship.

$$\sigma_h = \rho_m V_{\text{rim}}^2 \left[1 + \frac{R_b \rho}{2 t_b \rho_m} \left(1 - \frac{R_s^2}{R_b^2} \right) \right] \qquad (2.22)$$

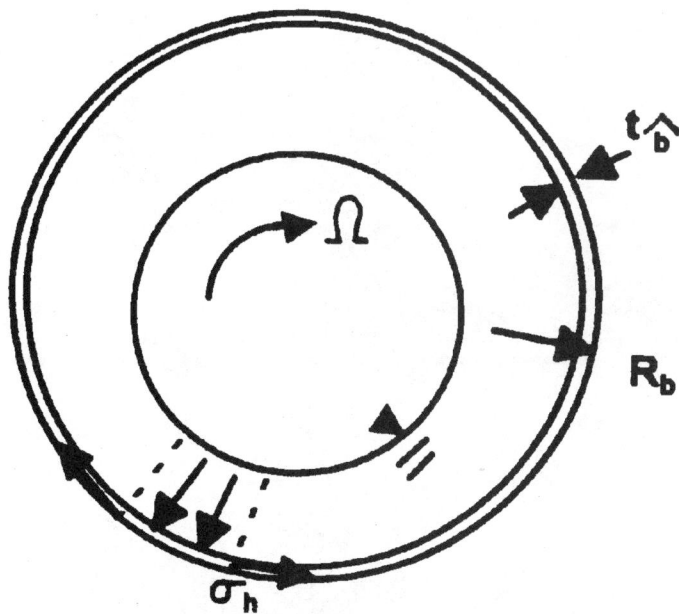

Figure 2.13 Hoop stress due to centrifugal load.

where $V_{rim} = \Omega R_b$ is the rim (or tangential) speed of the bowl. The term in square brackets is typically of order 1. If the maximum allowed σ_h is designed to be no more than 60% of the yield stress of the bowl material, which for steel is about 2.07×10^8 N/m² (30,000 lb$_f$/in²). Given $\rho_m = 7865$ kg/m³ (0.284 lb$_m$/in³) for stainless steel, then, in the absence of liquid load, $(V_{rim})_{max} = \{\sigma_h/\rho_m\}^{1/2} = 126$ m/s (412 ft/s). With additional liquid load, $\rho = 1000$ kg/m³ (0.0361 lb/in³), $R_b/t_b = 10$, and further assuming the worst case with liquid filling to the axis, the term in parentheses is 1.4237. Using Eq. (2.22), $(V_{rim})_{max} = 98$ m/s (346 ft/s). Indeed, almost all centrifuges are designed with top rim speeds about 91 m/s (300 ft/s). With special construction materials for the rotor, such as Ferralium or other duplex steel with higher yield stress approaching 5.5×10^8 N/m² (80,000 lb$_f$/in²), the maximum rim speed under full load can be much higher.

While Fig. 2.1a shows a useful plot of G/g versus bowl speed for a range of bowl sizes with diameters ranging from 0.076 m (3 in) to 1.37 m (54 in), it does not reveal the maximum G force a centrifuge bowl can attain. It could be misleading to bias toward the direction of a higher G force to make the separation without considering other limiting factors. Given $G = \Omega^2 R = V_{rim}\Omega$, a constant rim speed V_{rim} on a log G versus log Ω plot yields a straight line with unit slope. Figure 2.14 shows lines at

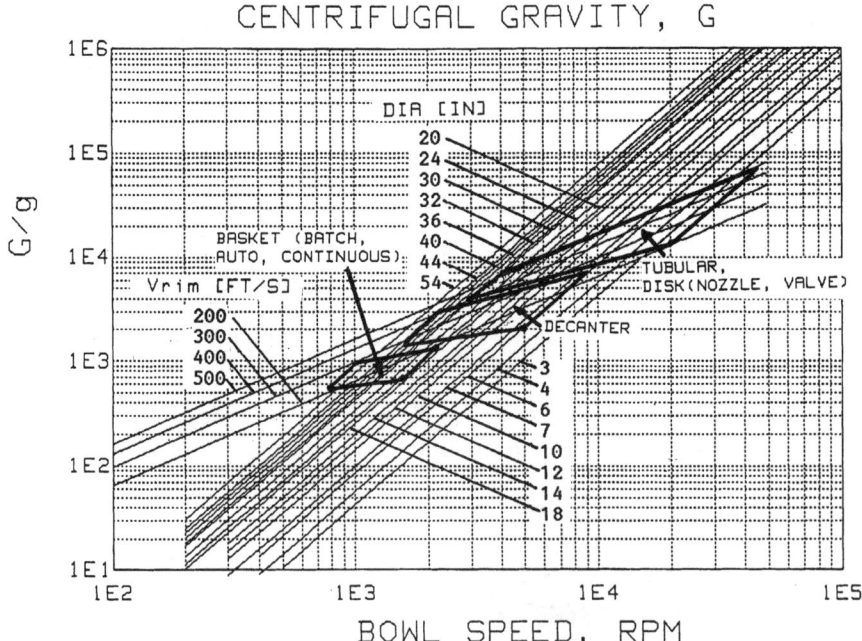

Figure 2.14 Classification of sedimenting and filtering centrifuges according to their G, speed, and size (1 in = 25.4 mm).

V_{rim} = 60, 90, 120, and 150 m/s (200, 300, 400, and 500 ft/s). For a given material of construction, V_{rim} is fixed, and this determines the maximum speed and G/g that the machine can achieve. For example, at V_{rim} = 90 m/s (300 ft/s) a 14-in-diameter bowl can achieve 4500g at 4700 rpm, whereas by changing the material of construction such that the rim speed can reach 122 m/s (400 ft/s), the bowl attains a maximum speed of 6500 rpm, which results in G = 8200g.

This shows clearly the trade-off between G force and machine size and the mechanical yield strength of the material of construction. The typical ranges of operation of the tubular or disk, decanter, and basket centrifuges are also mapped out in Fig. 2.14.

G Force versus Throughput

The G acceleration can be expressed as

$$G = \Omega^2 R_b = \frac{(V_{rim})^2_{max}}{R_b} \qquad (2.23)$$

Figure 2.15 shows the diameters of commercial centrifuges and the maximum G developed for each type. It demonstrates an inverse relationship between G and R_b at $V_{rim} = (V_{rim})_{max}$, which is fixed for a given material of construction of the rotor.

The throughput capacity of a machine, depending on the process need, is roughly proportional to the nth power of the bowl radius,

$$Q = c_1 R_b^n \qquad (2.24)$$

where n is normally between 2 and 3, depending on the process function—clarification, classification, degritting, thickening, or dewatering. Thus

$$G = c_2 (V_{rim})^2_{max} / Q^{1/n} \qquad (2.25)$$

Here c_1 and c_2 are constants. Consequently, large centrifuges can deliver high flow rates, yet separation is not as effective given it is done at a lower G force. Vice versa, smaller centrifuges deliver lower flow rates yet separation is more effective, given the higher G force. Also, by using a high-strength construction material for the rotating assembly, the centrifuge can attain higher rim speeds and G at a given rate for difficult separation applications.

Materials of Construction

Centrifuge bowls are made of almost every machinable alloy of reasonably high strength. Preference is given to those alloys having 1% elon-

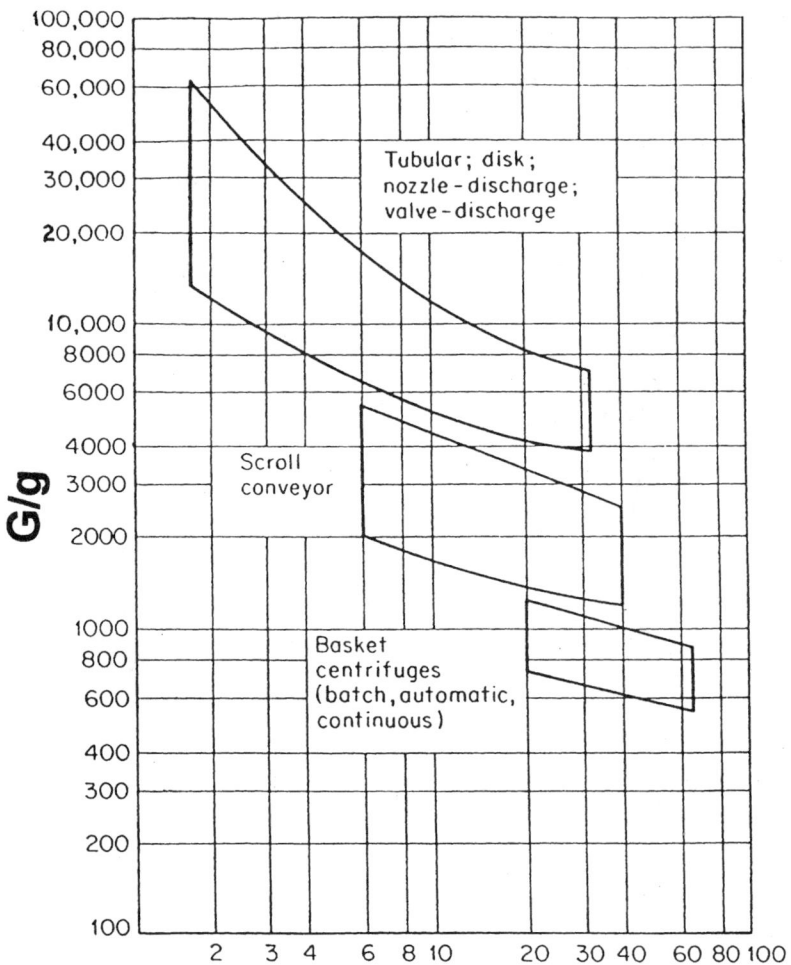

Figure 2.15 Variation of G with diameter in industrial centrifuges (1 in = 25.4 mm).

gation to minimize the risk of cracking at stress-concentration points. Typically, the list includes carbon steel, type 304, 316, and 317 stainless steels, duplex steel, alloy 20 (Carpenter) stainless steel, Monel metal, Inconel, nickel, Hastelloy B, titanium, and alloyed aluminum. Vertical-basket centrifuges are frequently constructed of carbon steel or stainless steel coated with rubber, neoprene, Penton, or Kynar. Casings and feed, rinse, and discharge lines that are stationary and lightly stressed may be constructed of any suitable rigid corrosion-resistant material. Wear-resistant materials—sintered tungsten carbide, silicone carbide, ceramic tiles, hard facing, and other coatings—are often used to protect the bare metal surfaces in high-wear areas, such as scroll blade tips, feed and discharge ports, bowl and case liners, bowl heads, screens, and nozzles.

Critical Speeds

In the design of any high-speed rotating machinery attention must be paid to the phenomenon of critical speed. This is the speed at which the frequency of rotation matches the natural frequency of the rotating part. At this speed any vibration induced by slight unbalance in the rotor is strongly reinforced, resulting in large deflections, high stresses, and even failure of the equipment. Speeds corresponding to harmonics of the natural frequency are also critical speeds but give relatively small deflections and are much less troublesome than the fundamental frequency. The critical speed of simple shapes may be calculated from the moment of inertia. With complex elements such as a loaded centrifuge bowl it is best determined through tests.

References

1. H. Greenspan, "Centrifugal Separation of a Mixture," *J. Fluid Mechan.*, 1983.
2. H. Greenspan, *The Theory of Rotating Fluids*, Cambridge University Press, London, 1968.
3. H. Lamb, *Hydrodynamics*, Dover, New York, 1945.
4. R. C. Emmett et al., "Sedimentation," in *Handbook of Separation Techniques for Chemical Engineers*, 3d ed., McGraw-Hill, New York, 1997, pp. 4–129.
5. J. F. Richardson and W. N. Zaki, "The Sedimentation of a Suspension of Uniform Spheres under Conditions of Viscous Flow," *Chem. Eng. Sci.*, 3 (1954), pp. 65–73.
6. H. Crosby, "Centrifuge with Torsional Vibration Sensing and Signaling," Canadian patent 1,077,450.

Chapter

3

Batch Sedimenting Centrifuges

Batch centrifuges are used to handle solid–liquid separation with a low amount of solids in the feed. The solids sediment stores temporarily in the bowl whereas the liquid is allowed to clarify in batch mode, as in test-tube and clinical centrifuges, or is discharged continuously in zonal, tubular, multibowl, or solid-wall basket centrifuges. The semicontinuous discharge of the liquid effluent is continued until the sediment builds up to a level at which it hampers the effluent quality. This is due to entrainment of the settled solids by the overlying fast-moving liquid layer. In this chapter the various designs of batch centrifuges are discussed, and an analytical solution is presented for transient centrifugal sedimentation.

In batch sedimenting centrifuges, the G level covers a wide spectrum from several hundred gs as in conventional basket or multibowl centrifuges, to 20,000g, as in tubular and high-G basket centrifuges. Up to half a million gs can be attained by ultracentrifuges. Different types of centrifuges are used, depending on the characteristics of the feed slurry that needs to be separated and on the process objectives.

Test-Tube and Clinical-Bottle Centrifuges

Centrifugal force is applied to the contents of a test tube, which rotates about a vertical axis. The tubes are contained in holders that are integral with the head assembly. The assembly is underdriven through a vertical shaft mounted on bearings.

Horizontal (also known as swinging-bucket) heads, as shown in Fig. 3.1a, accommodate cups and adapters designed to support a wide va-

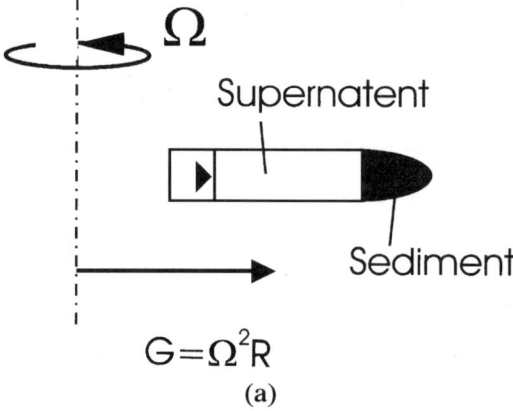

Figure 3.1 Spin tubes. (*a*) Horizontal head.

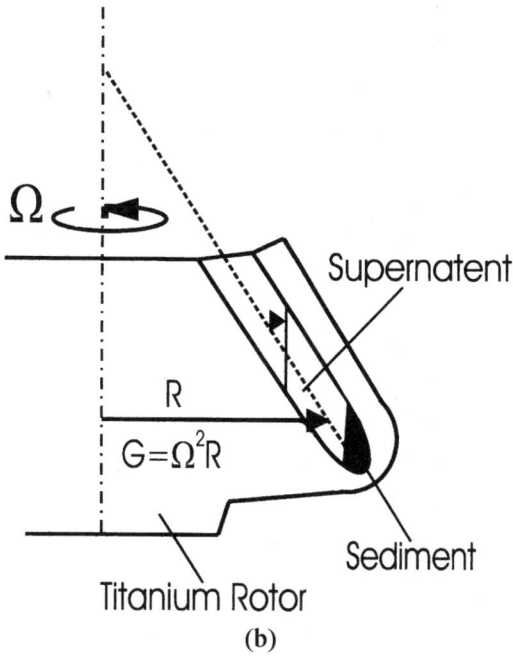

Figure 3.1 Spin tubes. (*b*) Angle head.

riety of containers for specific uses. These cups are mounted in trunnion rings with "shields," or buckets with ears hinged to to the rotor, which will allow the cups and their contents to swing from a vertical position while at rest to a horizontal position under rotation. The center of mass of the cup and its contents must be at a larger radius compared to that for the centerline of the trunnions or hinges for stability.

As shown in Fig. 3.1*b,* angle heads support the containers at a fixed angle, typically 40–48° with respect to the vertical axis of rotation. Using the inclined lamella sedimentation principle, it shortens the path at which particles have to settle before being captured by a solid surface. The walls of the tube also act to reduce any Coriolis effect, similar to disk centrifuges with azimuthal barriers, as will be discussed in Chap. 4.

Many test-tube centrifuges also accommodate basket heads, perforated for centrifugal filtration or imperforate for continuous-flow clarification. Some angle heads are also arranged for continuous liquid flow through the containers so that the quantity of solids accumulated per run is greatly increased.

Except in the smallest sizes, the rotating element is enclosed in a casing for safety. In some designs the temperature of the casing and its contents can be controlled through heating, while in others refrigeration is provided, both to permit subambient temperature operation and to absorb the heat generated by the windage of the rotor. The thermal effects of windage on ultrahigh-speed rotors are minimized by maintaining the casing under a vacuum.

Besides providing a convenient way of quickly settling and compacting small solid particles that settle or filter slowly, test-tube centrifuges are also used for the separation of refractory emulsions. Their use is specified in a number of ASTM standard petroleum and other tests, a variety of analytical methods, and clinical and public health procedures. When the volume of the separated phase is to be measured accurately, use of the swinging-bucket type is preferred. It gives an interface at right angles to the axis of the container. The containers are frequently shaped and calibrated to emphasize a small volume of the separated phase. Examples are small bore tips on pear-shaped bottles for the determination of small amounts of free moisture and sediment in oil or small bore tops on babcock test bottles for the determination of butterfat in milk products.

A large variety of containers is available for specific applications ranging in size from 175µL (4.6×10^{-5} gal), or even smaller, to 1 L (0.26 gal). Localized stressing of glass containers is minimized by supporting them on plastic cushions. Plastic bottles of polypropylene, polyallomer, and polycarbonate are less subject to breakage and do not require as careful handling as glass.

Horizontal heads provide from 2 to 16 places in which cups may be located. The number of small tubes handled per run can be increased to 296 microtubes (0.25 or 0.4 mL) with the use of adapters. Angle heads are available to centrifuges from four 600-mL blood bags to six 250-mL bottles or sixty 15-mL bottles. For the smaller samples, angle heads provide from four larger 50-mL tubes to 48 much smaller 0.5-mL microtubes in a single rotor head. New centrifugal membrane filters in 45° angle-head rotors are available with G forces up to 7000g, in which the transmembrane pressure for ultrafiltration reaches 100 lb$_f$/in^2 (6.8 atm). Shields and cushions are provided for 12 plastic sample tubes all in a single rotor. Each tube contains both the feed slurry compartment and the filtrate compartment, separated by the ultrafilter.

A large variety of drives is available to provide operating speeds up to 3000 rpm (1980g at the container tip) in the large sizes, to 23,400 rpm (38,000g) in smaller sizes and multispeed head attachments, to 65,000 rpm (429,000g) on ultrahigh-speed models. In every case the manufacturer's recommendations on speed limits to operating loads should be followed. Accessories include microprocessor control, speed control and indication, automatic timers, temperature control and indication, and safety interlocks.

Preparatory Centrifuges and Ultracentrifuges

Ultracentrifuges are commonly referred to as ultrahigh G-force centrifuges, and the original equipment, the Svedberg centrifuge, is known as the analytical or optical ultracentrifuge. Preparatory bottle ultracentrifuges may be of a fixed-angle or swinging-head type, as described in the preceding section. The spin tube in Fig. 3.1b represents a fixed-angle ultracentrifuge. They are most commonly applied in biological and biochemical techniques, such as the separation and isolation by selective classification of cells and macromolecules.

The analytical ultracentrifuge applies a high centrifugal force, up to 500,000g, to a small sample in a transparent cell. For example, a fixed-angle ultracentrifuge has the tube axis oriented at 20–25° from the vertical axis. The polycarbonate tube is typically about 16 mm in diameter by 76 mm long. At a rotational speed of 70,000 rpm in a titanium rotor, with minimum and maximum radii of 40.5 and 82 mm 1.59–3.23 in), respectively, it generates an average of 450,000g.

With schlieren or interference optics, the change in concentration can be measured as a function of time by means of the Svedberg equation.[1] This equation is given by

$$\frac{S}{D} = M \frac{1 - \rho_L/\rho_P}{\Re T} \qquad (3.1)$$

where the sedimentation coefficient $S = d^2(\rho_p - \rho_L)/18\mu$ and is measured in seconds. $\Re$ is the universal gas constant and T the temperature in kelvins. Molecular measurements typically give S in units of 10^{-13} s. In honor of the pioneering work by Svedberg, 1 svedberg is defined as 10^{-13} s. From measurement, molecular weight M, sedimentation coefficient S, diffusion coefficient D, and partial specific volume $1/\rho_p$ of the sample can be derived.

Separation Techniques in Preparative Ultracentrifuges

Differential centrifugation

In the analytical and preparatory centrifuges described in the preceding section the collected fraction reports to the bottom of the container as a pellet. This pellet is contaminated with larger and heavier particles as well as with smaller particles that started from a position proximate to the bottom of the container. The supernatant, after being decanted off, can be recentrifuged at higher speeds to obtain further purification, with the formation of a new pellet and supernatant. Likewise, the pellet from the former separation can be recentrifuged after resuspension in a small volume of a suitable solvent.

Density-gradient centrifugation

A density-gradient solution is introduced into the tube in layers through a special syringe such that the layers are at increasing densities from small to large radii. For example, sucrose solutions of 10, 20, 30, and 40% w/w are introduced into the tube to generate a density-gradient solution with densities of 1.0381, 1.0810, 1.1270, and 1.1764 g/cm^3 at 20°C. Over a period of time (on the order of 1 to 3 h), the solute diffuses out to form a more or less linear gradient along the radius. The rate of diffusion depends on the diffusion coefficient of the solute and the viscosity of the liquid, both of which are dependent on the temperature.

Settling-rate method. In this method a sample of solid particles (such as lysosome with a density 1.21 g/cm^3, plant virus with 1.3–1.45 g/cm^3, ribosome with 1.60–1.75 g/cm^3, glycogen with 1.7 g/cm^3) having a density greater than that of the liquid layers in the tube is introduced to the surface of the liquid. Settling is speeded up once the sample is under centrifugation. After some time, the solids in the sample settle in the tube at a velocity depending on the size and density of the solid. Light and smaller solids stay behind while heavier and larger solids settle faster, migrating closer toward the tube's outer radius. Segregation of the particles can therefore be made based on density and size.

Isopycnic method. Density-gradient solutions such as cesium chloride, which have a wider range of density differences are employed. A stock solution of cesium chloride at 65% w/w has a density of 1.91 g/cm^3. The sample can be mixed with the solution when it is introduced into the tube. At equilibrium the solids with their respective densities seek the level of the liquid layer with identical density. The technique is referred to as isopycnic banding and is used for the purification of many types of viruses.

Zonal Centrifuge

The zonal centrifuge is a cylindrical bowl with the interior divided into sector-shaped compartments by vanes attached to the core. The bowl is closed with a threaded lid. A rotating bowl assembly, which may be fixed or removable, allows fluid to be pumped in and out of the bowl while it is under rotation (Fig. 3.2a). In typical operation the zonal centrifuge is preloaded with a compatible solution, such as sucrose, in which a radial density gradient is established across the radius from light-density fluid near the center to heavy-density fluid at the periphery of the bowl (Fig. 3.2b). The feed sample is then introduced, either batchwise or continuously, and the solids of different densities in the sample locate the annular zones in the gradient corresponding to their respective densities at equilibrium. There are several ways of loading and unloading. The sequence of events during loading is illustrated in Fig. 3.2c–g. In Fig. 3.2f the separated sample is unloaded by injecting a dense solution at the outer radius of the rotor, displacing the fractions at center, whereas in Fig. 3.2g the fraction is unloaded at the outer radius by injecting water or buffer at the center.

Tubular-Bowl Centrifuge

The high-G tubular-bowl centrifuge in Fig. 3.3 consists of a hollow-cylinder rotating bowl, with lower and upper bowl heads, suspended from and driven by a spindle of designated flexibility. The spindle in turn is supported through radial and thrust bearings on a rigid frame. The drive may be direct, from a steam or air turbine mounted on the upper end of the spindle. More often the drive is through a flexible belt from an offset motor and pulley to a pulley that is mounted on the bearings and supports the spindle and rotor through a flexible coupling. A boss on the bottom bowl head fits into a sleeve bushing that provides controlled dampening during acceleration and deceleration and accommodates the excursion of the axis of rotation from the geometrical centerline as the speed of rotation passes above critical. The tubular-bowl centrifuge derives its stability above critical speed from its relatively large length-to-diameter aspect ratio, in the range of approximately 4.35 to 7.25.

The liquid to be clarified enters the bowl through a centrally located

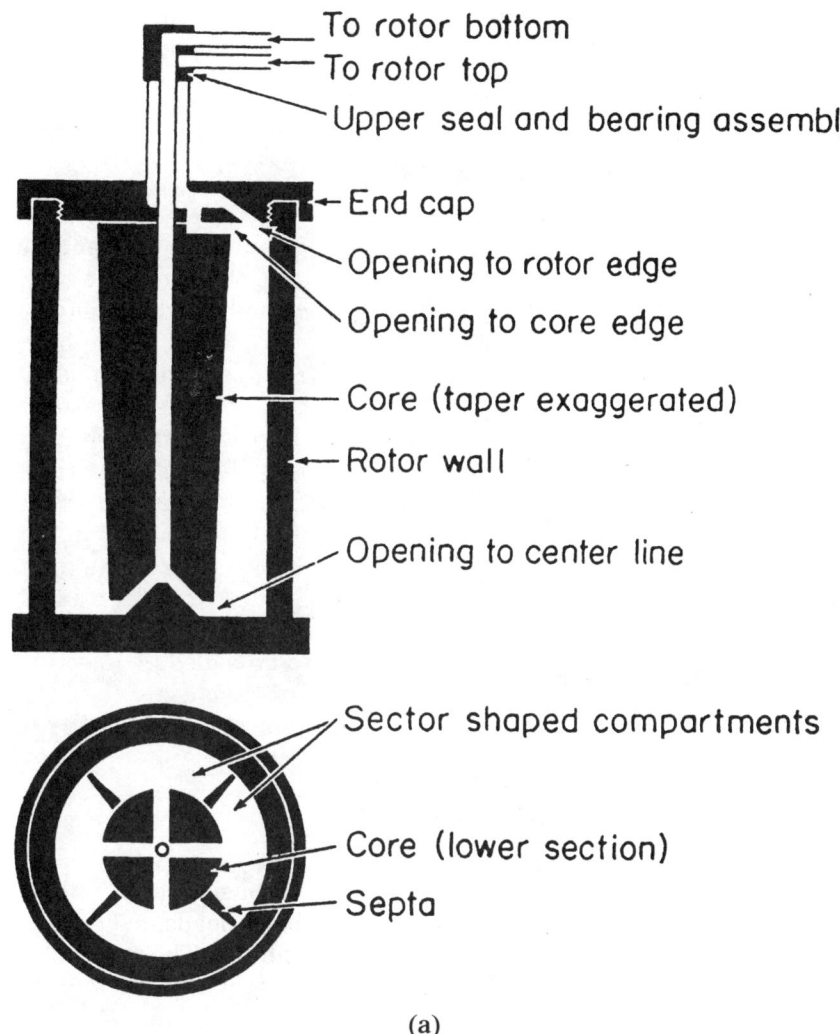

Figure 3.2 Zonal centrifuge. (*a*) Side and top views.

opening in the lower bowl head as a free-standing jet from a fixed feed nozzle. It is accelerated to the rotational velocity of the bowl by axial vanes (which also provide a stabilizing influence on deceleration) and flows upward along the inner bowl surface as an annular pool, where separation takes place under centrifugal force. Subsequently the clarified liquid overflows a ring weir, which is capped onto the upper bowl head. The radius of this opening establishes the pool depth within the bowl. When used to separate two liquids continuously,

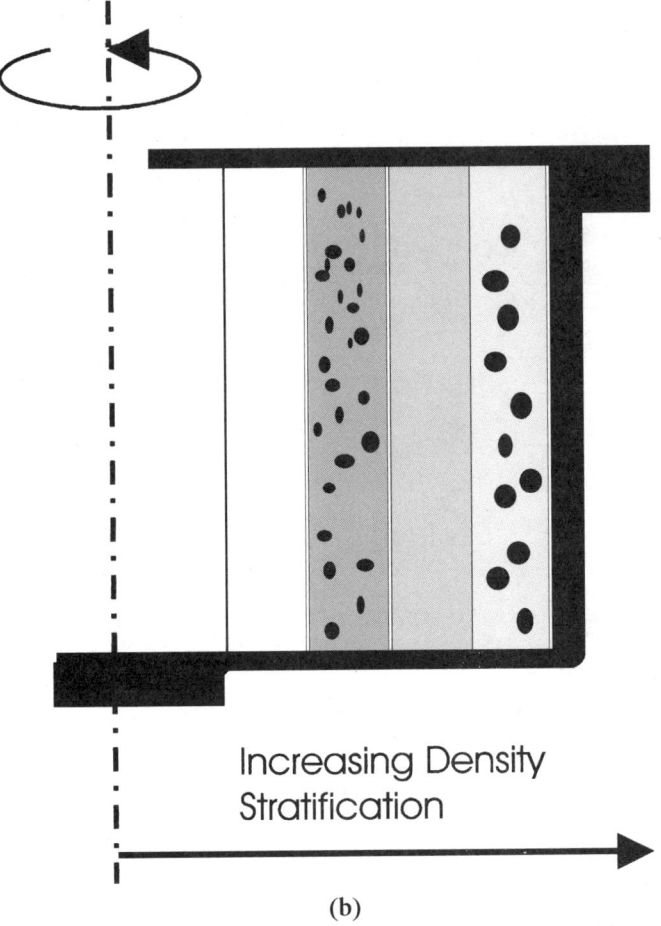

Figure 3.2 Zonal centrifuge. (*b*) Typical operation using isopycnic technique with particles of different densities locating their annular positions in fluid core having identical density.

an internal baffle provides a separate passage adjacent to the bowl wall to conduct the heavier liquid phase to a different discharge radius. The radial position of the liquid-liquid interface can be determined (see Chap. 4). It decreases as the heavy-phase discharge radius decreases, and vice versa. The discharged heavy phase is largely free from light-phase contamination. However, the discharged light phase may have an undesirable fraction of the heavy phase. Similarly, as the heavy-phase discharge radius increases, the discharged light phase becomes more purified whereas the discharged heavy phase may contain an undesirable fraction of the lighter liquid. Thus the centrifuge can be optimized, depending on process needs, by adjusting

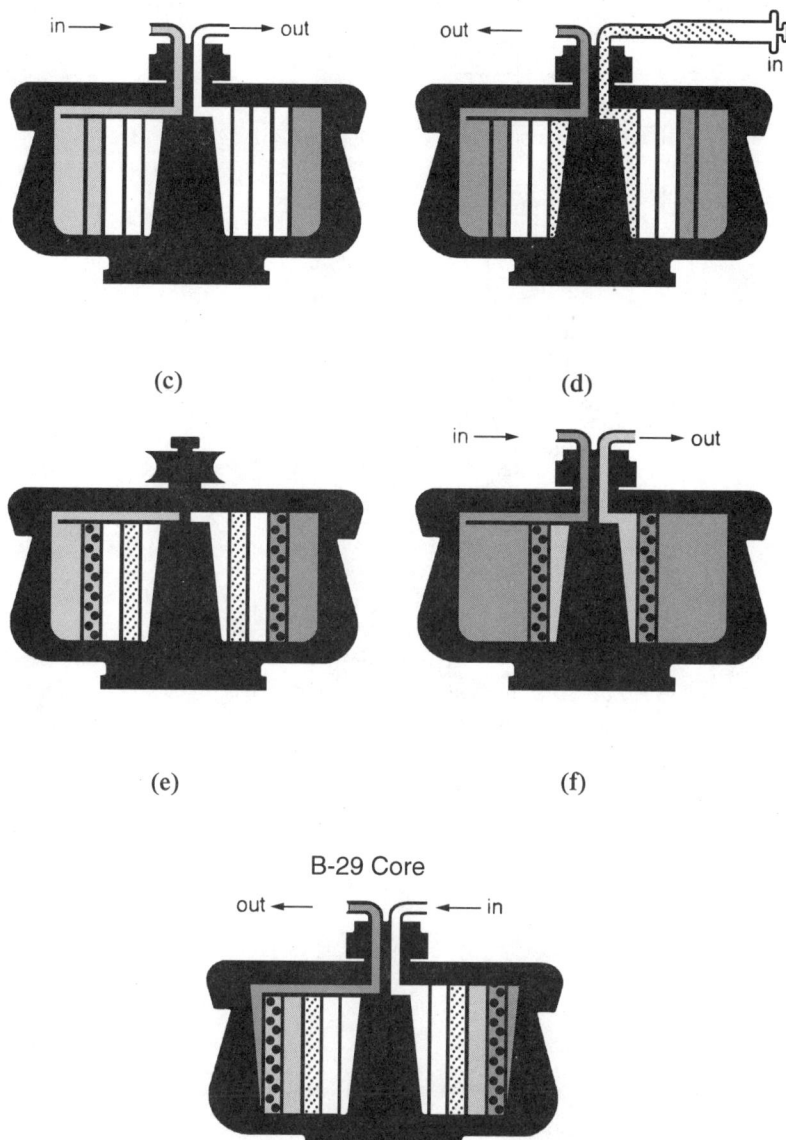

Figure 3.2 Zonal centrifuge. (c) Gradient loaded with rotor spinning at 2000 rpm. (d) Sample injected at 2000 rpm, followed by injection of overlay. (e) Particles separated with rotor at speed. (f) Contents unloaded by introducing dense solution at rotor edge, displacing fractions at center, standard core. (g) Contents unloaded at rotor edge by introducing water or buffer at center, B-29 core. (*Fig. 3.2c–g used with permission from Beckman Instruments, Inc.*)

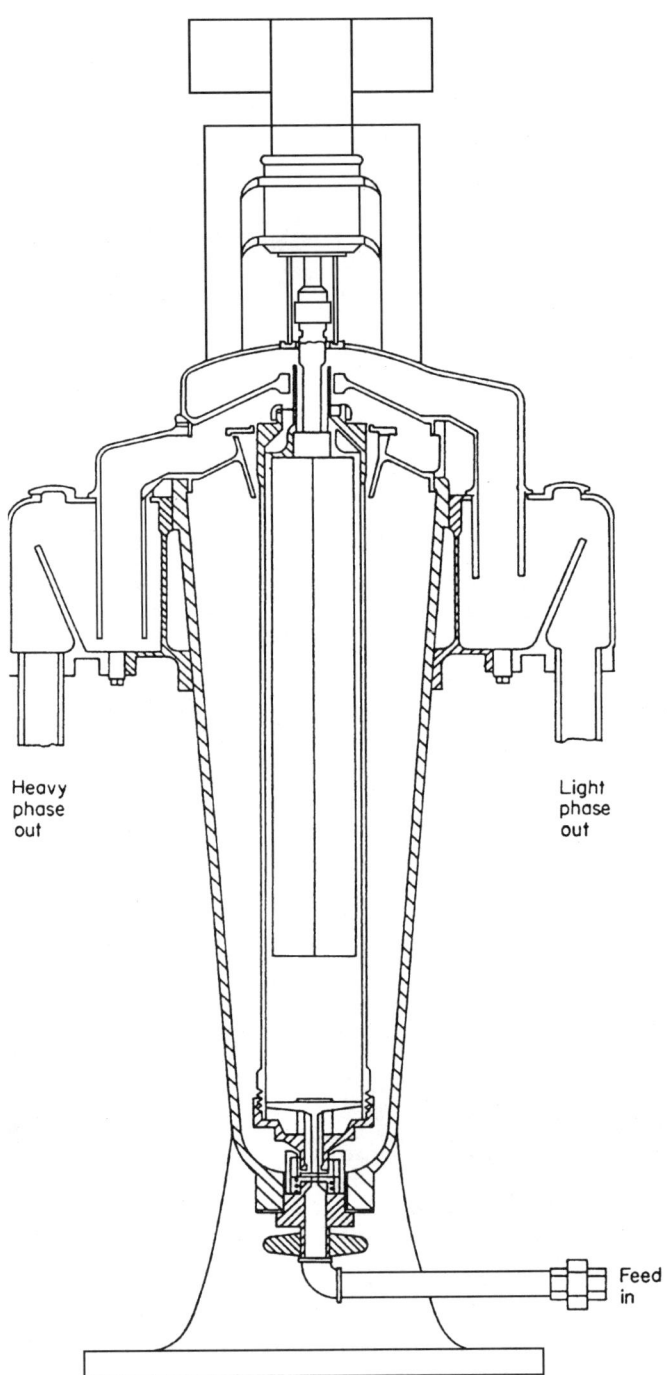

Figure 3.3 Tubular-bowl centrifuge.

the heavy-phase discharge radius. The foregoing discussion is typical for all liquid-liquid centrifugal separators.

The effluent liquids are captured in covers and carried to a convenient point for discharge. The liquid-handling capacity of the tubular-bowl centrifuge varies with use. The low end includes stripping small bacteria from a culture medium, the high end serves when purifying transformer oil and restoring its dielectric value. The solids are allowed to accumulate inside the bowl wall until the centrate clarity is affected. The accumulated solids are then recovered manually by a fast disassembly of the bowl and bottom head from the drive assembly and washing or scraping the solids out of the cylindrical bowl. This process typically takes less than 15 min. The solids-handling capacity of this centrifuge is limited to 5 kg (11 lb) or less. Typically, the feed-stream solids should be less than 1%.

The principal applications are for the separation of difficult-to-separate biological solids with small density differences between solids and the suspending aqueous medium, such as the harvesting of bacteria and other microorganisms and of precipitated blood protein fractions. The tubular bowl is also used for the clarification of viscous substances, such as molten chicle and nitrocellulose solution, and of essential oils, flavoring extracts, and other food products; the removal of microcrystalline wax from chilled lubricating oil in naphtha solution with continuous discharge of the separated wax on an aqueous carrier liquid; the recovery of fine metal particles such as silver resulting from the washing of scrap photographic film; the continuous separation of the soap stock formed by the alkali refining of vegetable and animal fats and oils; and the classification of pigment slurries.

Industrial models have bowls 102 to 127 mm (4 to 5 in) in diameter and 762 mm (30 in) long, which are capable of delivering up to 18,000g. The smallest size bowl [44 by 229 mm (1.75 by 9 in)] is a laboratory model capable of developing up to 65,000g. Other sizes are given in Table 3.1.

TABLE 3.1 Typical Tubular Centrifuges

Bowl diameter, inches (mm)	Length, inches (mm)	Maximum, rpm	Maximum, G/	Volumetric throughput, gal/min (m³/h)
1.75 (44)	9 (229)	50,000	62,400	0.0025–1 (0.011–4.4)
4.125 (105)	18–30 (457–762)	15,000	13,200	to 10 (to 44)
5* (127)	30 (457)	15,000	16,000	20 (88)

*A special model of the 5-in (127-mm) diameter×30-in (457-mm) long size (the KII zonal centrifuge) develops up to 87,000g at 35,000 rpm.

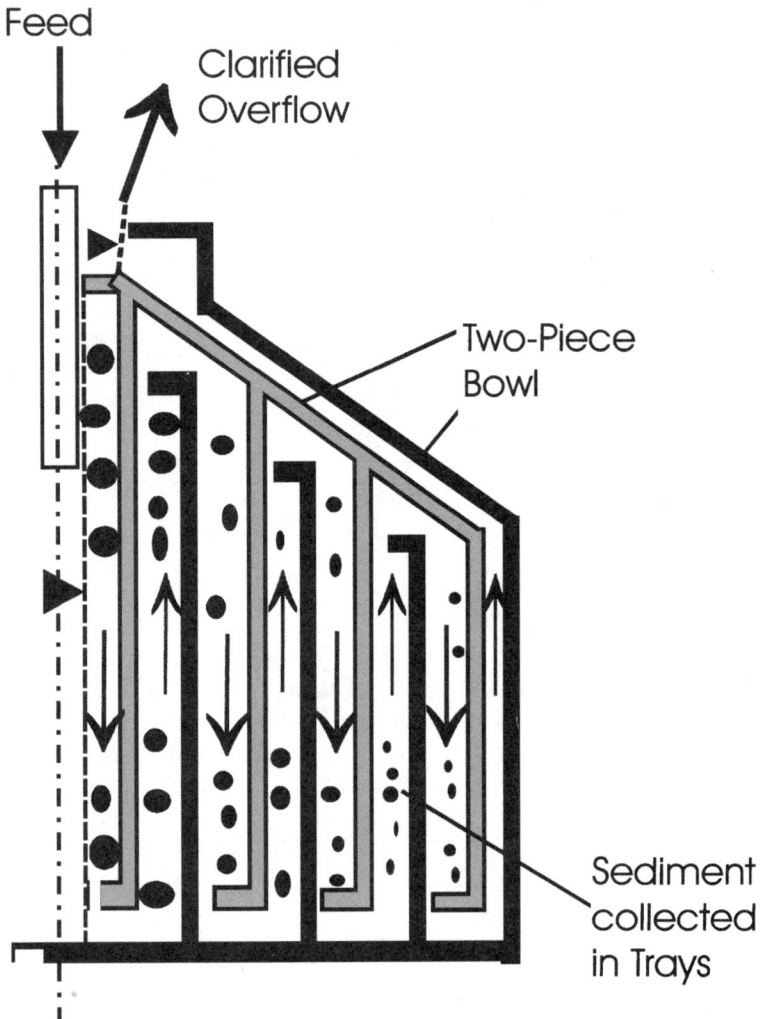

Figure 3.4 Multichamber bowl.

Multibowl Clarifier Centrifuge

The multibowl clarifier centrifuge shown in Fig. 3.4 functions as a series of interconnected tubular bowls, each of relatively short and constant length and stepwise increasing diameters, all in the same rotating assembly. The feed enters at the center, flows downward through the smallest diameter tube, then reverses direction to flow upward through the next larger diameter, and so on. There are up to six such annular passageways, the last being defined by the vertical bowl-shell wall. The tube diameters are selected so that each annulus (between adjacent tubes) has the same area and the velocity of flow through it remains constant. Since the flow at constant velocity is being subjected to zones of successively greater centrifugal force, the multibowl clarifier acts as a classifier, with the largest, heaviest particles being deposited in the first pass and the smallest in the last pass. The accumulated solids are removed manually after the bowl has been disassembled. The rotor is mounted on top of a drive assembly similar to that described for solid-wall centrifuges.

While the liquid discharge can be into open covers, most of the applications for this centrifuge will not tolerate exposure to air. In the hermetic design (discussed in greater depth in Chap. 4), the feed enters through a rotary joint on the bottom of the drive spindle. The effluent discharges at the top through a rotary face seal that may require supplementary cooling. In the centripetal pump design the effluent is led to an external chamber integral with the rotating bowl top. Set into this chamber is a stationary closed pump impeller. This converts the rotational-velocity energy of the contents of the chamber to pressure. The back pressure on the central discharge of this impeller controls its depth of submergence in the liquid in the chamber, and this in turn reduces the amount of entrained air and foaming in the discharged liquid to almost zero. The principal applications are for beverage and pigment varnish clarification.

Solid-Bowl Batch Centrifuge

These centrifuges, shown in Fig. 3.5, have many similarities with the tubular-bowl centrifuge, but their L/D ratio is much smaller, always less than 0.75 and frequently less than 0.6 effective. The rotor consists of a lip ring at the top, a cylindrical imperforate shell, and a bottom. The bottom may be solid, containing a central nave through which the rotor is driven. More often the nave is surrounded by an open spider (referred to as open-bottom design) through which the accumulated solids can be discharged while the rotor is at low speed or at rest. In this case the remainder of the bowl bottom is solid. The diameter of the opening of the spider must be less than the inner diameter of the lip ring, so the liquid being clarified will be directed to and overflow the lip ring at the top. In a modern configuration this rotor is mounted in the link-suspended casing and drive described in Chap. 8. The feed is directed to the bottom of the rotor through an accelerating cone or its equivalent, mounted on the nave; or through an offset J-shape feed pipe with the opening oriented along the circumferential direction. This provides the feed stream a tangential velocity component. Unfortunately the J-pipe arrangement does not provide a uniform distribution of feed along the axial length of the bowl, resulting in the loading

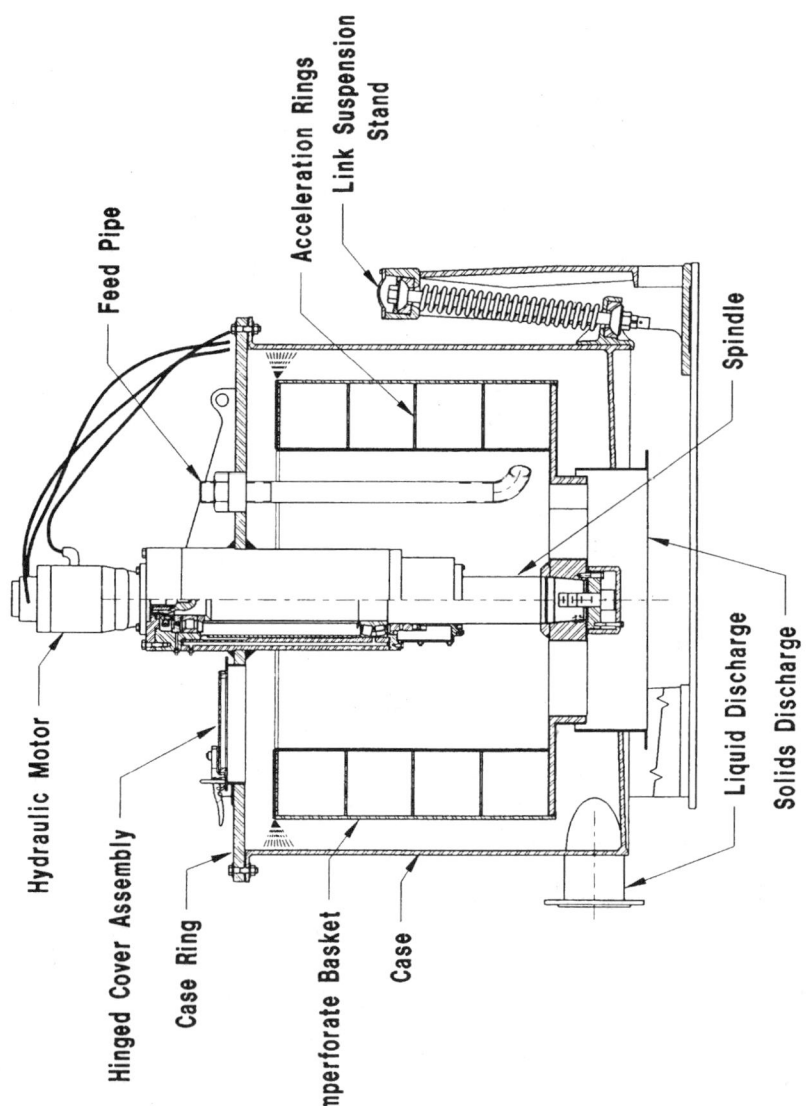

Figure 3.5 Solid-bowl basket. (*Courtesy of Bird/Ketema.*)

of more feed solids at the bottom and less at the top of the bowl. The cone accelerator is perhaps somewhat improved but still has this deficiency. A new feed-pipe design using two coaxial feed tubes, with the outer stationary tube having a series on in-line holes, while the inner rotating tube, a series of holes in a spiral pattern, has been shown to have better feed-slurry distribution along the bowl length.[2]

With all batch solid bowls the liquid being centrifuged travels upward as a thin annular layer, and about 60 to 80% of the volume under the lip ring is available for the accumulation of solids before the effectiveness of clarification is impaired. A cake detection system can be set to give a warning when the solids have reached a predetermined level, at which the feed cycle stops. A radially adjustable skimmer can be used to remove the free supernatant mother liquor from the surface of the cake. This same skimmer is also used to remove soft solids, such as metal hydrates or the biomass from activated sludge, without reducing the rotational speed. More refractory solids are removed manually from the bottom rotor while it is at rest. They can also be removed by an unloader plow that is moved into the solids layer while the bowl is rotating at low speed, causing the solids to drop through the openings in the bottom.

The system can be fully automated under timer control, with the cake detection system as overriding insurance for variations in feed solids concentration. The cycle typically consists of acceleration of basket feeding, cake compaction, skimming of pool liquid and soft solids, deceleration of basket, and cake unloading (Table 3.2). The gross flow rate during the feed part of the cycle is determined by the size and rotational speed of the centrifuge and the required degree of capture of the suspended solids. The net capacity of the centrifuge is this gross feed rate times the feed time per cycle divided by the total cycle time.

The cyclic solid-wall centrifuge is used primarily for the recovery and concentration of sludges and precipitates that remain plastic or fluid under a centrifugal force on the order of $1500g$. Typical are the metal hydroxide precipitates produced during electrochemical machining operations and the biomass of secondary waste-activated sludge from domestic and industrial sewage treatment.

TABLE 3.2   Solid-Bowl Basket Centrifuge—Operating Cycle

1. Accelerate
2. Feed (timer or cake detector to terminate)
3. Idle (optional if required to compact cake)
4. Skim Supernatant
5. Skim soft solids
6. Repeat 2, 3, and 4 (if hard-solids loading permits)
7. Decelerate
8. Unload hard solids (plow or manual unload), and repeat

The solid-bottom design is available in sizes ranging from 300 mm (12 in) in diameter by 150 mm (6 in) deep with a holding capacity under the lip ring of 0.006 m^3 (0.2 ft^3), to 1200 mm (48 in) in diameter by 750 mm (30 in) deep with a holding capacity under the lip ring of 0.45 m^3 (16 ft^3). The centrifugal force ranges from 1800g on the smaller sizes to 1300g on the largest. A bottom-discharge design is available in the larger sizes with the same parameters.

On a typical waste-activated sludge, gravity-settled to 1% dry solids concentration, the automated 1200-mm (48-in) by 750-mm (30-in) design will recover 90% of the suspended solids at 10% concentration at an instantaneous feed rate of 189 L/min (50 gal/min) and a net throughput rate of 170 L/min (45 gal/min).

Commercial centrifuges of this type have bowl diameters ranging from 0.3 to 2.4 m (12 to 96 in). The large sizes are used on heavy-duty applications such as coal dewatering and are limited by stress considerations to operate at 300g. The intermediate sizes, used for chemical process services, develop up to 1000g.

Fully Automatic High-G Basket Centrifuge

A recent innovation is the high-G sedimenting basket centrifuge (Fig. 3.6), which combines the operating strengths of both disk centrifuges (to be discussed in Chap. 4) and conventional basket designs. The self-aligning pendulum design and the use of titanium rotor construction permit operation up to 20,000g, resulting in excellent clarification and cake compaction all in a single unit. The clarifier bowl, with an aspect ratio of approximately 1:1, is further stiffened by internal circumferential baffle rings, which serve to minimize detrimental wave action along the axial direction. This is especially important during the acceleration phase of the operating sequence. Feed is introduced through a top-mounted stationary feed pipe and is accelerated to full rotational speed in the conical section. Clarified liquid overflows a bottom weir and drains off at the bottom of the bowl. Sediment compacts in the bowl until approximately 80% of the available volume has been filled. At this point timers or sensors slow the machine down to permit the drainage of surface liquid to a separate residuals compartment. Subsequently a low-speed drive brings the bowl up to 50 rpm, and a pivoting scraper arm advances into the settled solids, discharging them through a retractable solids gate and into a collection vessel. Following discharge, optional clean-in-place (CIP) or sanitary-in-place (SIP) cycles may be added to the program. The entire sequence of operations is fully automated and is controlled by an on-board programmable logic controller (PLC). This is a significant advancement compared to the conventional operation of the manual-discharge tubular centrifuge.

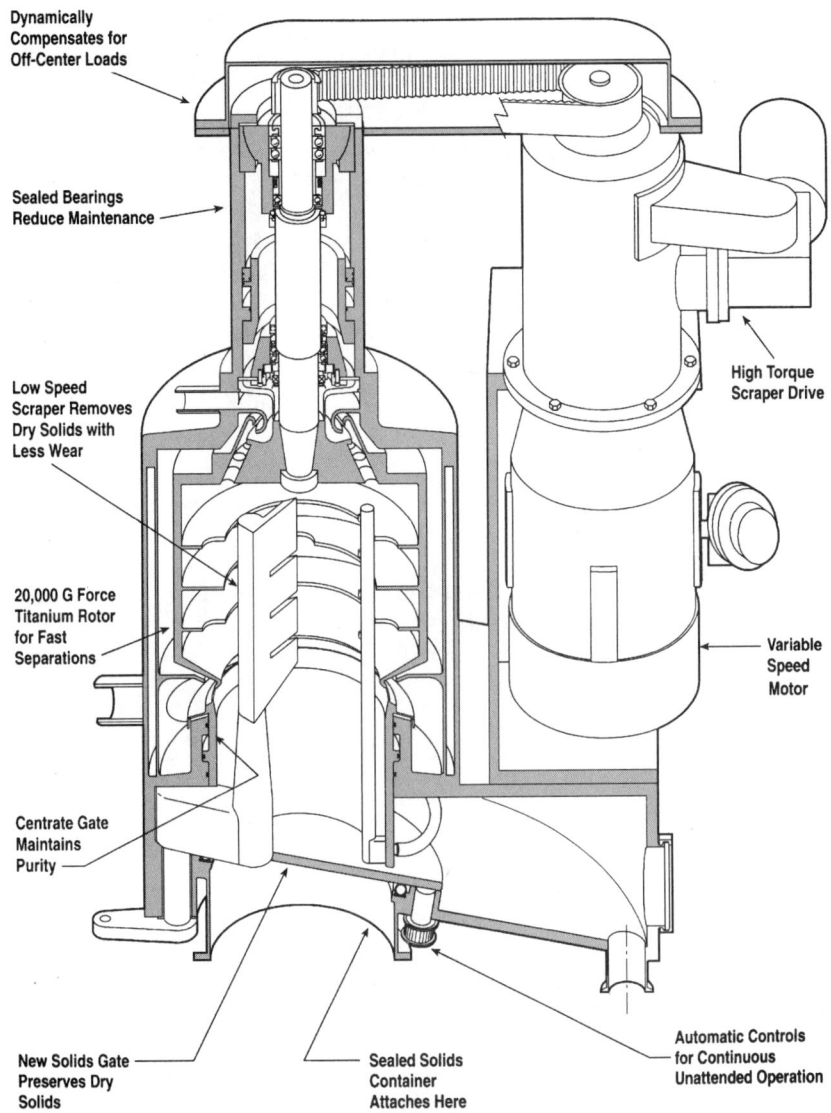

Figure 3.6 Fully automated high-G basket centrifuge. (*Courtesy of CARR Separations, Inc.*)

Transient Centrifugation Theory

It is appropriate in this context to review the theoretical analysis of transient centrifugal sedimentation.[3,4] As in gravitational sedimentation, there are three layers present during batch settling of a slurry in a centrifuge—a clarified liquid layer closest to the axis, a middle feed-slurry layer with suspended solids, and a cake layer adjacent to the bowl wall with concentrated solids. Unlike the constant gravity g, the centrifugal gravity G increases linearly with the radius. It is maximum near the bowl wall and zero at the axis of rotation. Also, the cylindrical surface area through which the particle has to settle increases linearly with the radius. Both of these factors give rise to some interesting, yet rather unexpected results.

Consider the simple initial condition $t = 0$ when the solids concentration ϕ_{s0} is constant across the entire slurry domain, $R_b \geq R \geq R_L$, where R_L and R_b are the radii of the slurry surface and the bowl, respectively. At a later time $t > 0$, three layers coexist—the top clarified layer, a middle slurry layer, and a bottom sediment layer. The air-liquid interface remains stationary at the pool surface radius, R_p, while the liquid-slurry interface with radius R_s expands radially outward with increasing time. R_s is given by the relationship

$$\frac{R_s}{R_p} = \sqrt{\frac{\phi_{s0}}{\phi_s}} \qquad (3.2)$$

Equation (3.2) can be derived from the principle of conservation of angular momentum as applied to the liquid-slurry interface.

Interestingly, the solids concentration in the slurry layer ϕ_s does not remain constant with time, as in gravitational sedimentation. Instead, ϕ_s decreases with time uniformly in the entire slurry layer, in accordance with

$$\frac{\phi_s}{\phi_{s\,max}} = 1 - \left(1 - \frac{\phi_{s0}}{\phi_{s\,max}}\right) e^{2\zeta} \qquad (3.3)$$

where ζ is a dimensionless time variable,

$$\zeta = \left(\frac{V_{g0} t}{R_b}\right) \frac{G}{g} \qquad (3.4)$$

In Eq. (3.3) under hindered settling and $1g$ the solids flux $\phi_s V_s$ is assumed to be a linear function of ϕ_s, decreasing at a rate of V_{g0}. Also, the solids flux is taken to be zero at the maximum solids concentration $\phi_{s\,max}$. As $G/g \gg 1$, this solids flux behavior based on $1g$ is assumed to be proportional to the ratio G/g.

Concurrent with the liquid-slurry interface moving radially outward, the cake layer builds up, with the cake-slurry interface moving radially inward and the radial position determined by

$$\frac{R_c}{R_b} = \sqrt{\frac{\varepsilon_s - \phi_{s0}}{\varepsilon_s - \phi_s}} \qquad (3.5)$$

where ε_s is a constant cake solids concentration. Sedimentation stops when the growing cake-slurry interface meets the decreasing slurry-liquid interface with $R_c = R_s$. This point is reached when $\phi_s = \phi_s^*$ at $t = t^*$. These variables (denoted by superscript *) can be determined from

$$\frac{1}{\phi_s^*} = \frac{1}{\varepsilon_s} + \left(\frac{R_b}{R_p}\right)^2 \left(\frac{1}{\phi_{s0}} - \frac{1}{\varepsilon_s}\right) \tag{3.6}$$

$$t^* = \frac{1}{2}\left(\frac{gR_b}{GV_{g0}}\right)\ln\left(\frac{\phi_{s\,max} - \phi_s^*}{\phi_{s\,max} - \phi_{s0}}\right) \tag{3.7}$$

Example: calcium carbonate–water slurry. A small sample of slurry with calcium carbonate in water is centrifuged in a batch solid bowl. The radius of the clarified liquid-feed interface as well as the feed-cake interface are calculated as a function of time using the above equations. The parameters of the example are given below:

G	$= 2667g$
V_{g0}	$= 1.31 \times 10^{-6}$ m/s $(5.16 \times 10^{-5}$ in/s$)$
$\phi_{s\,max}$	$= 0.26$ (with $\phi_s V_g = 0$)
ϕ_{s0}	$= 0.13$
R_p	$= 0.0508$ m (2 in)
R_b	$= 0.1016$ m (4 in)
ζ	$= 2667 \times 1.31 \times 10^{-6}(1/0.1016)t = 0.0344t$
ε_s	$= 0.52$

The results of the calculations are tabulated in Table 3.3 for various times t. As can be seen, the feed concentration as a ratio of the maximum concentration decreases from 0.5 to 0.16, during which the clarified liquid-feed interface moves radially outward whereas the feed-sediment interface grows radially inward. These two interfaces meet at $t = 7.6$ s, when the sedimentation process is completed.

TABLE 3.3 Results of Example Calculations on Transient Centrifugal Sedimentation

t, s	ζ	$\phi_s/\phi_{s\,max}$	R_s, m	R_c, m
0.0	0.0	0.50	0.051	0.102
1.0	0.034	0.46	0.053	0.1
5.0	0.173	0.29	0.067	0.095
7.6	0.261	0.16	0.091	0.091

References

1. T. Svedberg and Pedersen, "The Ultracentrifuge," 1940.
2. W. Wilkie and P. Blackburn., "Rotary Distribution Pipe Assembly," U.S. patent 5,582,742, Dec. 10, 1996.
3. H. Greenspan, "Separation of a Rotating Mixture," *J. Fluid Mech.*, vol. 127, no. 9, p. 91, 1983.
4. M. Sambuichi, H. Nakakura, K. Osasa, F. Tiller, "Theory of Batchwise Centrifugal Filtration," *AIChE J.*, vol. 33, no. 1, 1987.

Chapter

4

Continuous Sedimenting Centrifuges

The two predominant types of continuous centrifuges are disks and decanters.[1,2] They cover a wide range of applications with flow rates from less than 1 to more than 4000 L/min (0.26 to 1000+ gal/min), slurry particle sizes from 0.1 μm to 25 mm, solids concentrations of less than 0.2% to 70+% w/w, and Gs from 500 to over 15,000g. The disk centrifuge handles the small end of the size spectrum from 0.1 to 50-μm particles by using high G in the range of 4000 to 15,000g. On the other hand, the decanter handles the coarser end of the particle sizes from 1 μm to 10 mm (10,000 μm) using moderate to high G in the range of 500 to 10,000g. Also, the disk is typically used for dilute feed solids concentrations, whereas the decanter is used for both dilute as well as concentrated slurry. In the following, disk centrifuges will be discussed first, followed by decanters.

Disk Centrifuges

A popular clarification and separation centrifuge is the vertically mounted disk centrifuge shown in Fig. 4.1. The solid-liquid slurry is introduced via a stationary feed pipe along the axis of the machine, accelerated to tangential speed by a radial vane assembly and flowing through a stack of closely spaced conical disks (Fig. 4.2). Under centrifugal force the heavier solid settles against the underside of the disk surface in the disk stack and moves down toward the large end of the conical channel while the liquid phase by buoyancy flows up the conical channel and is taken out at a small radius. The solid is discharged either manually, intermittently, or continuously. Disk centrifuges are also frequently used for liquid-liquid-solid separation. In this case the

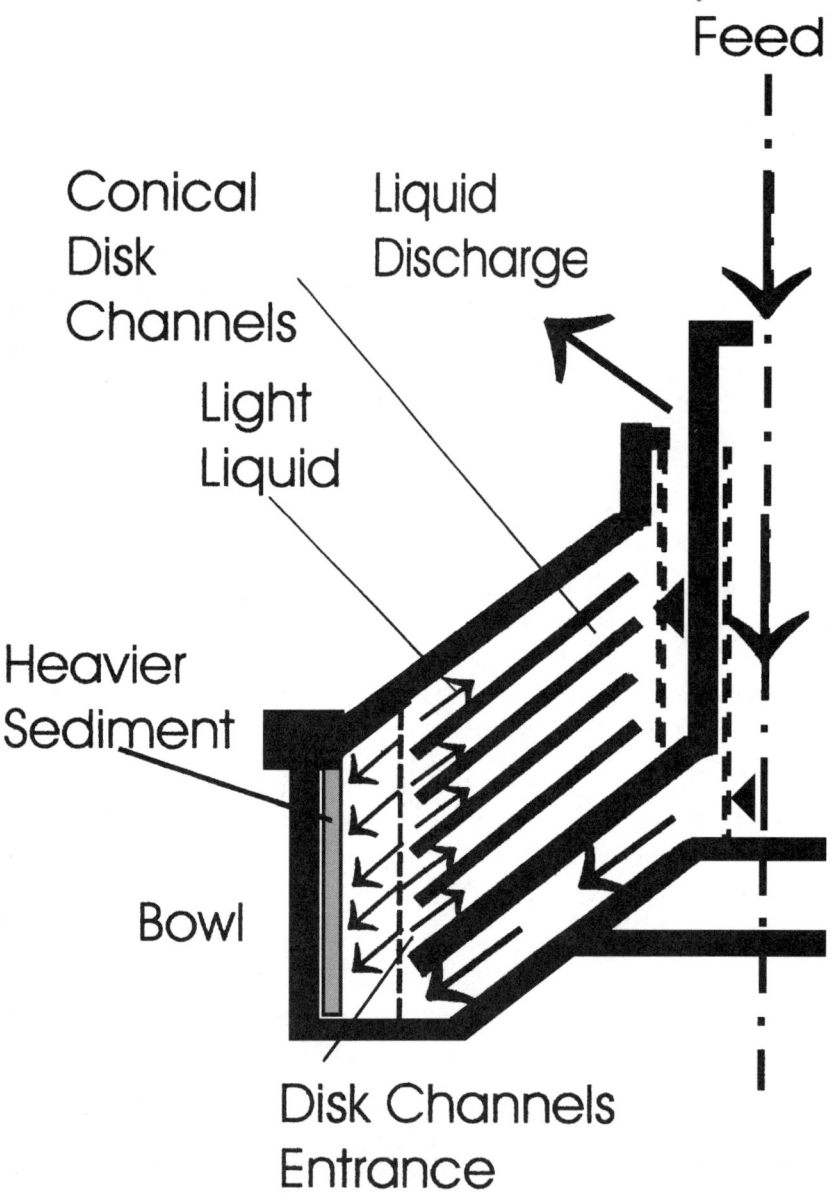

Figure 4.1 Schematic of disk for solid-liquid separation.

heavier liquid phase is discharged at a larger radius than the lighter phase, taken at a smaller discharge radius (see Fig. 4.3).

Disk geometry. The disks are truncated cones flanged at both their inner and outer diameters to provide strength and rigidity. Generally 50

Continuous Sedimenting Centrifuges 73

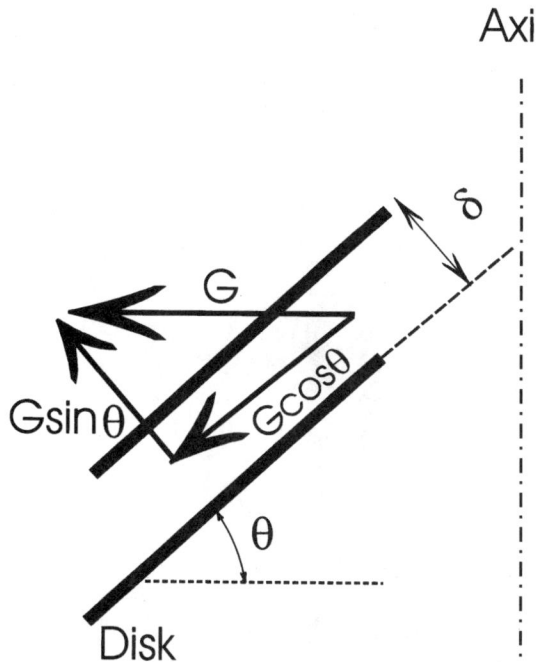

Figure 4.2 Centrifugal separation in disk stack.

to 150 disks are used, and the number of disks employed depends on the process requirements. The spacing δ between adjacent disks, measured normal to the disk surface, determines the number of parallel conical channels in a given disk stack. The closely spaced disks also reduce the maximum distance a suspended particle must settle by centrifugal force to reach the settling surface, that is, the underside surface of a disk. As illustrated in Fig. 4.2, δ is typically between 0.4 and 3 mm (0.015 and 0.125 in). This spacing also needs to be large enough to prevent oversized solids from clogging the channels. The driving force for sedimentation is $(\rho_s - \rho_L)G \sin \theta$, which is normal to the slanting disk surface and is greatly enhanced by the high G in the range of 4000–15,000g with 5000–7000g more typical. θ is the angle between the slanted disk surface and the horizontal. Another way of assessing the benefit of a disk separator is that the effective sedimentation area is greatly increased with the use of multiple disks having a total projected sedimentation area $(N - 1)A \sin \theta$, where N is the number of disks and A is the slanted area per disk.

Another important requirement for the disk is that the sediment needs to be removed continuously from the collecting surfaces by centrifugal force. As such, the sediment collection surfaces need to be tilted at a steep angle so that the sediment can slide down the inclined sur-

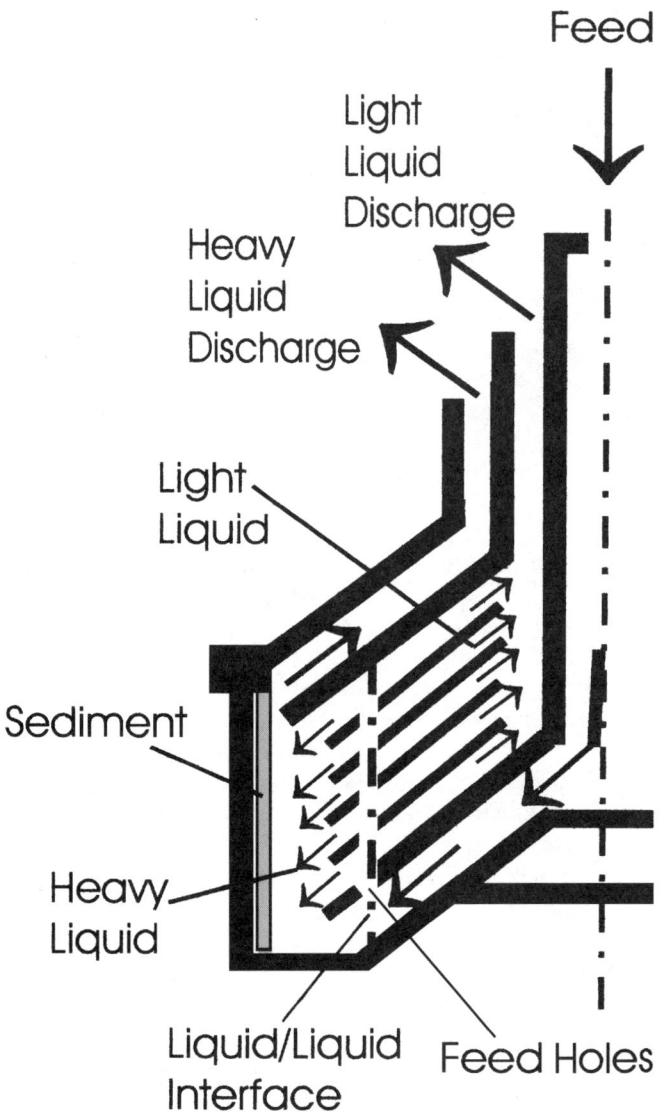

Figure 4.3 Schematic of disk equipped with feed holes in the disk stack for solid-liquid-liquid separation.

face under the component of the G force, $G \cos \theta$, and be removed continuously. This ensures the channel from being clogged by the sediment. (This is in contrast to the multibowl, which is a disk centrifuge but with $\theta = 90°$. Therefore cake solids cannot be removed and are allowed to accumulate until the channels are filled.) Based on these considerations, the angle θ made by conical disks with the horizontal is typically between 45 and 55° to ensure continuous conveyance of the sediment by the component $G \cos \theta$, while at the same time providing an increased sedimentation area $(N - 1)A \sin \theta$ (see Fig. 4.2).

For a solid-liquid-liquid separation, the feed slurry is introduced through a set of feed holes in the disk stack, as shown in Fig. 4.3. Each disk carries several feed holes spaced uniformly around the circumference. When the disks are assembled, the holes provide a continuous upward passage through which the feed slurry flows and distributes into the disk channels. For proper separation, the light–heavy-liquid interface should be located at the radius of the feed holes. The interface position is adjusted by changing the discharge diameter of the liquid phases or by adjusting the back pressure of the discharged liquid phases.

Flow pattern. It is reasonable to assume that the flow of liquid is radially inward toward the center while the flow of solids is radially outward away from the axis of rotation. This is due to sedimentation of the heavier solid at the underside of the disks, and by the component of G force the sediment slides down along the underside of the conical surface to the dirt-holding area at the periphery of the centrifuge. Concurrently the lighter liquid phase is displaced to the upper side of the conical surface by buoyancy and flows toward the exit located in the upper region of the centrifuge. It has been shown in testing that this assumed flow pattern is largely correct. However, there are further complications. A vortical flow pattern is observed in each sector between adjacent disk spacers due to the Coriolis effect. Assuming a clockwise rotation as the fluid moves radially inward relative to the rotating disk, a Coriolis force [see Eq. (2.4)] is induced on the fluid, directing it in the direction of rotation, yet perpendicular to the radially inward velocity. The flow path of the fluid thus curves inward in the direction of rotation (that is, prograde rotation), forming a vortex which reduces the effectiveness of sedimentation. This vortical flow pattern can be partially offset with the use of radial spacing bars. These azimuthal barriers counteract the Coriolis force and limit the size of the vortices.

Two-phase solid-liquid separation

In clarification or purification, the object is to obtain a pure effluent free from solids. As such, the entire length of the disk channel is utilized for the removal of solids so as to get a solid-free liquid effluent. Therefore the disk is fed at the outer diameter of the disk stack or through feed holes located at the outer periphery of the disks. This is illustrated in Fig. 4.1. The former can be achieved by blinding the feed hole of the first disk located at the bottom of the disk stack shown in Fig. 4.3.

In thickening or concentration of the underflow solids, the feed slurry is introduced through the feed holes located in the middle of the disk channel or closer to the axis (see Fig. 4.3) so as to allow ample time for the solids to concentrate or purify before discharging as underflow. This is applicable for difficult-to-compact cake. Given the feed holes are biased toward the axis of the machine, a balance is required to ensure that the loss of solids in the effluent is minimized as the clarification length of the disk channels is also reduced.

Three-phase solid-liquid-liquid separation

For three-phase separation, the solids together with two liquid phases, one light-liquid and one heavy-liquid phase, are fed to the centrifuge for separation. Under centrifugal force, each respective phase seeks its location within the bowl and conical channels to form a stable stratification with the light liquid residing closer to the axis, the heavier liquid in the middle, and the concentrated solid phase at the periphery of the bowl. There are two interfaces assuming that the two liquids are immiscible—a liquid-liquid interface and a liquid-cake interface (see Fig. 4.3). For proper operation, the liquid-liquid interface should be located at the feed holes in the conical channels. Otherwise improper separation will result, as illustrated in Fig. 4.4. A second interface is located outside the conical channels between the cake and the heavy-liquid phase. The discharge of the heavy-liquid phase is facilitated by the top disk of the disk stack, which has a slightly larger diameter to preclude feed slurry or separated light-liquid phase from entering the top conical channel (see Fig. 4.4). It also functions as a baffle for skimming only the heavy-liquid phase residing at the larger diameter. It is imperative that this channel be cleared from sediment blockage. Therefore the solid–heavy-liquid interface should be located away from the intake of this channel for proper operation with a continuous removal of the solid cake.

The feed holes are located near the axis for purification and clarification of the heavy-liquid phase. Likewise, if the object is to clarify the light phase, the holes should be located at the periphery of the disks. In both cases, the liquid-liquid interface is located at the feed holes and not at a radius, which is either too small or too large.

Liquid seal. For clarification of the light-liquid phase, the interface is positioned close to the outer diameter of the disks, but not outside the disks; otherwise the light liquid escapes with the heavy liquid in the discharge. This is further prevented by introducing to the bowl prior to feeding a sealing liquid which is compatible with the heavy liquid. This is to form an annular seal with a radius less than that of the top disk in the disk stack. Another requirement is that the sealing liquid be insoluble in the light liquid, yet soluble and compatible with the heavy phase,

Continuous Sedimenting Centrifuges 77

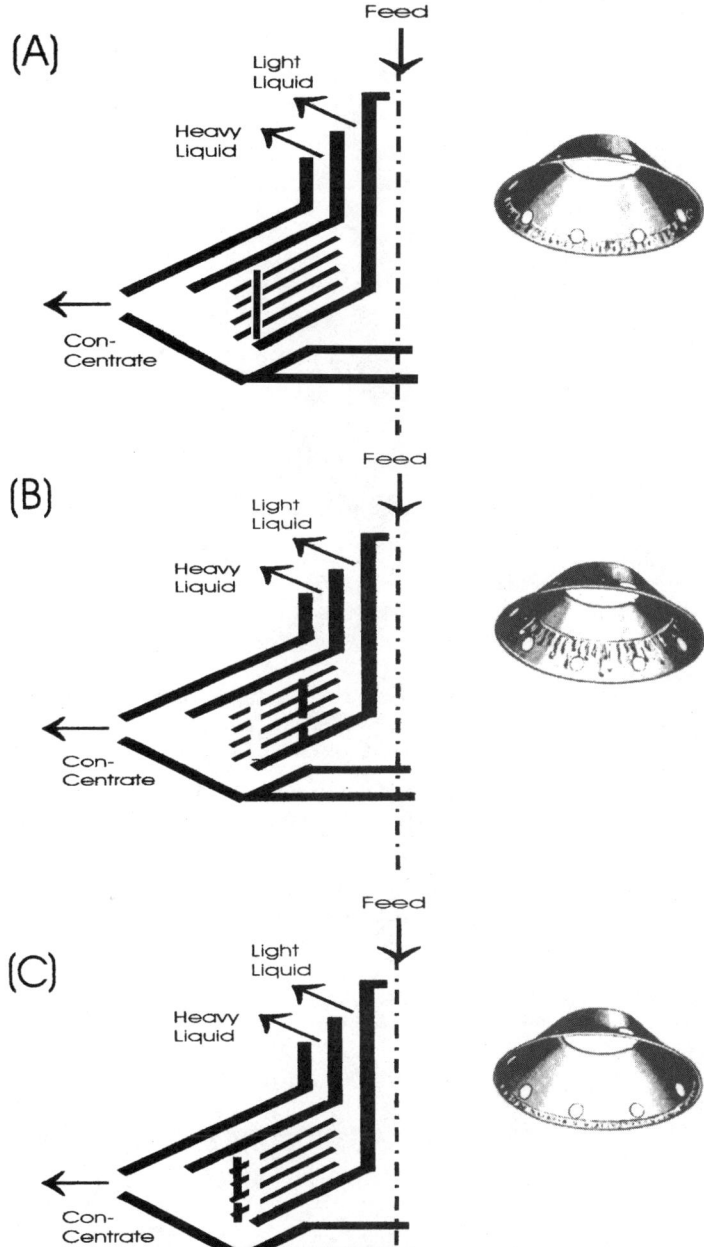

Figure 4.4 Interface setting on separation performance. (*a*) Interface correctly set at the radius of the feed holes. (*b*) Interface set at a smaller diameter compared to feed holes. The light liquid discharge contains a fraction of heavy phase. (*c*) Interface set at a larger diameter compared to feed holes. Discharged heavy phase contains light liquid phase.

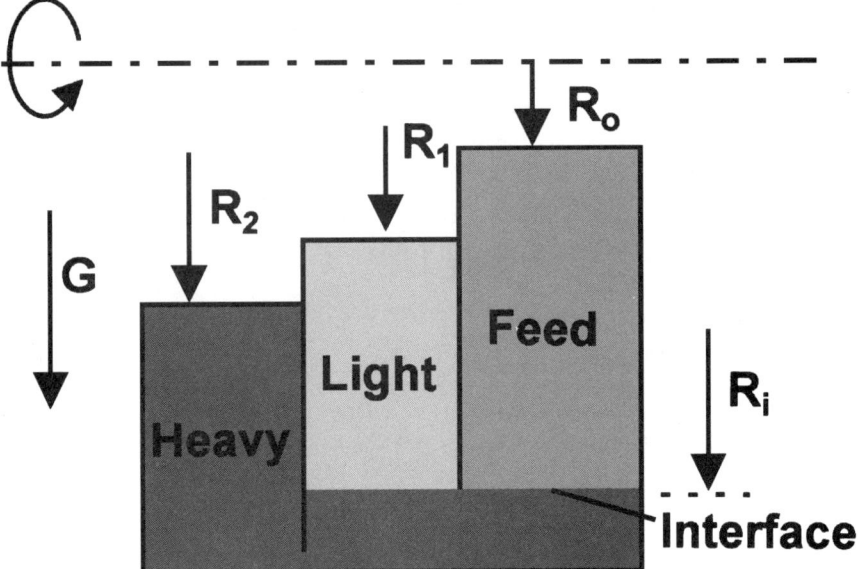

Figure 4.5 Schematic of liquid levels in liquid-liquid separator.

and preferably having a density between those of the light and heavy liquids to establish proper stratification inside the bowl.

Pressure loss. As shown by the schematic of Fig. 4.5 for the case of a two-phase liquid-liquid separation, there is a difference in pool levels between that of the feed pool at radius R_0 and that at the liquid discharge with radius R_1 (which is approximately the physical radius of the ring weir, with the exception of the dynamic liquid head needed to drive the flow over a dam). The difference in pool levels $R_1 - R_0$ represents a head loss due to flow through the disk stack, which depends on the flow rate, number and geometry of disks, and liquid viscosity. It is readily determined by $\Delta p_1 = \frac{1}{2}\Omega^2[\rho_f(R_i^2 - R_0^2) - \rho_1(R_i^2 - R_1^2)] \cong \frac{1}{2}\rho_1\Omega^2(R_1^2 - R_0^2)$, where ρ_1 is the light-liquid density, ρ_f is the feed density, and R_i is the interface between the feed and the light liquid. In other words, given R_1 is fixed by the light-liquid discharge radius, R_0 can be determined by the head loss relationship. Similarly for the heavy liquid discharged at radius R_2. With $R_2 > R_1 > R_0$ the head loss due to the heavy phase flowing through the top conical channel is then $\Delta p_2 = \frac{1}{2}\Omega^2[\rho_f(R_i^2 - R_0^2) - \rho_2(R_i^2 - R_2^2)]$. Here ρ_2 is the heavier liquid-phase density and R_i is the interface radius. Typically, the loss through the disk stack channels Δp_1 is obviously much greater than the loss through the top channel Δp_2.

Liquid discharge mechanisms

There are two liquid discharge mechanisms: (1) ring weirs and (2) centripetal pump and pairing disks.

Weirs. Circular weirs of different inner diameters are used for discharging liquids with the outer diameter held fixed by the geometry of the bowl (see Figs. 4.1 and 4.3). The discharge is over the entire circumference. The dynamic head, which is needed to drive the flow over the ring weir, is typically small under centrifugal force and increases with higher flow rates. This is by far the simplest arrangement with which liquid is discharged. With this arrangement, however, a full set of ring weirs is required to make discrete changes with rings of various inner diameters corresponding to different set pool levels.

Centripetal pump and pairing disks. The centripetal pump and the pairing disk used for disk centrifuges are shown in Fig. 4.6a and b. The centripetal pump has an impeller with finite curved conduits spaced evenly around the circumference of the pump. The opening of the conduits faces the incoming rotating flow, whereas the circumference of

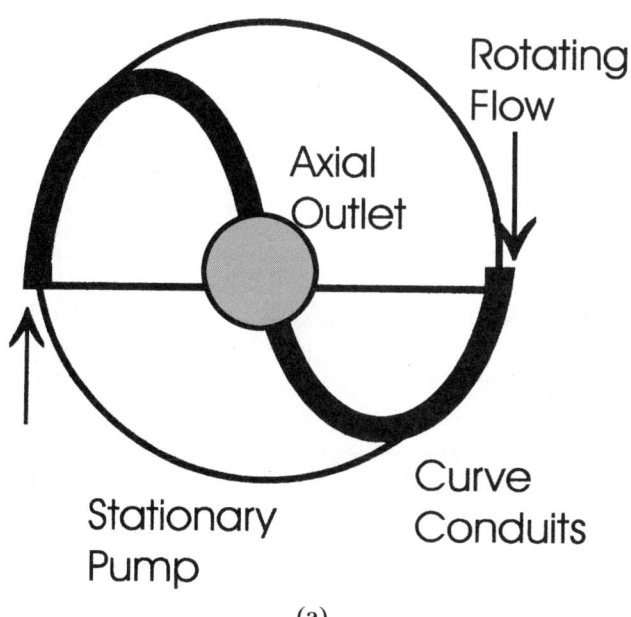

Figure 4.6 (a) Exploded view of centripetal pump showing a pair of conduits.

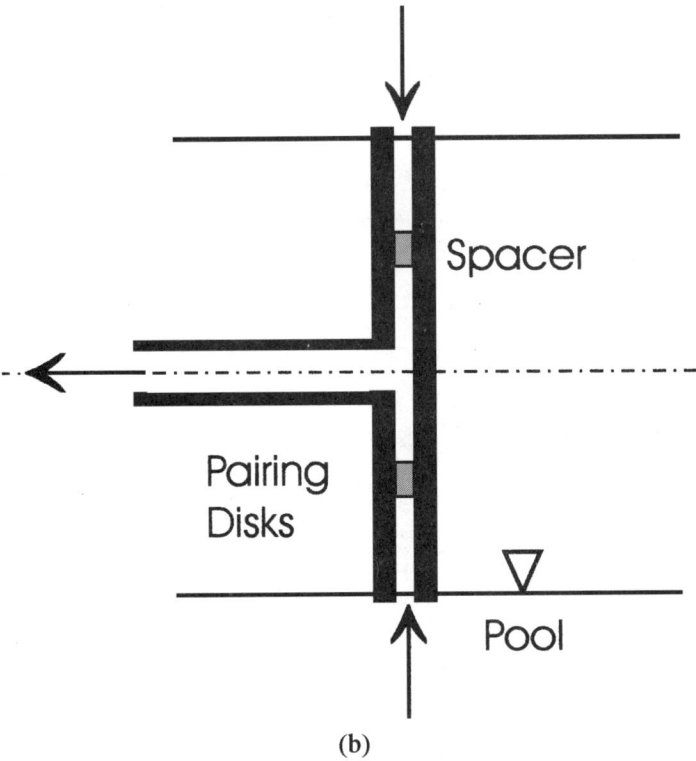

Figure 4.6 (b) Cross-sectional view of a stationary pairing disk in a rotating pool.

the pairing disk opens entirely to the flow without restriction. Both units are interconnected to an axially oriented shaft, which typically is mounted on the outer diameter of the stationary feed pipe. The rotating liquid flows tangentially into the openings of the pump or disk at their outer diameter and subsequently swirls freely or is guided into convoluted passageways in the disk or pump, which will ultimately turn into an axial direction. Simultaneously the cross-sectional area also increases. This allows the kinetic energy of the entering stream to be converted to a pressure head. This is particularly important for applications such as biological materials to avoid the high-velocity effluent liquid from foaming onto the wall of a stationary casing upon discharge. In addition to the single centripetal pump, a double centripetal design is often employed, as shown in Fig. 4.6c. This allows the light and heavy liquids to be removed at different radial and axial locations.

Skimming pipes. The skimming pipe is a variation to the systems just described. Here the pipe opening is directed to catch the rotating fluid flowing into the pipe through which a large part of the kinetic energy

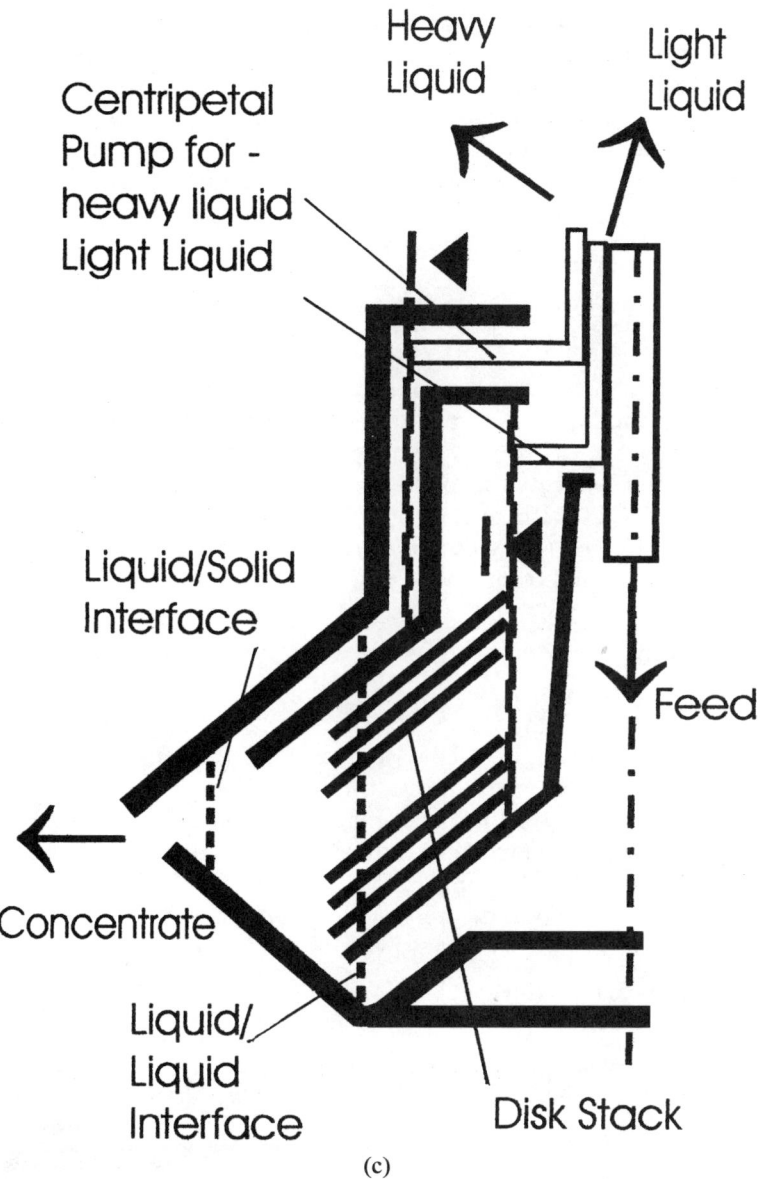

Figure 4.6 (c) Double centripetal pumps used for light- and heavy-liquid discharge.

is converted to pressure by choking the flow through the downstream back pressure valve. The effectiveness of the skimmer depends on the design which ranges from a simple standing pipe to a more streamline profiled skimming device with reduced drag on the rotating pool.

Consider a stationary skimming device such as a centripetal pump or a pipe with intake at radial location R_{intake} (Fig. 4.7). It is submerged in a pool of liquid with density ρ rotating at angular speed Ω. The pool surface at radius R_p is opened to atmospheric pressure p_{atm}. Downstream of the skimming device is a pressure gauge and a back-pressure valve which controls the flow velocity (and possible interface position). Applying the Bernoulli equation in the inertial laboratory reference frame to point 1 at radius R_{intake} in the rotating pool and also to point 2 in the flow inside the stationary pipe, we get

$$p_1 + \frac{1}{2}\rho V_1^2 = p_2 + \frac{1}{2}\rho V_2^2 + \frac{1}{2}\rho C_{loss} V_2^2 \qquad (4.1a)$$

where the pressure due to the liquid head in the stationary pipe is neglected. Because the liquid is in a solid-body rotation, hence $V_1 = \Omega R$. Also, $p_1 = 1/2 \rho \Omega^2 (R_{int}^2 - R_p^2) + p_{atm}$ where R_{int} is the intake radius and R_p the pool radius, $p_2 = p_6$ the back pressure at the valve, and the flow velocity $V_2 = Q/A$ with Q the total effluent flow rate and A the cross-sectional area at the valve. Rearranging, the back pressure (gage) can be determined from

$$p_b - p_{atm} + \frac{1}{2}\rho(\Omega R_{int})^2 - \frac{1}{2}\rho(Q/A)^2(1 + C_{loss}) + \frac{1}{2}\rho\Omega^2(R_{int}^2 - R_p^2) \qquad (4.1b)$$

The head between the skimming device and the pool surface (last term of Eq. 4.1b) is usually negligible. Therefore, the back pressure recovered is essentially the dynamic head due to the rotating pool $\frac{1}{2}\rho\Omega^2 R_{int}^2$ discounting the kinetic energy of the effluent stream, the head loss due to friction in flow through the pipe, and the head loss at intake to the skimming device, both of which are lumped in the C_{loss} term. Obviously, these head losses as well as the kinetic energy of the effluent should be made as small as possible.

Liquid-interface position

Figure 4.3 shows a schematic of the liquid-liquid interface in a disk centrifuge. Here the top disk has a larger diameter as compared to the disks in the stack. It forms a dedicated passage for skimming the heavy-liquid phase, which is located at a larger radius. Under equilibrium, the interface between the light and heavy liquids should be stationary, that is, time-invariant. Its position can be determined by balancing the hydrostatic pressure at the interface (see also Figure 4.5) assuming minimal "head loss" on flow of the light phase through the disk stack, that is,

$$p_i = \frac{1}{2}\rho_L \Omega^2 (R_i^2 - R_L^2) = \frac{1}{2}\rho_H \Omega^2 (R_i^2 - R_H^2) \qquad (4.2)$$

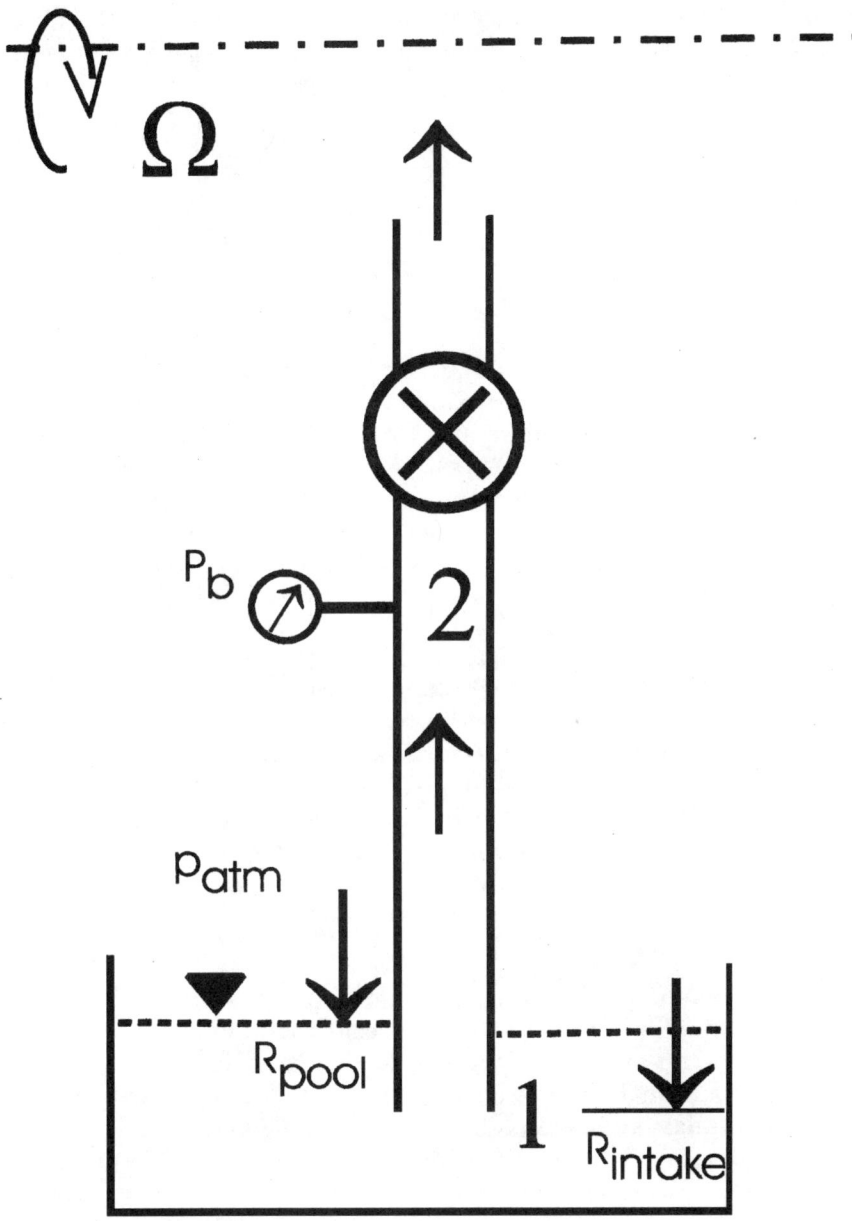

Figure 4.7 Principle of back-pressuring liquid discharge.

where R_i is the interface between the light and heavy phase, and R_L and R_H the discharge radius of the light and heavy phase, respectively.

From Equation 4.2,

$$\frac{R_i}{R_L} = \frac{\left(\frac{R_H}{R_L}\right)^2 - \frac{\rho_L}{\rho_H}}{1 - \frac{\rho_L}{\rho_H}} \qquad (4.3)$$

Figure 4.8 is a graph of Eq. (4.3), where the normalized interface position R_i/R_L is plotted against R_H/R_L with ρ_L/ρ_H as the parameter. Suppose the light-liquid-phase discharge radius is held fixed; then as the heavy-liquid discharge position is changed, the interface position is also changed accordingly. Of importance in Fig. 4.8 is that R_i increases and decreases simultaneously with R_H. A set of operating curves is also displayed in Fig. 4.8 for different ρ_L/ρ_H ratios. For example, when $\rho_L/\rho_H = 0.85$ and $R_H/R_L = 1.1$, $R_i/R_L = 1.5$. If R_i needs to be reduced to, say, 1.4 R_L, then R_H/R_L needs to be reduced to 1.07. Figure 4.8 further maps out the typical settings of the interface radius (normalized with respect to the light-liquid discharge radius) for the two important functions of the liquid-liquid separation: (1) concentration of the heavy-liquid phase and (2) purification of the light-liquid phase. The manner with which the liquid-liquid interface is determined as well as applied to practical applications is pertinent to all centrifuges used for the separation of two immiscible liquid phases. This includes the three-phase decanter to be discussed later in this chapter.

Hermetic liquid seal

Various "dynamic" sealing arrangements can be made with a stationary baffle attached conveniently to the centripetal pump and possibly a rotating baffle attached to the bowl. This is generally known as a hermetic liquid seal. Two designs are shown in Fig. 4.9a and b. In Fig. 4.9a, an annular baffle with a larger outer diameter is attached to the centripetal pump. This restricts liquid from overflowing the end plate and further forces the clarified liquid to be taken by the centripetal pump. In Fig. 4.9b, in addition to the stationary baffle ring, an annular ring is attached to the small end of the bowl and corotates with the bowl. Both the end plate and the annular ring, which rotate at the same speed, form an annular chamber with the liquid level maintained at the inner diameter of the end plate through continuous introduction and overflowing of the sealing liquid in the chamber. Furthermore, the stationary baffle, which is immersed in the pool of

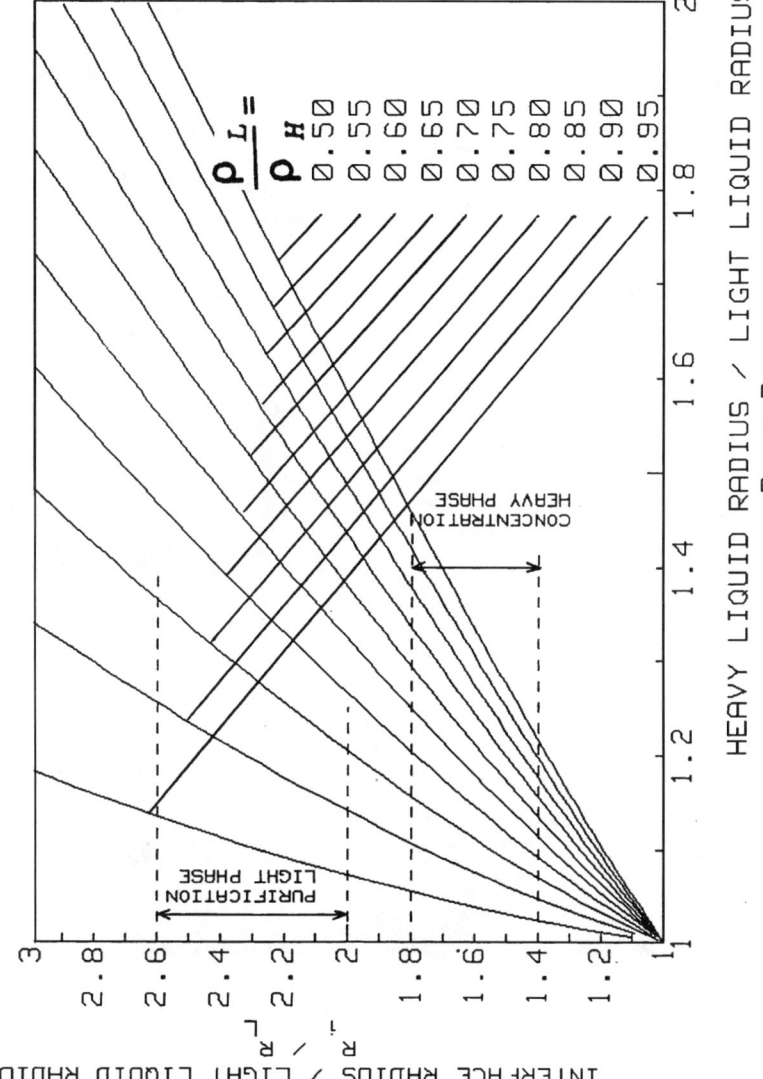

Figure 4.8 Interface as a function of light- and heavy-liquid radii.

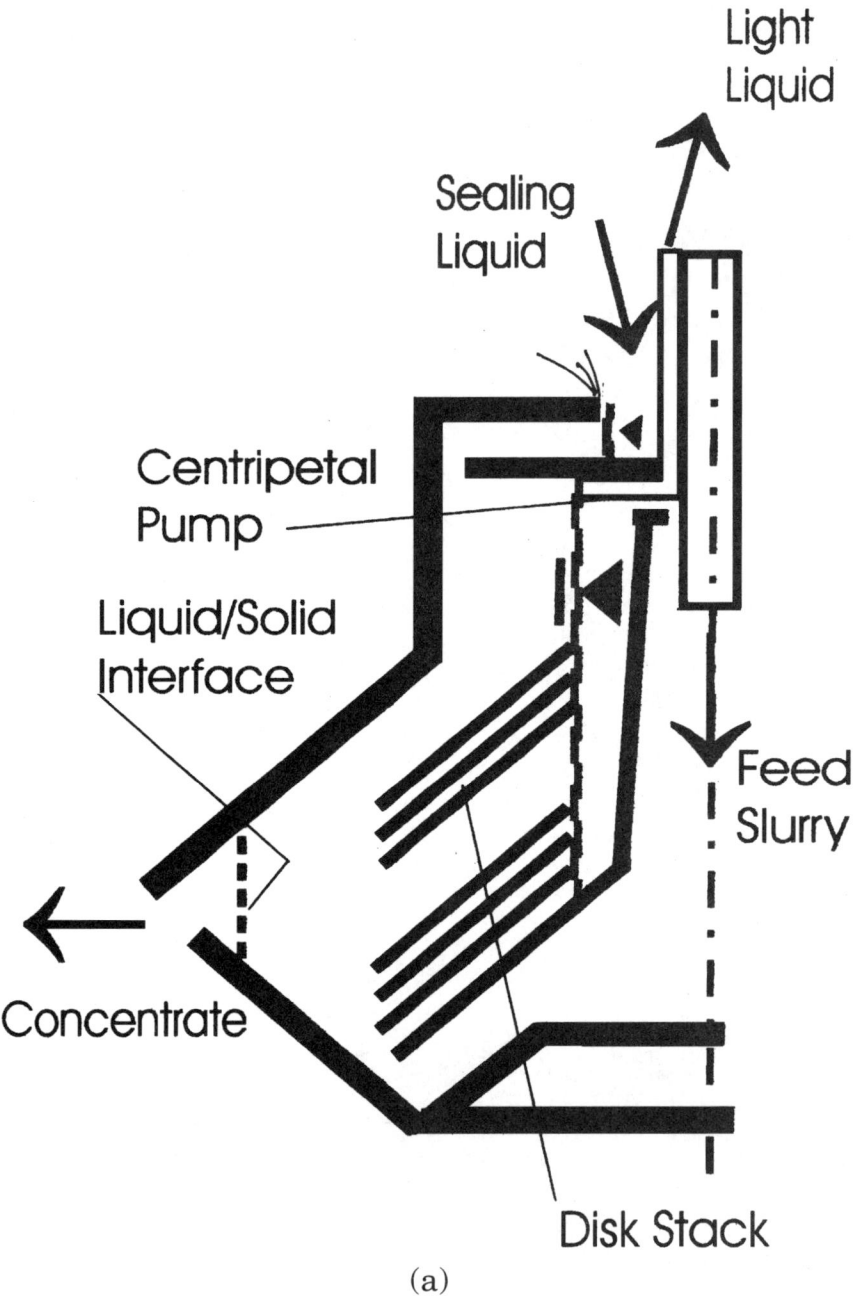

Figure 4.9 (*a*) Hermetic liquid seal design for centrate discharge.

Continuous Sedimenting Centrifuges 87

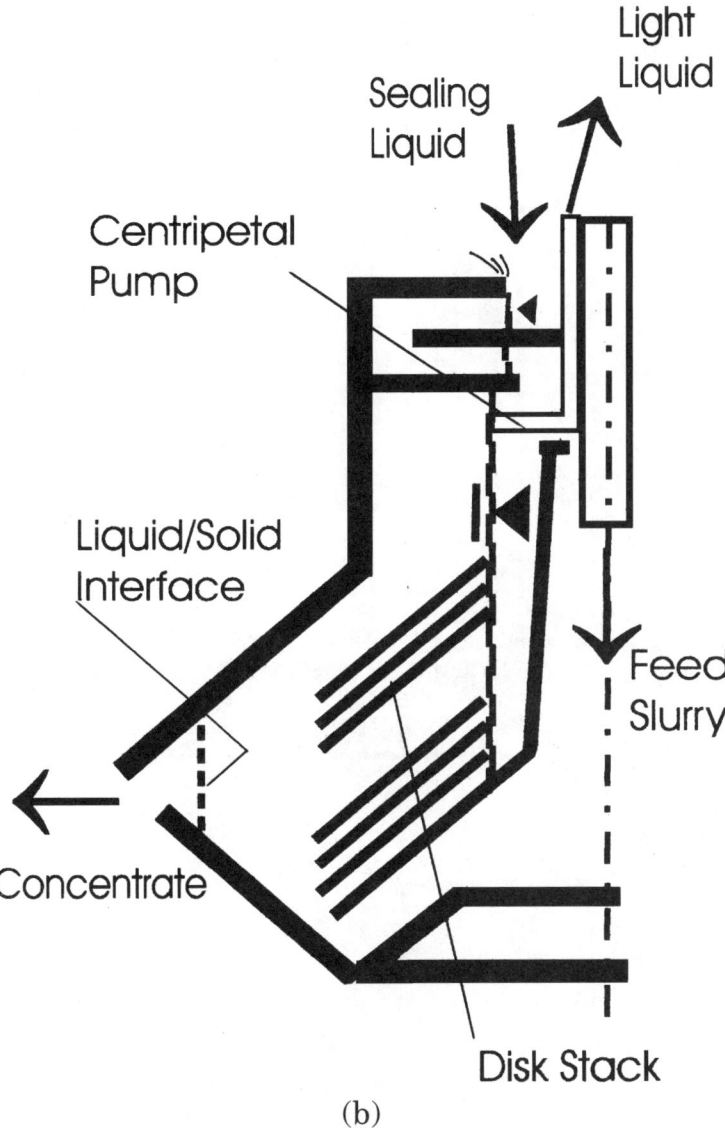

(b)

Figure 4.9 (b) Hermetic sealing chamber design for centrate discharge.

liquid in the chamber, provides a means of sealing the centrifuge interior from the ambient. In both designs the pressure of the air core inside the centrifuge can be maintained higher than that of the ambient. This is more readily attainable with the second design by virtue of the sealed chamber.

When the liquid discharge is dynamically sealed so that the pressure in the centrifuge does not communicate with the ambient, the

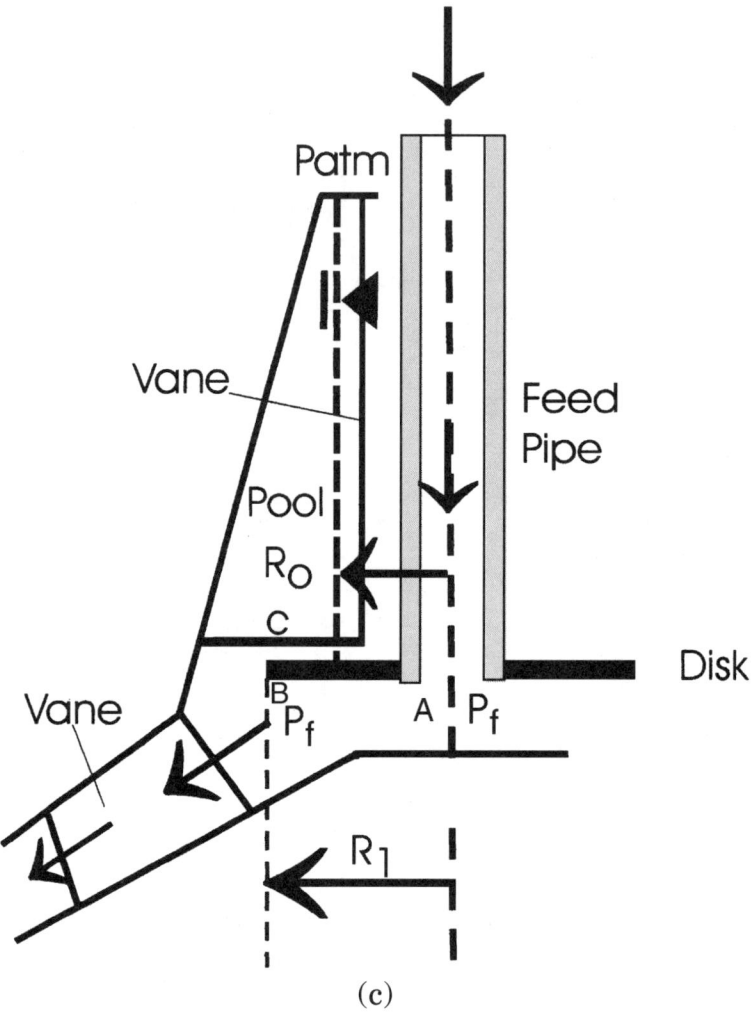

Figure 4.9 (c) Hermetic feed chamber design without mechanical seal.

feed zone can be pressurized and completely flooded with a feed-slurry pool. Thus the feed can accelerate from the axis at zero radius through a rotating feed tube with minimal shear. This requires a mechanical rotary seal placed between the stationary feed pipe and the downstream rotating feed tube which is integral with the harness of the disk stack to prevent short-circuiting of the feedback to the centrate discharge. Two advantages are evident. (1) The pressure of the feed to the centrifuge can be significantly higher compared to the ambient, which is suitable

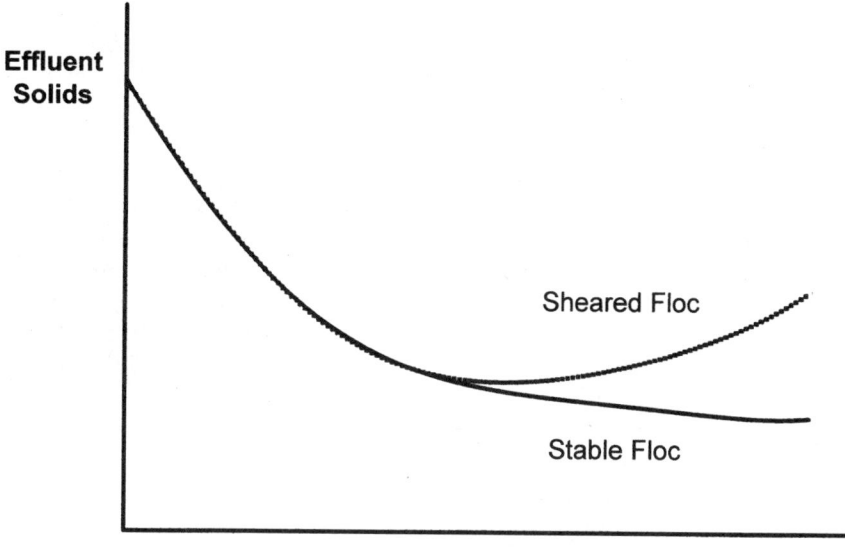

Figure 4.10 Centrate characteristics of disk with rotational speed.

for upstream pressurized processes. (2) Because the feed forms a pool at the feed zone, the feed accelerates virtually from zero radius, providing a more gentle acceleration. Otherwise unstable flocculated particles would break up by the shear action in the course of feed acceleration in high-G disk centrifuges where the underaccelerated feed slurry is abruptly accelerated to speed by the rotating vanes located further downstream of the axis. Figure 4.10 shows the improvement provided by the hermetic design over the conventional design for shear-sensitive slurry when high speed and thus high G is applied to the separation.

A hydrodynamics-based design without rotary seal is shown in Fig. 4.9c.[3] The feed is brought in from a stationary pipe. This pipe terminates at a stationary annular disk, which has a central opening interconnected to that of the feed pipe. The disk provides a baffle which shields the feed pressure p_f on one side at the feed-pipe discharge (point A) as well as at the large diameter of the disk (point B) from the opposite side of the disk. A set of longitudinal vanes is used to accelerate the annular pool to solid-body rotation such that the pressure at point C on the opposite side of the disk reaches the pressure from centrifugation $p = \frac{1}{2}\rho\Omega^2(R_1^2 - R_0^2) + p_{atm}$. Given that the feed has not acquired much angular rotational speed at point B as well as at point A, which is at the axis of the pipe, the pressure at both locations is approximately the same as that of the feed p_f. Also pressures at points B and C should be in equilibrium. Therefore $p_f = \frac{1}{2}\rho\Omega^2(R_1^2 - R_0^2) + p_{atm}$. This shows that the feed delivered to the centrifuge can be pressurized above the ambient by $\frac{1}{2}\rho\Omega^2(R_1^2 - R_0^2)$. The higher the rotation speed, the higher is this overpressure. Another means of explaining this is that in the zone with the vanes (point C) the pool is deliberately brought to speed so that the liquid has the "centrifugal weight" and therefore appears heavier. On the other hand, in the feed entrance zone under the disk there is not much angular acceleration as there are no vanes, and therefore the fluid appears lighter. In order to maintain hydrostatic equilibrium at points B and C, the lighter fluid column should be higher than the heavier fluid one. Therefore it is possible for the lighter fluid, which is the incoming feed, to fill up to the axis of the machine at point A. Further downstream of point B, the rotating vanes accelerate the feed to full tangential speed at the larger discharge radius (Fig. 4.9c). In this arrangement, as p_f increases, the pool becomes deeper with R_0 approaching the spill radius. Indeed, this sets the maximum feeding pressure at a given speed, which of course can increase further, with p_f increasing concurrently as the square of the speed until the maximum machine speed is reached.

Disk centrifuges can be classified according to the mechanism by which the cake solid is discharged. There are three discharge mechanisms for the cake:

1. *Manual discharge:* Solids accumulate in the bowl until they reach a maximum. Also when the centrate quality is impaired and the centrifuge stopped, the accumulated solids are removed manually. With this operation, the feed solids concentration should be no greater than 1% by bulk volume.

2. *Intermittent discharge or self-cleaning:* Solids are discharged periodically. Typical solids concentration can be as high as 10% by bulk volume. This method is used to discharge nonflowable cake, which cannot be discharged with nozzle disks.

3. *Continuous or nozzle discharge:* Solids are discharged continuously. Solids concentration is between 6 and 25% by bulk volume. This mechanism is used for cake that is flowable.

Manual discharge

In this configuration (see Figs. 4.1 and 4.3) the sediment is directed to an annular dirt-holding space between the outer diameter of the disk stack and the inner diameter of the bowl-shell wall, from where it can be removed manually, as required, after the rotor has been stopped and disassembled. For separating immiscible liquids, the disk stack is surmounted with a dividing conical baffle that includes a cylindrical extension at its small diameter. This leads the light-phase discharge to a different radial location than the heavy phase. The separated phases may be discharged into open covers, led to separate chambers in the rotor from which they can be removed by centripetal pumps, or taken off through rotary seals. The position of the liquid-liquid interface is usually controlled and optimized by adjusting the heavy-phase discharge radius, the value of R_H in Eq. (4.3) (see also Fig. 4.8), but in sealed-discharge models this may be controlled externally by adjusting the back pressure of the discharge lines connected upstream to the centripetal pumps or pairing disks.

The solid-wall disk rotor is supported and driven by a stiff rotating spindle. The spindle is mounted in appropriate thrust and radial bearings, and the assembly is supported and aligned in a rigid frame on which the covers are also mounted. The upper spindle bearing, nearest the rotor, is preloaded radially with springs or elastomeric cushions to provide the required degree of flexibility and dampening. The drive is usually through step-up right-angle gearing from a horizontal-shaft electric motor. This drive train is almost exclusively limited to underdriven centrifuges. It is splash-lubricated and consists of a bearing-grade bronze gear on the motor shaft driving a steel worm shell on the spindle. It is limited to the transmission of 19 kW (25 hp). On larger sizes, a V-belt drive from a vertical motor is used.

By far the largest single application of the solid-wall disk centrifuge is the separation of cream from milk at flow rates of up to 22,500 kg/h (50,000 lb/h). Other major uses are for the purification of used lubricating oils, a variety of used industrial oils, and new and used insulating oils (to raise their dielectric value); and for the continuous separation of soap stock in the process of refining vegetable and animal fats and oils.

The sizes of the solid-wall disk centrifuges range over a broad spectrum, with disk stacks from less than 150 mm (6 in) to over 500 mm (20 in) in diameter and with upward of 100 or more disks in a stack. The corresponding effective capacity for separation or clarification covers the range from a small fraction of a liter/min (0.26 gal/min) to upward of 760 L/min (200 gal/min). The volume available for holding settled solids is defined by the difference between the outer radius of the disks (or stack) R_3 and the inner radius of the bowl shell R_2 times the height of the stack L, $\pi(R_2^2 - R_3^2)L$. Since the rotor must be stopped, disassembled, and cleaned manually when this volume has been filled, the use of the solid-wall disk centrifuge is limited to applications in which the content of settled solids is low. Increased dirt-holding capacity can be obtained by:

1. Decreasing the disk stack diameter with a corresponding decrease in the area parameter Σ to be discussed

2. Increasing the bowl-wall diameter with a corresponding decrease in speed of rotation Ω to maintain the same stress in the shell
3. Increasing the stack and bowl-wall height L, limited by decreasing rotational stability above critical speed as the ellipsoid of inertia approaches a circle

The small L/D ratio of the disk centrifuge makes it feasible to slope the bowl walls to direct the solids sediment to a centrally located zone, from which their discharge can be effected, without greatly increasing the diameter and consequently the stress level in the bowl shell at a given speed of rotation. It can be shown that when the height and the diameter of the disk stack are controlled by the sloping sidewalls of the bowl shell, the sedimentation capacity of the assembly is optimized when the radius of the inner diameter of the bowl shell is

$$R_2 = \frac{4(R_3^3 - R_1^3)}{3 R_3^2} \tag{4.4}$$

where R_1 is the inner radius of the disks. Because the accumulated solids must be removed manually on a periodical basis, similar to the tubular-bowl centrifuge, this requires stopping, disassembling the bowl, and removing the disk stack. Although the individual disks rarely require cleaning, manual removal of solids is economical only when the fraction of solids in the feed is very small.

Intermittent discharge

For the intermittent-discharge disk centrifuge, solids discharge through valves or ports at the periphery of the bowl with the valves open on a timed cycle while the bowl is at speed. With the split-bowl design, the bottom of the bowl drops, exposing an annular opening where accumulated solids are discharged. The bowl closes and the cycle repeats.

Valve discharge. These centrifuges permit the accumulation of sedimented solids in the holding space between the disk stack and the bowl shell. At controlled intervals, radially disposed ports around the periphery of the largest diameter of the shell are opened by mechanical or hydraulic force acting on the valve tips that normally seal each port. This unloading action may be under timed-cycle control or triggered by a change in the clarity of the effluent stream. In one version the valve is held closed by an internal flow of supernatant liquid. When this flow is interrupted by the accumulation of sedimented solids, the valve-operating liquor drains off through a leak hole and the internal hydrostatic pressure causes the valve to open automatically.

The satisfactory operation of valve-discharge centrifuges is highly dependent on the nature of the sediment. The externally controlled valve bowls are usually applied with relatively soft fluid solids. The automatic valve bowls require the sediment to be plastic and of substantially higher density than the liquid phase. They have the further requirement that the feed flow rate be sufficiently greater than the flow rate through the open ports to restore the valve-operating liquor circuit when the solids have been cleared from the bowl.

Valve-discharge disk centrifuges are used for the control of the pulp content of fruit and vegetable juices, the separation of wool grease from wool scouring liquor, the removal of solid fats and waxes from chilled hydrocarbon solutions of vegetable and animal triglycerides and mineral oils, and the polishing of rendered animal fats. To a large extent they have been replaced by the split-bowl desludger centrifuges and the continuous nozzle-discharge centrifuges.

Split bowl. In the split-bowl configuration shown in Fig. 4.11, also commonly referred to as desludger disk or dropping bottom, a space is provided between the outer diameter of the disk stack and the sloping walls of the bowl shell for the accumulation of sedimented solids. This space is sealed by an elastomeric ring in the upper part held against the lip ring of a movable piston or sleeve in the lower part. The seal is maintained differently in various designs:

1. By spring pressure
2. By the pressure from a continuous flow of an auxiliary operating liquid (usually water) that is continuously being drained at a lower rate through peripheral leak holes
3. By a residual pool of operating liquid below the piston that serves the same function as the springs

When a suitable load of solids has accumulated, the upper and lower sections of the bowl are caused to separate, exposing an annular slot through which the solids, if of proper plasticity, discharge under centrifugal force assisted by the head of liquid in the bowl. The operating means in design 2 is an interruption in the flow of the operating liquid; in designs 1 and 3 it is a brief application of the operating liquid.

The triggering mechanism may be (1) under time-cycle control, (2) the response to a change in effluent clarity, or (3) the response to the rejection of a small recycle stream of effluent caused by the buildup of solids between the bowl wall and the outer diameter of the equivalent of a dividing cone. Both clarifier and liquid-liquid separator models are available.

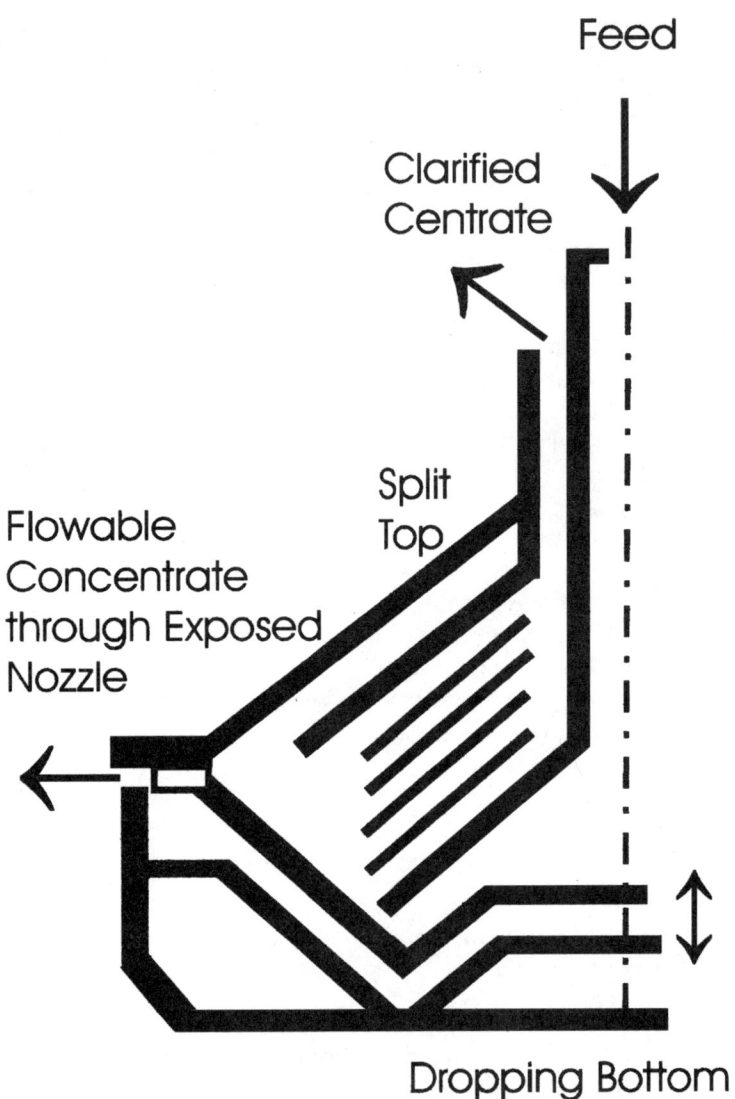

Figure 4.11 Dropping bottom intermittent-discharge disk.

The liquid discharge can be into open ring weirs, through rotary seals, or through centripetal pumps, as selected. The smaller sizes are driven by a horizontal motor through a right-angle gear train; the larger sizes, with hydraulic capacities of up to approximately 1000 L/min (264 gal/min), are underdriven by belts from an offset vertical shaft motor.

The sludge-holding space between the outer diameter of the disk stack and the bowl shell ranges from a fraction of a liter (0.25 gal) in the smallest size to approximately 25 L (6.6 gal) in the largest. The desludging action is quite rapid, requiring no more than 1 or 2 s. It is recommended that the desludging action should occur no more frequently than once a minute. This factor, combined with sludge-holding volume, volumetric feed rate, and volume fraction of solids in the feed and sediment should be considered in sizing a centrifuge for a given application. Stainless steels of the AISI 300 series, or their European equivalents, are the usual materials of construction.

Desludging centrifuges are used for the clarification and pulp content control of fruit and vegetable juices, the clarification of extracts such as coffee, the clarification of animal fats, the recovery and concentration of vegetable, milk, and single-cell proteins, and numerous other applications where the solids to be separated are soft, plastic, and not abrasive. They are frequently used in applications in which a high degree of sanitation is needed to avoid bacterial buildup and contamination. At intervals, as required, the feed is interrupted and a sanitizing solution passed through the assembly with several desludging cycles. The time for manual disassembly and cleaning is replaced by a few minutes of off-line time, without shutting down the centrifuge.

The maximum holdup time t_{max} can be roughly calculated by $t_{max} = V_{dirt}\varepsilon_s/Q_f\phi_f$, where V_{dirt} is the dirt holdup volume of the centrifuge, ε_s and ϕ_f are the solids volume fraction in the sediment and the feed, respectively, and Q_f is the feed volumetric flow rate. As an example, if $\varepsilon_s = 5\%$, $\phi_f = 1\%$, $V_{dirt} = 2$ L, and $Q_f = 200$ L/h, then $t_{max} = 3$ min.

Continuous or nozzle discharge

The mechanical complexities of the valve-discharge and split-bowl desludging centrifuges can be reduced by discharging the sediment solids continuously from the annular band through ports or nozzles in the outer perimeter of the bowl shell. Figure 4.12a shows a schematic of a solid-liquid nozzle disk, Fig. 4.12b that of a solid-liquid-liquid nozzle disk. Under steady state, the volumetric cake flow rate Q_s at discharge is determined by

$$Q_s = \left(\frac{\phi_f - \phi_e}{\varepsilon_s - \phi_e}\right)Q_f \qquad (4.5)$$

where ϕ_f, ϕ_e, and ϵ_s are the volume fraction of solids in the feed, effluent and cake, respectively, and Q_f is the volumetric feed rate.

Example: Given $\phi_f = 1\%$, $\phi_e = 0.1\%$ and $\epsilon_s = 5\%$, $Q_f = 757$ L/min (200 gal/min), number of $N_{noz} = 12$ nozzles each having 0.14 mm diameter, therefore $Q_s = 15.4$ L/min (4.1 gal/min) and the discharge velocity v $= Q_s/N_{noz}A_{noz} = 13.9$ m/s (44ft/s) where A_{noz} represents the cross-sectional area of the nozzle.

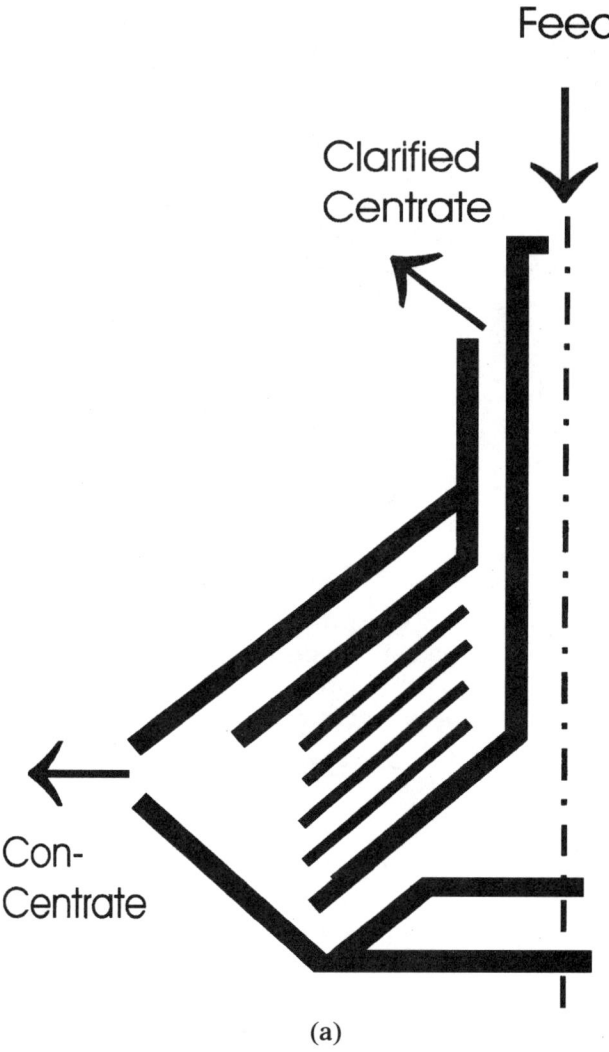

Figure 4.12 Nozzle disk. (*a*) As clarifier.

When the nozzles wear out over time such that the worn nozzles have larger effective diameter of d_{noz} = 2.0 and 2.5 mm, the flow rates of the cake increases. The flow rate is extremely sensitive to the nozzle opening. The downstream cake line needs to be back-pressured, reducing the cake flow to compensate for the worn nozzles. If necessary the worn nozzles should be replaced in accordance with the manufacturer's recommendations.

The nozzles range in size from 0.5-mm (0.02-in)-diameter openings of

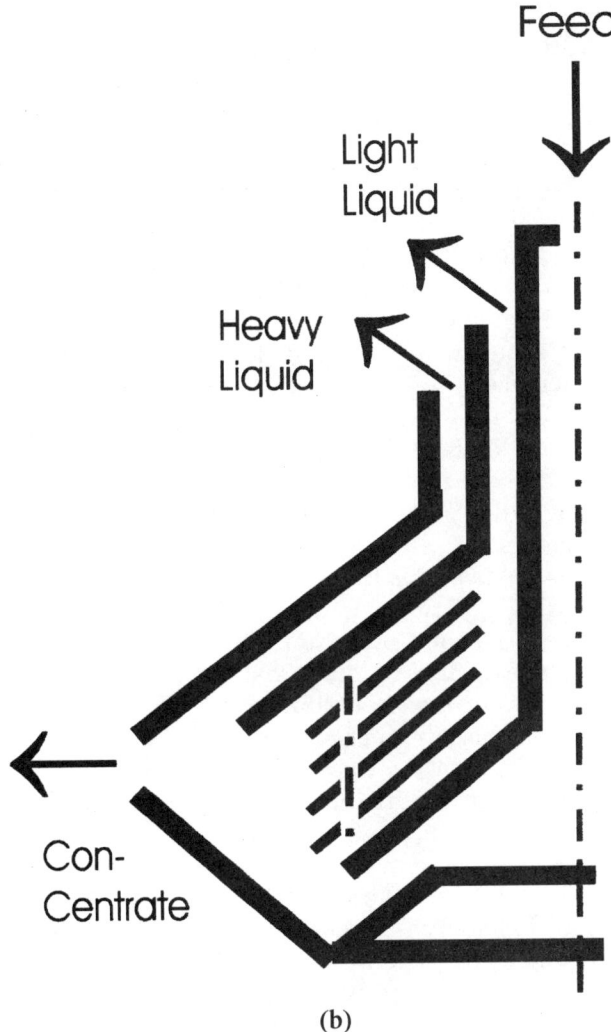

Figure 4.12 Nozzle disk. (b) As two-liquid-phase separator.

use on the smaller centrifuges to 3.2 mm (0.125 in) on the larger sizes. The design number of nozzles per centrifuge, depending on its size, ranges typically from 12 to 24. For satisfactory operation, the minimum allowable nozzle size is at least twice the diameter of the largest particle to be discharged through it. Large particles must be removed by pretreatment such as screening. The number of nozzles is controlled by the angle of repose of the sedimenting solids and must be selected so that the accumulation of solids between adjacent nozzles will not build into the disk stack and interfere with its clarification effectiveness. For once-through operation, the underflow concentration is

$$W_s = \frac{W_f m_f - W_e m_e}{m_s} \qquad (4.6)$$

where s indicates cake solids underflow, f feed, and e effluent. W denotes the solids weight fraction and m the combined (solid and liquid) mass throughput rate in kg/h (lb/h). The underflow concentration can be increased by recycling a portion of it back through the feed, that is, increasing the quantity W_f. This method necessitates a reduction in capacity to maintain the same amount of solids lost to overflow W_e or, conversely, causes an increase in W_e at the same throughput rate. In light of these concerns, better performance is obtained by recycling a portion of the nozzle discharge directly back to the dirt-holding space upstream of the nozzles instead of mixing it with the feed. In effect, this preloads the nozzle with underflow that has already passed through it and reduces the net m_s taken away from the system. (Similar material balance holds if m is taken to be the volumetric flow rate and W the volume fraction of the solids in each stream.)

When used as a liquid-liquid-solid separator, the conventional nozzle-discharge centrifuge shown in Fig. 4.12b has one characteristic that is different from the other disk centrifuges described. The volume of the separated heavy-phase liquid must be great enough to satisfy the flow requirement through the fixed size and number of nozzles, with sufficient excess to maintain the hydrostatic balance between the liquid phases. One modification permits the addition of a supplementary stream of the heavy phase behind the ring dam that controls the liquid-liquid interface position inside the bowl. Any excess beyond that required to satisfy the nozzles and maintain this interface position is automatically rejected.

To conserve the energy required to drive the nozzle-discharge centrifuge, the nozzles are directed "away" from the direction of rotation of the bowl. The flow from them is approximately tangential to the outer perimeter of the bowl shell. If the relative velocity of discharge with respect to the rotating bowl is comparable in magnitude to the peripheral velocity of the bowl, much of the energy required to accelerate the nozzle-discharge liquid to the peripheral velocity of the bowl is recovered as in a reaction turbine. The discharge absolute velocity of the cake and thus the kinetic energy are reduced to a minimum. This further reduces foaming for a liquid effluent with low surface tension as in food application.

A cross section through the nozzles in the plane perpendicular to the axis of the machine reveals that the cake solids typically form an angle of repose also laterally. This creates pockets of dead space. To avoid buildup of cake solids in these dead spaces, which if not properly distributed could lead to unbalance, wall inserts are placed in between nozzles to fill these dead spaces.

Nozzle-discharge centrifuges are usually constructed of AISI 300 series stainless steels. The discharge nozzles, being subject to erosion from the very high velocity through them (on the order of up to 150 m/s [492 ft/s]), are constructed of hard abrasion-resistant materials, such as tungsten carbide in an appropriate matrix, fused boron carbide, or fused Al_2O_3 (synthetic sapphire), and are easily replaceable. When the nozzles are worn, depending on the process, there is excessive underflow rate, resulting in poor cake quality. When used for the classification of kaolin, with worn nozzles the recovery of the fine products of 1 and 2 μm (in liquid effluent) is reduced significantly as the product also escapes with the coarser reject sediment through the nozzles. In operation, the condition of the nozzles is closely monitored by mechanical and process measurements, and a reduction of solids loading upstream, where possible, will alleviate the problem.

The relative separation or clarification capacity of the nozzle-discharge centrifuge can be calculated from the parameters of its disk stack in accordance with equations developed in Chap. 6. The resultant area function applies only to discharge overflow in the case of a clarification application. The gross feed rate is the sum of this overflow and the underflow through the nozzles.

Disk centrifuge with light-phase skimmer

In certain separation applications, the separated light phase is too viscous and gummy to flow over the annular weir at a discharge radius R_L. The discharge can be facilitated on disk-centrifuge separators by incorporating a stationary light-phase skimmer or centripetal pump into the design.

Typical applications are the concentration and recovery of high-butterfat-content cream cheese from hot mix, the separation of crystalline waxes and sterol from chlorinated hydrocarbon solutions, and the separation of unhydrolyzed fat and protein from dextrose (corn syrup). Since these materials tend to plug the disks, it is conventional to establish the phase interface inside the inner radius R_1 of a special stack.

Continuous Solid-Bowl Decanter Centrifuges

The solid-bowl decanter consists of a solid-wall bowl inside of which a helical screw (or scroll) conveyor rotates at a slightly different speed (the differential speed, commonly known as Δ) either faster or slower. An assembly schematic is shown in Fig. 4.13a. The rotor is conical or conical-cylindrical in shape, as illustrated in Fig. 4.13b, for solid-liquid

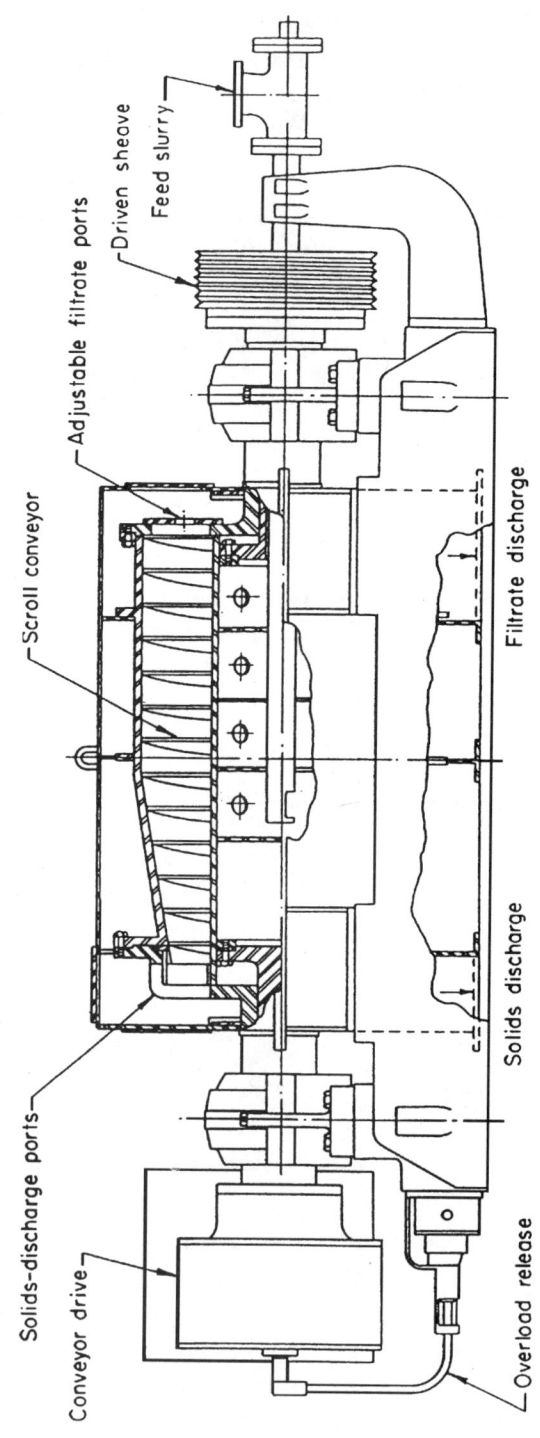

Figure 4.13 Decanter solid-bowl layout. (*a*) Assembly. (*Courtesy of Bird Machine Company.*)

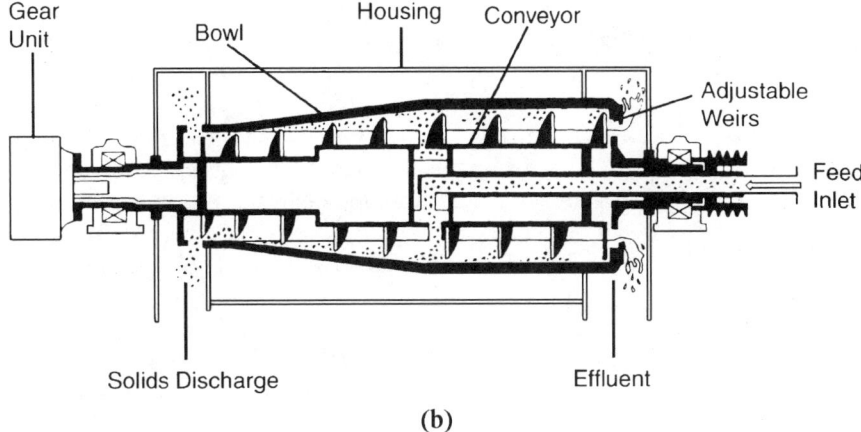

(b)

Figure 4.13 Decanter solid-bowl layout. (b) Rotor layout. (*Courtesy of Bird Machine Company.*)

separation. While Fig. 4.13 shows the decanter design for the separation of solids from liquid, the modified version shown in Fig. 4.14 allows for a three-phase separation (two liquid and one solid phases), where the light- and heavy-liquid phases are taken off at two different discharge radii. In either case the assembly is rotated about a horizontal axis (in limited applications on a vertical axis) to develop the required centrifugal force. In some special applications the centrifuge is mounted vertically with the weight of the rotating assembly supported by a single bearing at the bottom, or with the entire machine suspended from the top. The former provides a good sealing surface at the bearing for high-pressure applications.

The slurry to be centrifuged is delivered by pump or liquid head

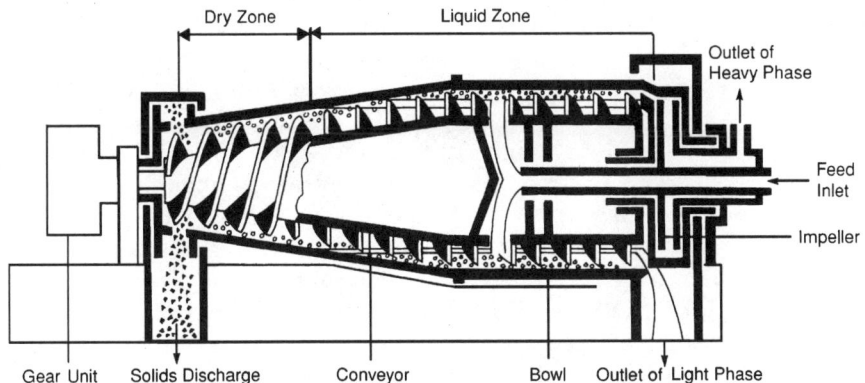

Figure 4.14 Three-phase decanter. (*Courtesy of Bird Machine Company.*)

through a stationary feed tube to the rotating feed chamber (Fig. 4.13b), where it is accelerated before discharge through a set of feed ports to the separating pool. As it travels along the helix of the conveyor toward the liquid discharge, the slurry is subject to centrifugal force where the solids settle out against the bowl wall. The cake adjacent to the bowl wall is transported by the differential speed from the cylinder up the cone, also known as the beach. The half-cone or beach angle is generally between 5 and 20°. In the cylindrical clarifier and at the beginning of the beach section, the cake is immersed in the liquid pool. Liquid buoyancy reduces the weight of the cake under centrifugal gravity, resulting in a lower effective weight and consequently also a lower conveyance torque. Further up the beach, the cake emerges above the liquid pool and moves along the dry beach. Here the buoyancy force is absent, resulting in higher conveyance torque from the full weight of the cake. However, it is also in this section that the expressed liquid from the cake is drained back to the pool. While the centrifugal force helps dewater the cake, it also hinders the transport of the cake in the dry beach. Therefore a balance in cake conveyance and cake dewatering is the key in setting the pool and the G force for a given application. Also, clarification requirements of the pool may further influence this decision.

The clarified liquid overflows at the cylindrical end of the machine, similar to flow over a dam. In general the feed slurry is introduced near the intersection of the cylindrical and conical sections of the rotor. The effective clarifying length L is the distance from the liquid overflow discharge ports to the point where the pool intersects the dry beach. Conservatively one can take the distance of the cylinder or the distance from the feed ports to the overflow ports.

Conveyor drive and assembly

The differential speed between the bowl and the conveyor is obtained through a two-stage planetary gear box or a cyclo gear box. Typically the conveyor rotates more slowly than the bowl with a two-stage gear box (it rotates faster with a three-stage planetary gear box, and slower with a four-stage box, and so on). In contrast the conveyor rotates faster with a cyclo gear box. The input shaft, the pinion, of the gear or cyclo gear box may be held stationary, in which case the differential is fixed. In other applications the speed is variable, with the pinion to the planetary gear or cyclo box connected to an electric (variable-frequency ac or dc) motor or a hydraulic motor. Alternatively, a rotating hydraulic motor directly drives the conveyor after receiving pressurized hydraulic fluid through a rotating fluid coupling. With any of these arrangements, the resulting differential speed Δ is typically less than 2–5% of the bowl speed Ω_b. (For example, with a locked pinion and a gear ratio gr = 20:1

the differential speed is 5% of the bowl speed given $\Delta/\Omega_b = 1/\text{gr.}$) In some applications where the retention time of the cake is desired, the differential speed can be as low as a small fraction of a percent of the bowl speed down to 0.2 rpm. Very high conveyance torque and possible dirty centrate are expected under this low differential operation.

With reference to Fig. 4.13b, the stationary feed pipe passes through the driven pulley and the bore of the hub to deliver the feed slurry into a feed zone inside the conveyor. The feed pipe can be brought in from either the conical end or the cylindrical end of the machine. The rotating assembly is housed in a casing, with separate hoppers taking the respective separated phases to the appropriate discharge. Tongue-and-groove seals are often used as case seals to prevent cross flow of the already separated phases. It is also important that the hoppers be properly vented to avoid pressure gradient from building up inside the casing, which might trigger flow of the separated phases. A rigid frame serves to support and align the assembly between appropriate bearings. The drive is from an offset motor and pulley to the driven pulley through six to eight V-belts. For best performance the horizontal model should be mounted on vibration isolators in coiled spring modules confined by adjustable side snubber plates or visco dampers. The vertical model includes its self-contained isolators, which work on similar principles.

Capacity and G

Table 4.1 shows some typical data of decanters with their sizes, rotation speeds, Gs, liquid and solids throughputs, and motor sizes. As shown in Chap. 2, the smaller the decanter, the lower the throughput yet the higher the G to meet the separation process requirements, and vice versa.

TABLE 4.1 Typical Decanter (Solid-Bowl) Centrifuges

Bowl diameter, inches (mm)	Speed, rpm	Maximum, G/g	Maximum liquid throughput, gal/min (m³/h)	Solids throughput (dry basis), tons/h (t/h)	Typical motor size, hp (kW)
6 (150)	8000	5500	10(44)	0.03–0.25 (0.03–0.23)	5 (4)
14 (350)	4000	3200	75(330)	0.5–3.5 (0.45–3.2)	30–75 (22–56)
18 (450)	4000	4000	100(440)	0.5–5 (0.45–4.5)	30–100 (22–75)
24 (600)	3000	3100	250(1100)	2.5–12 (2.3–11)	60–125 (45–93)
30 (750)	2700	3100	350(1540)	3–15 (2.7–14)	125–300 (93–224)
36 (900)	1650	1400	500(2200)	10–25 (9–23)	200–500 (149–373)
44 (1100)	1600	1600	700(3080)	15–50 (14–45)	200–500 (149–373)
54 (1350)	700	370	800(3520)	50–120 (45–110)	150–400 (112–300)

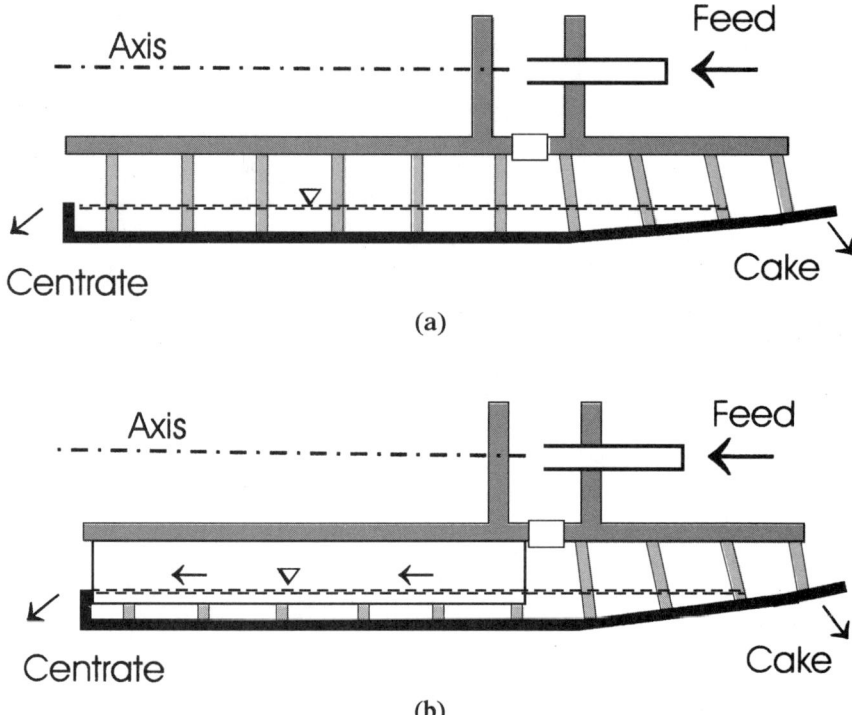

Figure 4.15 (a) Solid-blade decanter. (b) Ribbon-blade decanter.

Solid-blade and ribbon-blade designs

With the solid-blade design shown in Fig. 4.15a the clarified liquid flow follows the helical channel of the conveyor in the cylinder toward the effluent discharged while the cake solid is conveyed countercurrently toward the beach, again following the helix. When the solid throughput is high, the cake thickness is also deep, resulting in possible entrainment of the already settled solid by the fast-moving effluent stream. This is especially pronounced with a shallow-pool operation. With the ribbon-blade design, the conveyor blade height should be as tall as the thickest sediment, and this ribbon blade is supported by either radial poles or axial vanes (Fig. 4.15b). The clarified liquid is allowed to flow axially along the surface of the pool from the feed point to the overflow ports. This reduces entrainment with the cake solids as they both flow through separate paths. Given the more open cross-sectional area, the effluent liquid velocity is also reduced significantly. The ratio of the average velocity u of the solid-blade conveyor design to that of the ribbon-blade design can be shown to be

$$\frac{u_{\text{Sb}}}{u_{\text{Rb}}} = \frac{R_b}{R_p} \frac{L_{\text{lead}}}{2\pi R_b} = \frac{R_b}{R_p} \frac{1}{\tan \alpha}$$

where R_p is the pool surface radius, R_b the radius of the bowl wall, L_{lead} the lead, and α the helix angle. The latter two variables are defined in the next section. This velocity ratio is typically between 5 and 10.

The ribbon-blade design is particularly useful with high flow rates, or for small machines where the velocity of the clarified liquid is significant, or for multiple-lead arrangements where the helical cross section is reduced and effluent liquid velocity is high. Independent of the blade design, or even for the case without a conveyor, it has been shown from testing and flow observation that the liquid flow is confined in a small boundary layer adjacent to the pool surface so that the thickness and velocity of this layer depends on the total volumetric rate. The total available pool depth from the pool surface to the bowl wall must be deep enough to accommodate this moving layer and the loading of solids being conveyed in the opposite direction without any interaction or obstruction at their interface. Solid blade designs have also been modified to provide axial flow cut-out windows on the blade surface near the conveyor hub.

Lead design

The conveyor lead is defined geometrically by the pitch, which is also loosely referred to as the lead length, or simply lead L_{lead}. The helix angle α for a "developed" or "unwrapped" conveyor is then

$$\alpha = \tan^{-1} \frac{L_{\text{lead}}}{\pi D} \tag{4.7}$$

where D is the bowl diameter. The helix angle is constant for the cylindrical bowl section, but increases in the beach section as the bowl diameter decreases toward the cake discharge for a fixed lead design.

Multiple leads, that is, two (double), three (triple), or four (quadruple) leads, are used for granular cake solids which exhibit an angle of repose. The use of multiple leads cuts down the cake height and thereby reduces the distance through which the liquid moisture has to drain, resulting in drier cake. For fluid or pastelike cake, which does not exhibit this behavior, multiple leads are useless as the cake surface flattens out under G. Multiple leads only increase the viscous frictional drag on the conveyor from the cake. Therefore single leads should be used for fluidlike cake with negligible angle of repose and multiple leads for granular cake that has a finite angle of repose.

There are three possible lead designs and operating conditions which dictate cake conveyance along one direction versus the other. It

TABLE 4.2 Cake Conveyance Mechanics

Correct setup: cake conveys toward beach solid discharge end	Incorrect setup: cake conveys toward liquid discharge end
Right-hand lead Bowl faster Rotation clockwise from liquid discharge end	Left-hand lead Bowl faster Rotation clockwise from liquid discharge end
Right-hand lead Bowl slower Rotation anticlockwise from liquid discharge end	Left-hand lead Bowl slower Rotation anticlockwise from liquid discharge end
Left-hand lead Bowl slower Rotation clockwise from liquid discharge end	Right-hand lead Bowl slower Rotation clockwise from liquid discharge end
Left-hand lead Bowl faster Rotation anticlockwise from liquid discharge end	Right-hand lead Bowl faster Rotation anticlockwise from liquid discharge end

is obvious that the cake needs to move toward the beach for proper operation, and failure of meeting this requirement results in no cake discharge, gradual accumulation of solids in the bowl, discharge of solids in the overflow, or "peak out" on the conveyance torque of the machine. As such it is useful to know the lead design and the associate operation to ensure that the cake is conveyed in the proper direction. The three influential factors are (1) left-hand versus right-hand lead, (2) conveyor slower than bowl versus conveyor faster than bowl, and (3) direction of rotation, clockwise versus anticlockwise.

Each design variable has two possibilities, and therefore there are a total of eight permutations, four of which allow the cake to convey in a given direction and the remaining four in the opposite direction. The design and process engineer needs to be aware that if the choices are selected incorrectly, the cake will not convey toward the beach, resulting in no cake discharge and possibly other serious consequences. The possible combinations for cake conveyance are summarized in Table 4.2.

Beach

Dry beach. For most applications the liquid-discharge radius is greater than the solid-discharge radius, so that the cake is conveyed over a dry beach in the conical section of the bowl. Here the moisture of the cake can further drain under the centrifugal gravity G, or the cake can be washed on a limited basis before reaching the solids discharge ports. When it becomes difficult for the cake to convey from the below-pool to the above-pool region (dry beach) in the beach section, it may be desirable to raise the pool level closer to, or at times even above the spillover (that is, the solids discharge level). This provides a buoyancy effect on the cake solids to facilitate the "lift" from the bowl wall to the conical beach discharge against the opposing G force. Unfortunately, raising the pool reduces the dry beach with consequence of wetter cake. Another passive means is to reduce the G force so that the resistance force against cake conveyance is reduced. This has a negative impact on the separation process since it reduces (1) the clarification efficiency and (2) the liquid drainage from the cake back to the pool.

Mechanical conveyance. The cake is not transported directly along the beach, but climbs on a much shallower incline along the helix, similar to a car negotiating up a spiral mountain road. The effective incline angle is known as the climb angle γ, which is very much less than the beach angle ß. From Fig. 4.16a, it can be shown that

$$\gamma = \alpha ß = \frac{L_{\text{lead}}}{\pi D} ß \qquad (4.8)$$

Here α, ß, and γ are expressed in radians for $\gamma = \alpha ß$; γ and ß are either in degrees or in radians for $\gamma = (L/\pi D)ß$. For example, when $L_{\text{lead}} = 4$ in (102 mm), $D = 18$ in (457 mm), and ß = 10°, then $\gamma = 0.71°$. If the lead is 6 in (152 mm), then the climb angle increases to 1.06°. The resistance force to cake conveyance is due to the component of the G force, $\rho_{\text{eff}} G \sin ß$ (Fig. 4.16b), where $\rho_{\text{eff}} = \rho_c - \rho$ when the cake is below the

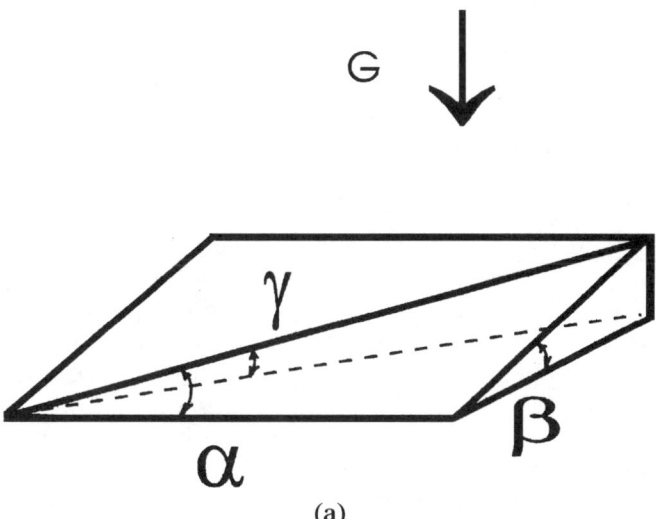

Figure 4.16 (a) Climb angle.

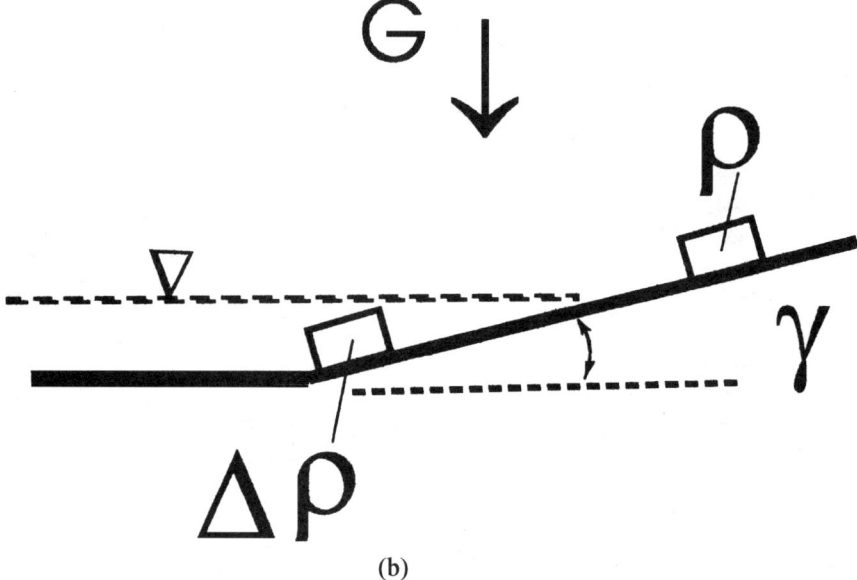

Figure 4.16 (b) Pool effect on cake conveyance.

pool and $\rho_{\text{eff}} = \rho_c$ when the cake is above the pool. (Note ρ is the liquid density and ρ_c the bulk density of the cake.) Therefore if the G force cannot be compromised because of the reasons discussed, the climb angle should be decreased, which effectively requires a smaller lead or a smaller beach angle. Often for difficult-to-convey cake solids the beach angle is made small, typically 5–8°. Unfortunately, with reduced beach angle the beach takes up a significant fraction of the overall length of the decanter for a given spillover of the beach.

Alternatively, instead of altering the design to a small climb angle during operation, the differential speed can be increased to enhance conveyance, leading to smaller cake height.

Hydraulic assist. The foregoing discussion pertains to material that requires a dry beach for liquid drainage from the cake. This is typical of incompactible drainable cake. For compactible and nondrainable cake, where cake dryness is achieved by compaction even under the pool level, circular dip weirs (or baffles) are often used, where the dip weir allows a differential height of the liquid level across it, a deeper pool level upstream of the weir, and a slightly shallower pool downstream. This is shown in Fig. 4.17a for a circular dip weir, in Fig. 4.17b for a conical dip weir, and in Fig. 4.17c for a cake baffle. Typically, the pool is brought above spillover downstream of the weir to facilitate discharge. The actual pool upstream of the weir can be 3

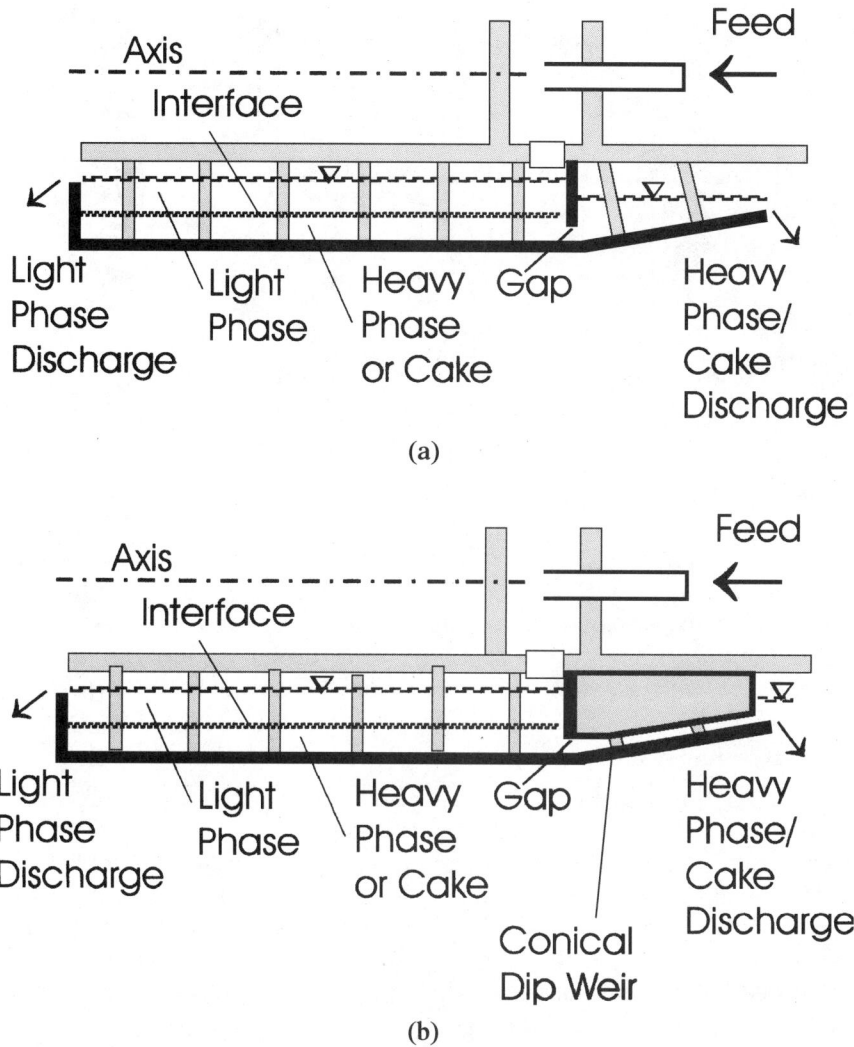

Figure 4.17 (a) Circular dip weir. (b) Conical dip weir.

to 25 mm (0.13 to 1 in) above the physical spillover. This provides a large driving force due to the dynamic head under the centrifugal field. This hydraulic force is used together with the differential speed to convey the cake solids up the beach, overcoming the resistance due to the dip weir and the component of the G force acting against cake conveyance. A delicate balance is in order such that the driving force, especially that due to the hydraulic head, does not overpower the combined resistance which otherwise may lead to "washout" of the

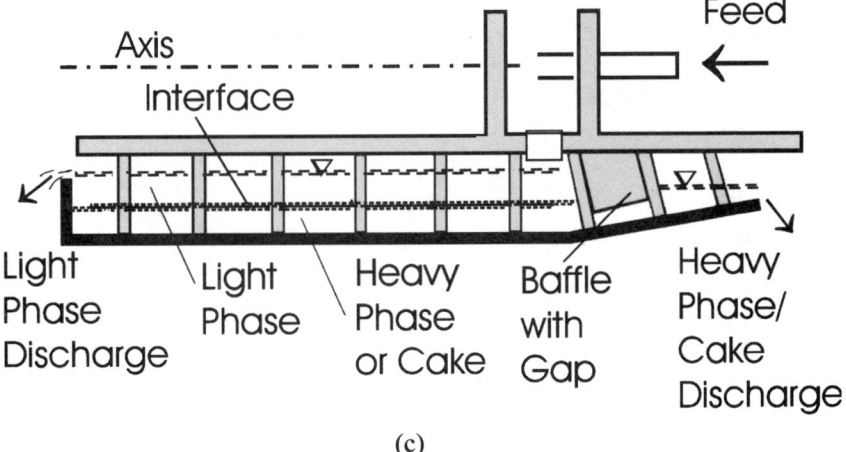

Figure 4.17 (c) Axial baffle.

cake at the beach discharge. The gap under the weir can be anywhere between 6 and 50 mm (0.25 and 2.0 in), depending on the flow rate and the nature of the cake solids. It must be noted that once a dip weir is installed, moisture expressed from the cake downstream of the weir in the beach section cannot be returned back to the pool.

Countercurrent and concurrent designs

In certain applications the time required to compact the solids to equilibrium concentration at a given centrifugal force may be relatively long compared with the time required to remove them from the moving liquid layer. One design introduces the feed next to the hub furthest from the solids discharge ports, that is, at the end of the cylindrical section of the machine. Therefore the entire bowl length is available for the compaction of the cake, and the clarified liquid layer is taken off by a set of effluent tubes spaced out circumferentially at the center diameter of the conveyor hub with opening located near the cylindrical and conical intersection (Fig. 4.18b). This concurrent flow design, where feed and settled solids are moving in the same direction, is in contrast with the previous discussion so far where the feed stream flows in the direction opposite to the movement of the cake, establishing a countercurrent flow (Fig. 4.18a).

Operating variables

For a solid-bowl decanter the operating variables are G force, differential speed, and pool level. The following summarizes the trend one would expect when these variables are changed independently.

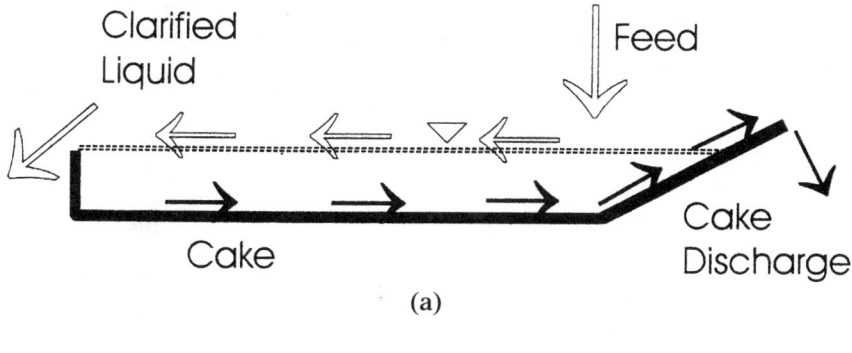

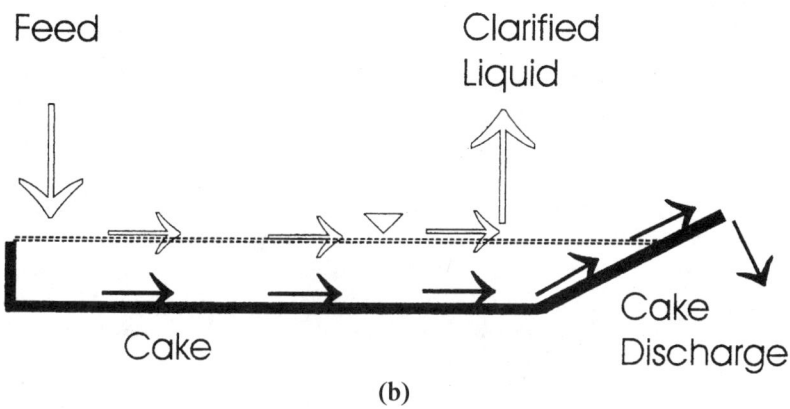

Figure 4.18 Flow decanters. (a) Countercurrent. (b) Concurrent.

Effects of G force. Increasing G force often results in the following effects with the first three being beneficial and the last a limitation.

1. Improved settling velocity, separation efficiency, and centrate clarity
2. Faster cake dewatering via cake compaction and subsequent expression and drainage mechanism
3. Faster return of expressed liquid back to the pool
4. Increased difficulty of cake conveyance up the beach

Effect of differential speed Δ. The speed at which the cake transports is controlled by the differential speed. High differential speed facilitates high solids throughput, where the cake thickness is kept to a minimum so as not to impair the quality of the centrate due to entrainment of fine solids. Also, cake dewatering is improved due to a reduction in the drainage path with smaller cake height. However, this is offset by the fact that higher differential speed also reduces cake residence time, especially in the dry beach. Vice versa, the opposite holds for low differential speed. Therefore an optimal differential speed is required to balance centrate clarity and cake dryness. The following summarizes the effects as Δ increases:

1. Reduced cake height, reduced conveyance torque, improved cake dewatering by reducing the path with which the moisture needs to travel, resulting in drier cake
2. Reduced cake dewatering time, resulting in wetter cake
3. Improved solids transport, which is favorable for high solids throughput
4. Possibly generating secondary flow in the clarifier at high Δ with subsequent resuspension of sediment and dirtier centrate
5. Reduced difficulty associated with cake conveyance up the beach

Increasing Δ results in opposing effects, as seen in items 1 and 2. For reasonable Δ the effect of retention time typically dominates that of cake height, resulting in a wetter cake. Δ can be changed with a variable-speed backdrive hydraulic motor, a variable-speed ac or dc drive, or, for the case of a gear box, by changing the gear ratio (gr). For the latter, increasing the gear ratio reduces Δ and, likewise, decreasing the gear ratio increases Δ. Δ is related to the speed of the bowl and the pinion in the gear box arrangement via

$$\Delta = \frac{\Omega_b - \Omega_p}{\text{gr}} \qquad (4.9)$$

When the pinion speed is fixed, $\Omega_p = 0$, the minimum gr is about 20 and the maximum about 300. This gives $\Delta = 0.3$ to 5% of Ω_b. For the backdrive arrangement, Δ is infinitely adjustable, and with special multi-stage gear designs it can be reduced to as low as 0.2 rpm.

Pool effect. The cylindrical section provides clarification under high centrifugal gravity. In some cases the pool should be shallow to maximize the G force for separation. In other cases, when the cake layer is too thick inside the cylinder, the settled solids, especially the finer particles at the cake surface, get entrained into the fast-moving liquid stream above, which eventually ends up in the centrate. A slightly deeper pool becomes beneficial in these cases, because there is a thicker buffer liquid layer to ensure settling of any resuspended solids. This can be at the expense of cake dryness due to a reduction of the dry beach. It is apparent that centrate clarity is often compromised with cake dryness. Another reason for such trade-off is that in losing fine solids to the centrate resulting in poor solids recovery (that is, classification) the cake, with larger particles having less surface-to-volume ratio, can dewater to higher dryness. Given all these considerations, it is best to determine the optimal pool for a given application through testing.

As the pool level is increased, the following can be expected:

1. Lower conveyance torque by reason of buoyancy on the cake mass from the liquid pool.
2. Better centrate quality due to more pool volume, keeping sediment away from the fast-moving clarifier layer close to the pool surface, which reduces solids entrainment.
3. Pool located closer to the point where the incoming slurry discharges off the feed accelerator, thus reducing mismatch in the feed and pool velocity due to "radial drop" (to be discussed in Chap. 11).
4. Wetter cake due to a reduction of "dry beach length," in particular for incompactible cake. (For compactible cake, which dewaters under the pool, this is indifferent or perhaps has a small benefit.)
5. Reduced G surface area, resulting in slightly poorer centrate quality.

Again items 3 and 5 have contrasting effects; 3 seems to dominate over 5 in most applications. The slurry pool can be changed by adjusting the radial position of the weir openings, which take the form of circular holes, crescent-shaped slots, or rectangular windows. The opening of the stationery skimmer can be adjusted radially to change the pool depth while the machine is in operation.

Materials of construction

The most common construction material is stainless steel, followed by carbon steel. Titanium and Hastelloy C construction can be used when the process environment is highly corrosive. Ferralium is used for the high-speed rotors. For high-wear applications all the wear-prone areas, such as feed ports, conveyor blade tips, liquid overflow weirs, cake discharge plows, and solid-bowl heads, are protected by either tungsten carbide, silicone carbide, or simply hard-facing. The conical beach is lined with ceramic tiles for abrasive cake.

New decanter technologies

Rapid technological advances[4] have been made in recent years, some of which are briefly discussed below.

Clarification enhancement. For fine particles with sizes in the micrometer to submicrometer range or particles with small density differences as compared to the liquid (such as biological materials), clarification of the slurry becomes a challenge. Higher G forces of up to 10,000g can be attained for small decanters[5] designed with special "floating" bearing support. The solid throughput for these decanters is low because of their small size as the rim speed limit is still the overriding factor and so is the torque. Large industrial decanters of 750–1000 mm (30–40 in) in diameter have pushed their Gs up to 3000–4000g by means of special construction materials, allowing higher yield strengths to overcome the rim speed limit. At the same time these machines have longer clarifiers, with the overall length of the machine extending up to 3000–4000 mm (120–160 in). Smaller gear boxes are designed at high speed and high G to reduce the overhung load and moment on the rotating assembly (bowl, conveyor, heads, trunnions, and gear). In contrast, these smaller gear boxes are rated for higher torque capacity and high speed to meet the stringent process requirements.

In addition to high Gs, new technologies have been developed with a view to increase the settling area, reduce the entrainment of the already settled solid, and provide prompt acceleration of the feed slurry to the specified G force with reduced slipping in the separation pool in the clarifier.

Disk-stack decanter. The disk-stack decanter has a combined geometry of a decanter and a disk. A series of lamella-plate-like cones[6] is installed in the cylindrical section. With this geometry the G-field vector, which directs radially outward, has to intercept all the settling surfaces. This implies either a series of coaxial conical surfaces or a series of inclined surfaces at an acute angle to the radius and spaced out circumferentially around the annular space between the outer diameter of the conveyor hub and the inner bowl wall.

In the clarifier, the conveyor has a ribbon blade with:

1. A coaxial disk stack[7] in a cage with the frame of the cage supporting the ribbon, as shown in Fig. 4.19.
2. Disks stretched out to form helical fins running parallel to the conveyor screw.[8]
3. Angled fins emanting from and spread out circumferentially and radially around the conveyor hub. (This design came out in the 1950–1960s and has been offered by several manufacturers.)

In all cases the increased sedimentation surfaces of these three designs are such that they are at an angle to the G force so as to get the benefit of the enhanced lamella sedimentation effect in increased surface area or reduced sedimentation distance. The limitation with the

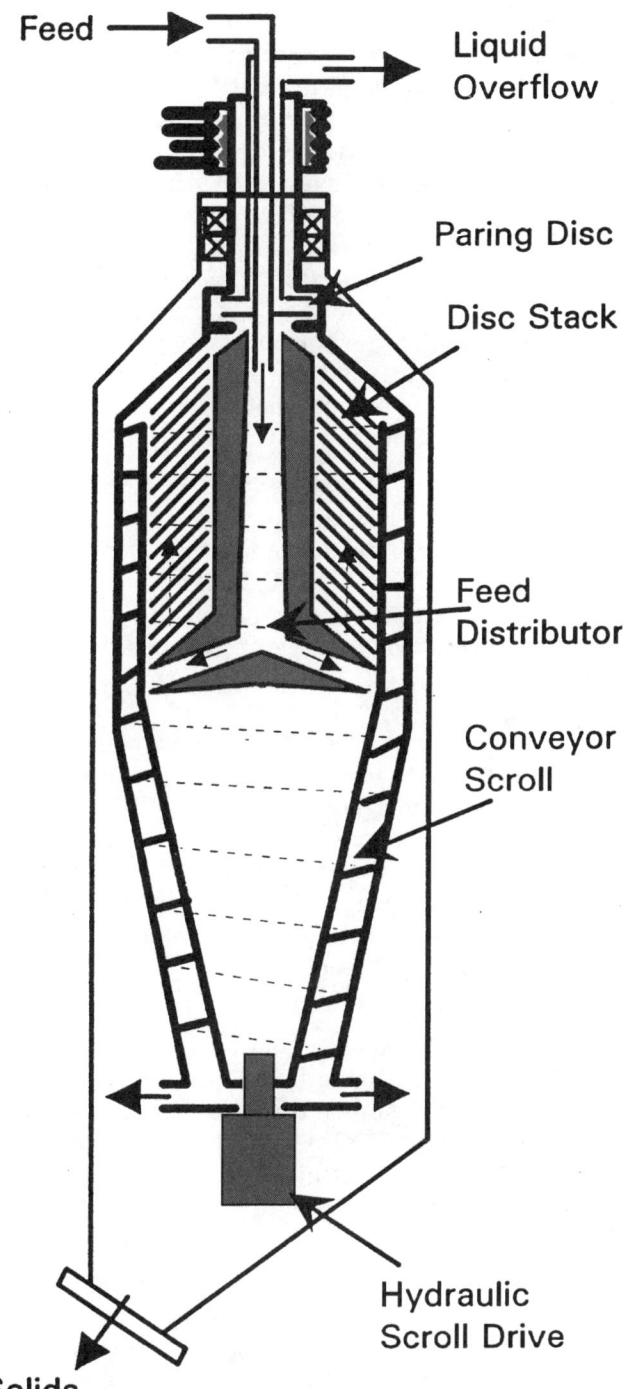

Figure 4.19 Disk-stack decanter. (*After Lee,*[7] *reprinted with permission from the American Filtration and Separations Society.*)

disk-stack decanter is that it cannot process high feed solids concentration or oversized solids which might plug up the disk stack. This seems to be inherent with all disk-stack geometries. In the former, the feed solids concentration is higher than in a regular disk centrifuge because the conveyor screw effectively conveys the cake away from the disk-stack area. Still, it is important to avoid cake solids accumulating in the clarifier which will interfere with the disk stack. Along that direction, longer lead can be used in the clarifier and smaller lead in the beach so that a moderate Δ can be used in operation without compromising the functions of the clarifier and the beach. Also to prevent clogging, larger spacings between adjacent disks, such as 10–20 mm (0.4–0.8 in), and a steeper inclination angle with respect to the axis, say, 45° or higher, should be used to ensure self-cleaning and that no solids accumulate in the disk channels.

Sloping clarifier. Instead of a straight-wall bowl in the clarifier, the bowl wall is slightly angled with a descent angle $\theta = 2$–$5°$ tilted toward a larger radius from the effluent discharge end to the beach-cylinder junction[9] (see Fig. 4.20). This helps the cake to convey by a component of its body force $(\rho_c - \rho_L)G \sin \theta$. (Recall that this is the very same force that resists cake conveyance up the beach whereby the climb angle γ can be as steep as 2°.) In this case this force helps cake conveyance, and the conveyor blade can be eliminated, or the pitch can be made very large. This reduces the secondary flow generated by the differential speed which becomes more pronounced as the conveyor blades come close together, such as with multiple-lead or close-pitch conveyor design. The secondary flow stirs up settled solids, especially the fine particles on the cake surface. The sloping clarifier in conjunction with a wide lead, or even without the conveyor lead but replaced with a raking mechanism, in the clarifier reduces solids entrainment. A design with similar features has been offered[10] with G up to 6000–8000g.

Prompt acceleration. The improved feed acceleration technology for continuous-feed sedimenting centrifuges helps accelerate the feed promptly to speed, eliminating any slip as the feed is introduced into the pool. The commercial technology XL • PLUS®* has been developed for both sedimenting and filtering centrifuges. The discussion of feed acceleration is taken up in greater depth in Chap. 11.

Dewatering enhancement

Fluid cake. When the cake behaves like a fluid, it flows under a pressure difference or driving gradient and the angle of repose is vir-

---

*XL • PLUS® is a registered trademark of Bird Machine Company.

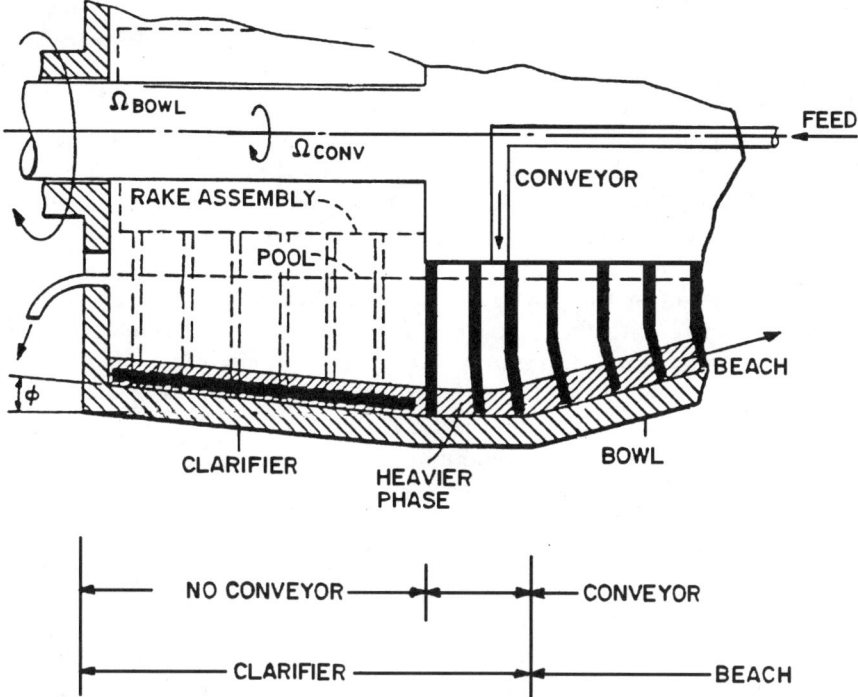

Figure 4.20 Sloping clarifier design. (*After Shapiro,[3] courtesy of Bird Machine Company.*)

tually zero, especially under G force. It is difficult to convey the cake up the beach, as it tends to flow back to the pool under G. Therefore the climb angle is deliberately made very small, typically under 0.5°. Also the blade tips should have close clearance from the bowl wall in the beach to prevent leakage under the blade.

A new cake-solids control technology has been developed recently. When the cake gets the benefit of dewatering in the dry beach, an adjustable exit restriction [11,12] located in the proximity of the beach exit can be used whereby the cake is impeded by the restriction at discharge. Such an impedance causes a buildup of the cake thickness upstream. Upon reaching a certain critical thickness, the surface of the cake flows backward toward the pool, thereby carrying expressed liquid (from the cake which percolates up to the cake surface) back to the liquid pool, while the cake layer adjacent to the bowl wall still conveys forward, passes the opening of the restriction, and discharges out of the machine. Because the cake flow is throttled by the restriction, the cake rate is reduced and the solid has to stay longer in the dry beach before it is discharged out of the machine, thereby increasing under control the dewatering time and consequently the cake dryness on the dry beach. Due to cake compaction and expression, the cake adjacent to the bowl wall is the driest and wettest at the cake surface. The exit restriction skims off the driest cake, rejecting the wetter cake back upstream for further dewatering. The drawback of the technology is the reduced flow rate, which can amount to 10–30% of the unimpeded rate, depending on the application. The restriction can be adjusted[11,12] to the appropriate opening suitable for a given application to yield the desired cake dryness. For optimal operation, the exit restriction is reduced for dilute fluidlike cake and increased for thick viscous cake material.

For the fluidlike cake, a dip weir has been used in the past. It is typically located at the junction of the beach and the cylinder. Thus the expressed liquid from dewatering downstream of the weir gets trapped by the weir and cannot return back to the pool. This is not the case with the adjustable exit restriction as it is positioned near the beach exit with minimal waste of the beach length.

Another arrangement is a compound beach design with an exit restriction.[13] Unlike the single beach with exit gate as described in the preceding section, this design does not compromise solid throughput with cake dryness by virtue of the design. The first beach has a steep angle whereas the second beach has a very small angle which is approximately parallel to the machine axis (that is, 0°). The pool is typically set at the intersection of the two beach sections. An adjustable restriction is installed near the beach exit to control the retention time of the cake in the second beach, the amount of cake discharge, and, most importantly, the increased dryness of the cake discharged. The design of the restriction is such that only the driest cake is allowed to be discharged out of the machine. This new design incorporates the adjustable-restriction concept while relaxing the limitation on throughput due to high torque because the second beach angle is very small, approaching 0°. This results in minimal torque and conveyance difficulty. Consequently, both high-solids throughput and superior cake dryness are obtained with this new innovative technology.

Granular cake. With granular cake, the cake filters and desaturates in the dry beach, as discussed in Chaps. 2 and 7. Multiple leads are used to reduce the cake height and maximize the dry beach length. For coarse particles without fine fraction, close-pitch multiple-lead design provides an increase in retention time and reduced cake height. For moderate-size particles with a fine solids fraction, a more open-pitch and multiple-lead design proves optimal.

Solid-solid-liquid separation. A double-cone double-blade decanter, CENSOR®* has been used to separate two solids in a liquid medium—one solid floats the other sinks in the liquid.[14] In the design

---

*CENSOR® and Centrisizer® are registered trademarks of Bird Humboldt.

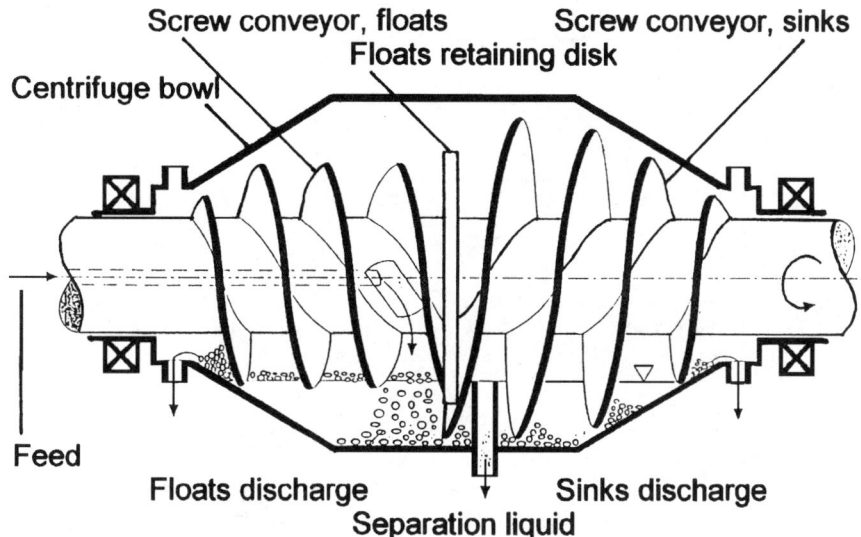

Figure 4.21 Double-cone double-blade decanter—CENSOR®. (*Courtesy of Bird Humboldt.*)

shown in Fig. 4.21 the feed is introduced near the center of the machine. The heavier solid settles out against the bowl wall and is conveyed by a set of screw blades toward one beach, whereas the lighter solid floats and is conveyed by different screw blades, directing the floating solid to discharge at the opposite beach. The clarified liquid is skimmed off near the middle of the clarifier, similar to the concurrent centrifuge. This design has been commonly applied to sort out recycled plastic particles or to separate food products of different densities, with one floating and the other sinking in the suspending liquid.

Nozzle decanter. It has been found beneficial in the classification of kaolin to discharge the heavy medium before scrolling the cake up the beach thinly reducing the conveyance torque and wear on the blade tips. In a Centrisizer®, nozzles are installed[15] at the junction of the cylindrical clarifier and the beach to discharge the grit and heavy media while the lighter materials are allowed to overflow and discharge up the beach. Given the feed is introduced at the larger end of the machine and the product overflow is withdrawn at the small end (concurrent configuration), therefore the full length of the machine is used for classification. In the beach, the conveyor flights are reversed to convey settled solids down the beach toward the nozzles (see Fig. 4.22).

Variable-speed drive. The variable-frequency drive (VFD) has become more attractive as an alternative drive for centrifuges due to its reduced cost and its versatility. The silicon-controlled rectifier of the VFD

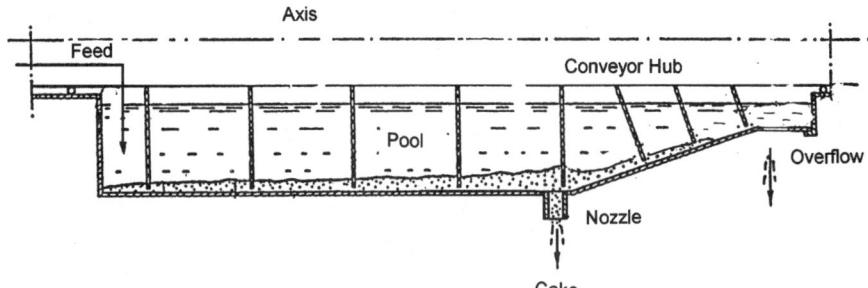

Figure 4.22 Nozzle decanter—Centrisizer®. (*Courtesy of Bird Humboldt.*)

takes in a three-phase power input and converts it into a dc output to a dc bus, which stores in a filter capacitor. The latter, driven by a microprocessor-based regulator, discharges to an inverter, where the dc voltage is converted to a controlled sinusoidal ac voltage with a three-phase output at a desired frequency. The speed of the motor can be adjusted through the frequency and the torque through the voltage. The VFD package, when used for driving the centrifuge, allows the machine speed to be changed while it is in operation. Certain products can reap additional benefits, such as better clarification, sharper classification, and drier cake, when the speed and thus the G force are tailored to the process needs. With other products, where the benefit is only marginal, the speed can be reduced to increase the longevity of the mechanical components due to lower wear and tear than under high-speed operation and to save on power consumption. The VFD eliminates the burden of changing the sheave size, either driven or driver, in order to change the speed as on a constant-speed drive motor. Also, it is a cleaner drive compared to the hydraulic drive which, operating at high pressure (such as 2.07×10^7 N/M² or 3000 lb/in²), has a notorious problem with oil leakage. The VFD is becoming easier to service, which sometimes may amount to changing a circuit board in contrast to disassembling and reassembling a hydraulic drive. More recent innovations in vector-field oriented control technology provide independent control of the flux-producing and torque-producing currents in the motor. This allows very accurate torque and power control, which is comparable to and frequently surpasses dc drives.

When used as a back drive, the VFD can act as a brake to maintain the differential speed for the decanter. In that capacity, power is regenerated and fed back to the main ac drive system during braking instead of being wasted as heat. Also, when the machine is flushed with wash liquid, it is useful to rotate the machine at low speed such that the contents of the rotating bowl cannot sustain a solid-body rotation. This tumbling and slashing motion of the wash liquid provides effective cleaning by scouring off any deposited solids left on the bowl wall and the conveyor. Also, high differential speed is maintained during cleaning to convey the residual solids out of the machine in a decanter equipped with a VFD backdrive. In most plants where the VFD is in place this cleaning procedure is carried out automatically by computer control on a routine basis.

In this chapter we have reviewed the basic designs of disks and decanters as well as the variations of each to address special needs and applications. Applications are dealt with in more detail in the next chapter.

References

1. W. W. F. Leung, "Centrifuges," in Perry and Green (eds.), *Chemical Engineers' Handbook,* 7th ed., McGraw-Hill, New York, 1997.
2. W. W. F. Leung, "Centrifugation," in P. Schweitzer (ed.), *Handbook of Separation Techniques for Chemical Engineers,* 3d ed., McGraw-Hill, New York, 1997, pp. 4-78 to 4-86.
3. H. Lehmann, "New Developments in Liquid/Liquid/Solid Centrifugal Separation," in *Proc. Am. Filt. Sep. Soc. Ann. Conf.* (Nashville, TN, Apr. 23–26), vol. 9, K. Choi (ed.), pp. 503–529.
4. W. W. F. Leung, "Advances in Industrial Centrifuges for Solid/Liquid Separation," presented at the AIChE/CIESC Conference (Beijing, China, May 1997).
5. A. Letki, "Development of the Multi-Purpose 10,000g Decanter Centrifuge," in *Proc. Am. Filt. Sep. Soc. Ann. Conf.* (Chicago, IL, May 10–14, 1992), vol. 6, R. Peters (ed.), pp. 21–24.
6. W. W. F. Leung, "Lamella Settlers: Part 1—Model; Part 2—Flow Stability," *I&EC Process Design and Dev.,* pp. 60–73, 1982.
7. A. Lee, "Do We Need a New Type of Centrifuge," in *Proc. Am. Filt. Sep. Soc. Ann. Conf.* (Valley Forge, PA, April 21–24, 1996), vol. 10, B. Scheiner (ed.), pp. 248–253.
8. S. Suzuki, "Sedimenting Centrifuge with Helical Fins Mounted on the Screw Conveyor," U.S. patent 5,314,399, May 24, 1994.
9. A. Shapiro, "Conveyorless Clarifier," U.S. patent 5,067,939, Nov. 26, 1991.
10. A. Karolis and E. Krautlein, "Flottweg Sedicanter—A New Centrifuge for Separation of Suspension with Poor Settling and Conveying Characteristics," in *Proc. Am. Filt. Sep. Soc. Ann. Conf.* (Valley Forge, PA, Apr 21–24, 1996) vol. 10, B. Scheiner (ed.), pp. 200–204.
11. W. W. F. Leung, A. Shapiro, and R. Yarnell, "Decanter Centrifuge with Adjustable Gate Control," U.S. patent 5,643,169, July 1, 1997.
12. W. W. F. Leung, "Decanter Centrifuge with Discharge Opening Adjustment Control and Associated Method of Operating," U.S. patent 5,653,674, August 5, 1997.
13. W. W. F. Leung, and A. Shapiro, "Decanter Centrifuge and Associated Method of Producing Cake with Reduced Cake Moisture Content and High Throughput," U.S. patent 5,695,442, December 9, 1997.
14. M. Wuensch, K. H. Unkelbach, and G. Arhelger, "CENSOR—The Decanter with Two Cones and Two Flights Washes—Separates and Dewaters Plastics for Recycling," in *Proc. Am. Filt. Sep. Soc. Ann. Conf.* (Chicago, IL, May 3–6, 1993), vol. 7, W. F. Leung (ed.), pp. 267–270.
15. F. Muller, J. Kompe, and R. Kluge, "Centrifugal Classifying with the Centrisizer in a Range of 1 Micron," in *Aufbereitungs Technik (Mineral Processing),* Sonderdruck aus reprint from the journal *Aufbereitungs-Technik/Mineral Processing,* no. 12, December 1993, pp. 611–618.

Chapter 5

Applications of Sedimenting Centrifuges

Sedimenting centrifuges utilize centrifugal force that acts on the density difference between the phases to effect separation. They are typically used for one or more of the following process functions. In clarification or purification, sedimenting centrifuges aim at producing centrate with acceptable clarity and minimal solids. In classification applications, particles of certain sizes are grouped based on their settling rates, which depend on their respective sizes and densities. In degritting, foreign or oversized particles are removed from the product centrate by the application of centrifugal force. In thickening and concentrating, solid-free liquid is removed with the goal of concentrating the feed stream for downstream processing. In dewatering, cake dryness is the objective of the separation. In general the choice of using sedimenting centrifuges versus filtering centrifuges is dictated by one of the following: the solid materials to be separated out must have a considerable fraction below 45 μm, the concentrate sediment (that is, the cake) is highly compactible, rendering filtering centrifuges inapplicable, or the liquid in the porous cake does not drain. In separation and repulping, sedimenting centrifuges are used to remove contaminants dissolved in mother liquid by separation and redilution in several stages. In extraction, reslurrying tanks and sedimenting centrifuges are set up in countercurrent stages to extract the valuables dissolved in the feed liquid with an immiscible solvent liquid. In this chapter several applications are discussed to reflect the typical functional aspects of sedimenting centrifuges. Both three-phase and two-phase separation applications are discussed.

Three-Phase Separation

In industrial applications, food processing, waste cleanup, oil refining, and so on, which involve handling of oil and grease, the intermediate product or waste often requires three-phase separation. If the solid in the feed stream is already very high, it can be removed first by a two-phase decanter followed by a two-phase liquid-liquid separation using a disk separator at high Gs. In the alternative, a three-phase decanter can be used to achieve the separation of all the phases at 4000–6000g. In most applications chemical treatment and heat (the temperature of the feed slurry is increased via passing through a heat exchanger) are applied upstream to facilitate this separation. Likewise, a three-phase nozzle disk can also be used for the case when the feed solids are low (a few percent of the solid) and nonabrasive. In this section a few examples are used to illustrate the application characteristics of a three-phase separation by centrifugation.

Oil recovery

Solid-bowl centrifuges are used to process the slop oil or off-spec oil from tank bottoms in oil refining and petroleum production. Solid bowls 457–610 mm (18–24 in) in diameter are used with operating flow rates of between 80 and 480 L/min (20 and 125 gal/min). The oily slurry from tank bottoms, lagoons, and slop oil contains by weight 60% solids, 25% oil, and 15% water. It first passes through a screen where the coarse-size materials are removed as tailings. The screen drain, containing finer solids with about 8% solids, 43% oil, and a balance of 49% water is then fed to a heat exchanger, increasing the temperature of the slurry to reduce the viscosity. Subsequently it is further treated with emulsion-breaking chemicals to separate oil-water emulsion and possibly with chemical flocculants to agglomerate the fines before feeding the slurry to a three-phase decanter. The decanter separates the slurry into three streams—a cake phase enriched in 55% solids, 5% oil, and 40% water; a solid-free water phase with 2% oil and 98% water; and an oil phase containing 98% oil, 1% solids, and 1% water, which can be sent to a disk centrifuge for further polishing. The disk further reduces the solids content in the oil to less than 100 ppm (0.01%) and the water content to less than 500 ppm (0.05%). This separation process converts a waste material to a higher-quality valuable product. The water stream from the centrifuge, from which the oil has been stripped off, can be reused for injection wells, or further treated using dissolved air flotation or membrane filtration for other uses. The flow sheet for the oil-sludge separation and reclamation is shown in Fig. 5.1.

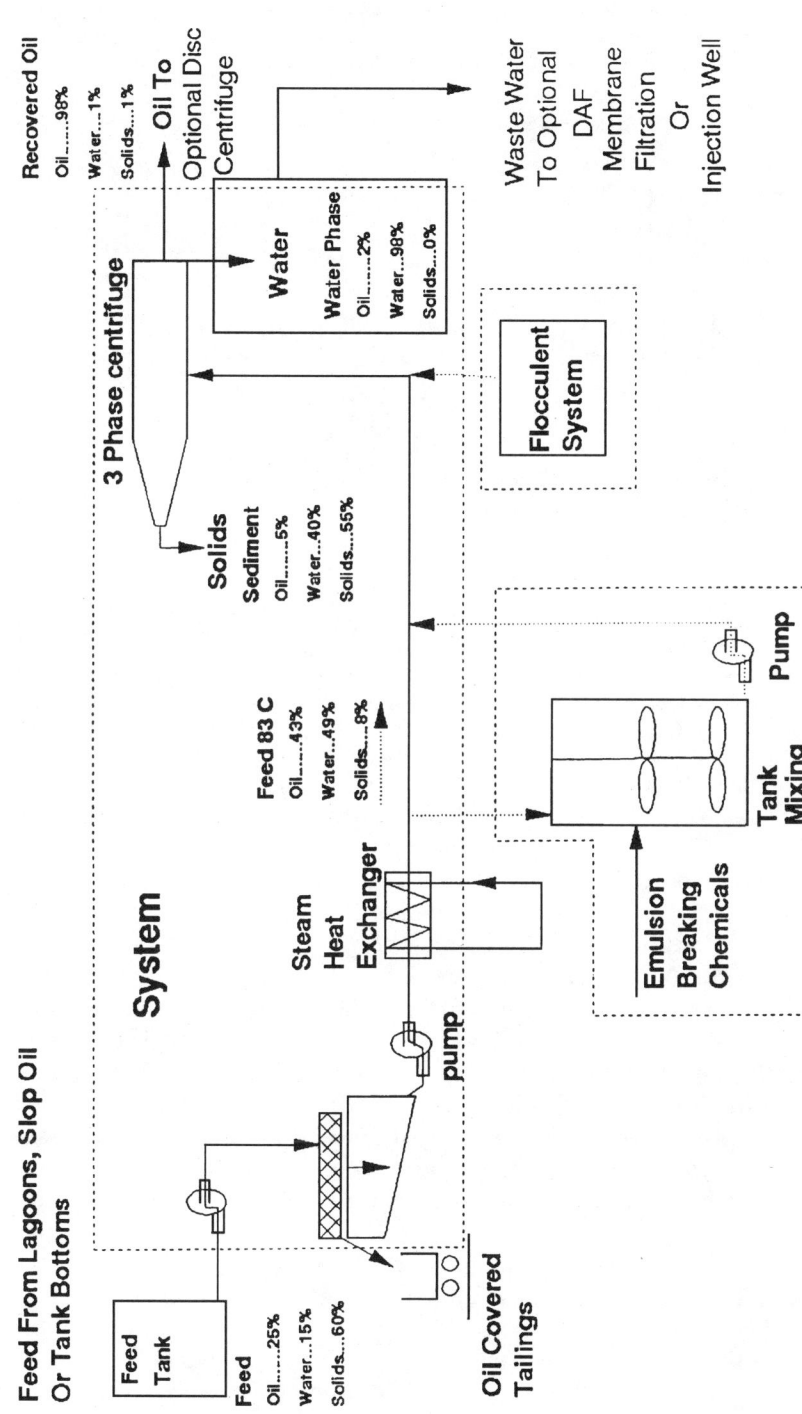

Figure 5.1 Oil-sludge separation and reclamation circuit.

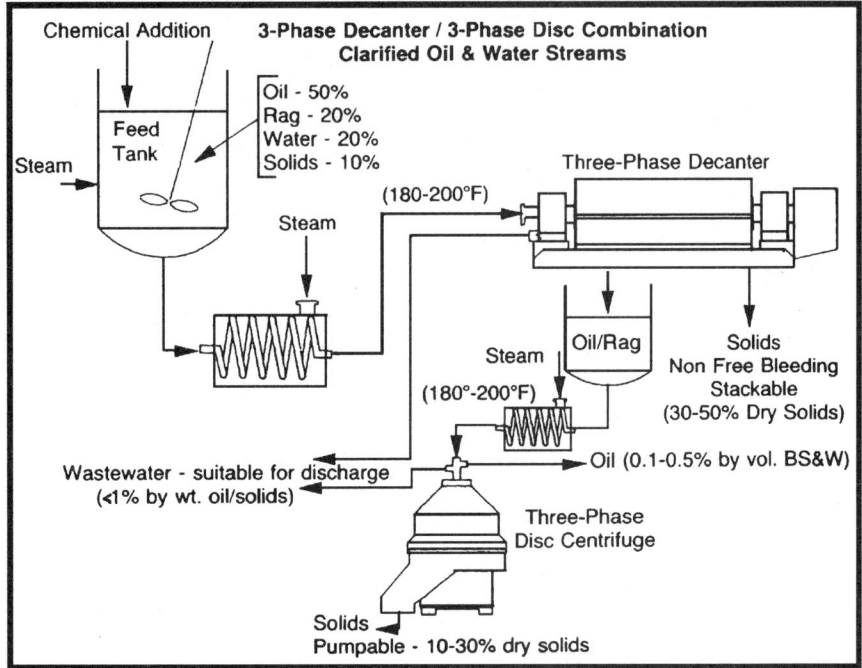

Figure 5.2 Three-phase decanter-disk combination for clarifying oil and water streams.

Figure 5.2 shows an industrial application flow sheet of oil-water-solids separation. The objective is to clean out waste from upstream processing so that the processed streams can be safely discharged or reused. The feed contains by weight 50% oil, 20% water, 20% water-in-oil emulsion (or rag), and 10% solids. It is passed through a heat exchanger where excess steam from the plant is used to heat up the oil to 82°–93°C (180–200°F). A solid bowl is used to carry out the three-phase separation. Exiting the centrifuge, the light-phase liquid contains oil and emulsion with minimal water and solids; the heavier water phase contains less than 1% oil and solids; and the cake phase contains 30–50% w/w of solids, which are nonbreeding and "stackable" for ease of handling. The light-liquid phase subsequently goes through the polishing step. First the cooled-off process stream passes through the heat exchanger again to ensure that the process temperature is maintained prior to separation in a high-speed nozzle-disk centrifuge. Three streams are obtained from centrifugation—the product oil phase contains 0.1–0.5% of residual solids and water, the water phase contains less than 1% w/w oil and solids, and the cake phase contains 10–30% w/w solids, which are "pumpable" condition. This water phase from the polisher is combined with the water phase from the decanter for discharge.

Alternatively, the disk can be eliminated, and the three-phase decanter can be operated with an oil-water interface set close to the bowl wall, leaving a large pool volume for clarification of the oil phase. Thus a cleaner oil is obtained, which is comparable to that obtained from the disk. This benefit is unfortunately at the expense of a dirtier processed water phase contaminated with oil. In some applications the wastewater can be disposed in this form; whereas in other applications, when it is important to strip the residual oil from the wastewater, further treatments such as chemicals, heat, or membrane filtration are used. It can be seen that the operational change around a centrifuge affects the required ancillary equipment downstream or upstream to complete the separation process.

Palm-oil separation

In a typical palm-oil separation plant, the palm slurry after passing through the digester is taken to a screw press, where the oil is squeezed out. The slurry is subsequently sent to a settling tank for degritting the coarse sand, and the overflow liquid is taken to a vibrating screen, where the oversized solids are further removed. Prior to feeding to a three-phase decanter, the screen drain is heated by pumping it through a heat exchanger so as to reduce the viscosity of the oil-water slurry to effect better separation. The feed to the centrifuge contains 29% oil, 67% water, and 4% solids. The product of the separation step is the oil phase and should be free from both water and solids, even if it is at the expense of a trace amount of oil in the water stream. Typically, the oil phase is of high purity with 99.78% oil and a balance of 0.22% water and solid sludge. The palm oil leaving the centrifuge is sent to a collection tank for downstream purification. On the other hand, the centrifuge discharged water contains 1.8% oil, 2.7% solids and 95.5% water. Fortunately, the discharged water phase from the decanter can be returned back to the buffer tank, which provides make-up water for the digester. Given that it is a closed loop, the oil loss in the water stream is not an issue. Finally the cake, which contains 3% oil, 77% water, and 20% solids, is taken either to a trailer for disposal or to a dryer, where the dried material can be used as a by-product.

Rendering

A typical rendering flow sheet is shown in Fig. 5.3a. In a low-energy rendering process the raw materials from cattle processing are sent to a preheater and subsequently to a screw press. The cake from the press is sent to a dryer, where after the moisture evaporates, the solid is bagged as product dry meal; the liquid slurry from the screw press is sent to a three-phase decanter (also known as tricanter). There the

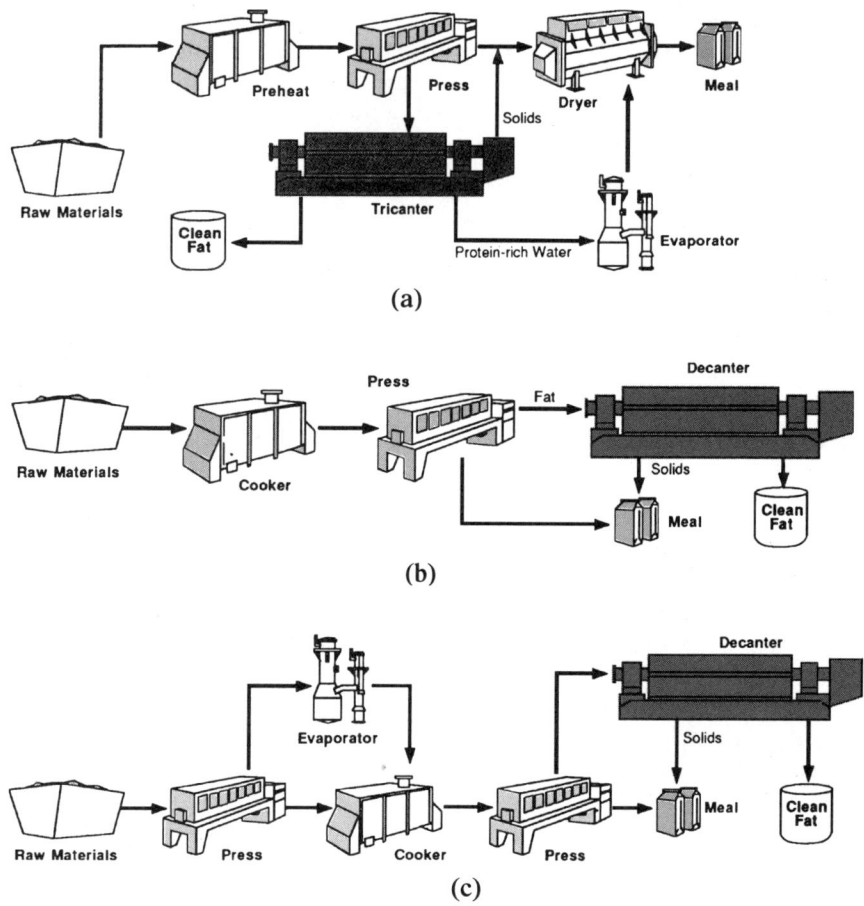

Figure 5.3 (a) Low-energy rendering. (b) Conventional dry rendering. (c) Retrofit dry rendering.

stream is further separated into a light oil or fat liquid, a protein-enriched water which is sent to an evaporator to remove the water, and a solid cake. The latter is blended with the solids from the screw press and those from the evaporator before being sent to the dryer.

Cream separation

By far the most common application of disk centrifuges is in the dairy industry, where the cream is separated from the milk. As shown in Fig. 5.4, an intermittent discharge disk is generally used. Milk is fed

Applications of Sedimenting Centrifuges   129

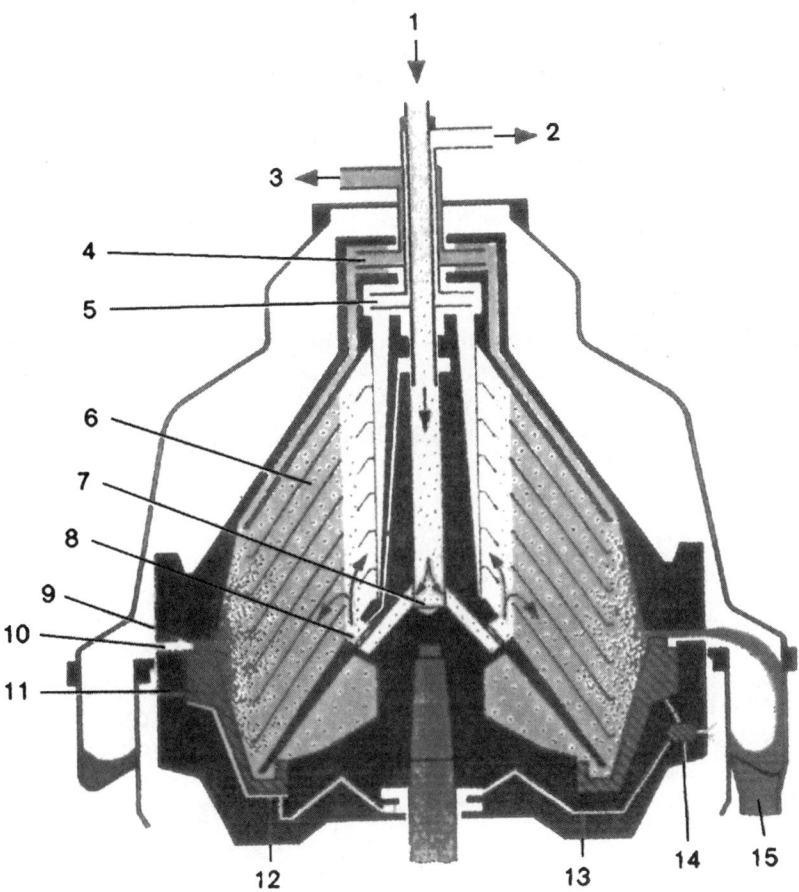

1 Feed
2 Cream discharge
3 Discharge, separated whey
4 Centr. pump for sep. whey
5 Centr. pump for cream
6 Disc stack
7 Soft-stream inlet
8 Rising channels
9 Solids space
10 Solids ejection ports
11 Sliding piston
12 Closing-water chamber
13 Opening-water channel
14 Piston valve
15 Solids discharge

Figure 5.4 Cream separator. (*Courtesy of Westfalia.*)

through the feed pipe into the machine and distributed and accelerated by a vane assembly from which it is distributed to the disks through the rising channels of the disk stack. After separation in the channels formed between adjacent disks, the lighter cream travels upward by buoyancy toward the axis and is discharged under pressure through the lower centripetal pump; whereas the heavier skim milk flows outward under G and is discharged into the outer periphery of the bowl, where it is picked up by the upper centripetal pump. Hermetic designs are used to avoid shear on the product, with feed acceleration starting from the axis. In cream separation, the low-fat skim milk is the key product and cream the by-product. Therefore the interface should be set such that a significant fraction of the disk channel is used to separate the cream from the skim milk in order to ensure that the fat content of the milk stays within specification. Consequently the rising channels are typically located closer to the axis of the machine.

Fish meal processing

A small 13-in-diameter decanter equipped with a deep-pool design and compound beach is used to separate sardine oil in a fish meal plant.[1] A hot feed stream of 50 L/min (13.2 gal/min) with fat and oil (16–32% v/v), solids (3–12% v/v), and water (remaining balance) at 90°C is fed to the decanter operating at 3000g. This stream emanates from an upstream screw press followed by screening. The three-phase decanter separates the raw feed into three streams, with the lighter liquid the product fish oil consisting of 99.85% w/w fat and oil, 0.15% w/w (or < 0.21% v/v) water, and trace amount of solids. The heavy liquid has 0.1–0.15% w/w oil, 99+% water, and 0.3% v/v solids. This goes into a concentrated protein tank. On the other hand, the cake solids consist of less than 1.8% w/w fat and oil, 65% w/w water, and 33.2% w/w solids. The residual liquid in the cake is removed by the downstream dryer. (Note that v/v cited here refers to the test-tube spin-down bulk volume fraction.) In some other applications where the solids content in the product oil is high, say 1–2% v/v, it is further polished by a disk centrifuge. Therefore the primary objective in the fish meal separation is for the decanter to obtain product purity (that is, the light-liquid phase) in oil and fat, with a secondary objective to dewater the cake to high dryness in order to reduce thermal drying costs. As such, the pool level is set quite high to achieve good liquid-liquid separation, even at the expense of cake dryness.

Miscellaneous liquid-liquid separation

By far the majority of the applications are for oil-water separation, such as for shipboard or marine use. In addition, the liquid-liquid disk centrifuges are used for the recovery of citrus oils from peel press liquid, wool grease from wool scouring liquor, fish oil from fish press liquid, and dehydration and purification of heavy fuel oils after water washing. The liquid phases are discharged under pressure to avoid foaming at discharge.

Two-Phase Separation

Rendering

Unlike low-energy rendering, a cooker is used in the circuit, as shown in Fig. 5.3b and c. Given that the moisture has already been evaporated in the cooker, the feed slurry contains solids and liquid fat squeezed out from the press. Therefore a two-phase decanter can be used in this two-phase separation. As an objective, the liquid fat in the cake should be minimized.

Calcium carbonate

Mined from stone quarries and pulverized into a powder, or precipitated out of a chemical reaction, calcium carbonate enhances the quality of an incredible range of consumer products from plastic and coated papers to medicines and foods.

Calcium carbonate is ground [ground calcium carbonate (GCC)] in a ball mill in the wet processing circuit. The stream leaving the ball mill is classified by hydrocyclones or decanter centrifuges to remove the oversized particles, which are returned back to the mill. The slurry that has a finer size is sent to a thickener, where it is concentrated to 30% or higher before the thickened slurry is dewatered by a decanter. The cake leaving the dewatering centrifuge has solids in the range of 70%–78% w/w. Simultaneously it is important to keep a high solids recovery of 95–99% to reduce the loss of valuable product. The median particle size of the feed slurry presented to the dewatering centrifuge ranges typically between 5 and 20 μm. Capturing the particles becomes a challenge as they get to finer sizes. It is important to keep the fine fraction (say, the smallest 10%) above 1 μm to ensure good capture by the dewatering centrifuge. The solids recovery curve illustrated in Fig. 2.9b is atypical for this application. The result depends critically on the fine fraction of solids present and the chemical conditioning of the fine particles such as the presence of leftover frothing chemicals from the upstream flotation thickening process.

Decanters have also been used to dewater precipitated calcium carbonate (PCC) with median particle sizes between 0.3 and 3 μm. Some special applications even call for ultrafine separation in the 0.02–0.05-μm range. Given that the particle-size distribution is narrow, the separation characteristics by centrifugation are very similar to those of the monodispersed slurry discussed in Chap. 2. The recovery stays close to 100% at low feed rates until it reaches a critical feed rate, where the recovery drops off sharply (see Fig. 2.9a in Chap. 2). This corresponds to the condition where the sedimentation rate of the particles is offset by the convection rate of the liquid effluent toward the liquid discharge ports. The cake solids concentration obtained from PCC depends on the shape and morphology of the solid. It generally ranges between 40 and 60+% w/w.

In the final step of processing, the concentrated slurry with 70+% w/w solids is sent for degritting, where oversized particles (larger than 25 or 45 μm) or foreign particles (from the grinding medium for GCC) are dropped out under centrifugation, leaving the finer product in suspension in the centrate. Usually a relatively medium speed of operation is all that is needed with higher differential speeds between the bowl and conveyor and shallower pool to discourage sedimentation of finer particles. It is important to maintain a high product quality, that is, low grit level, while maintaining a low loss of valuable product from sedimentation with the grit particles. In operating a degritting centrifuge, rotational speed, differential speed, and pool depth are tuned to attain optimal results.

Coal processing

Decanters have been used extensively for dewatering fine coal with a significant fraction of minus (less than) 325 mesh or minus 45 μm. Some applications are aimed at good clean coal whereas others are for waste refuse coal provided the economics of the latter are justified. It depends on the heating value Joules (Btu) of the coal and the amount of unwanted ash (nonhydrocarbon) which typically resides in the finer sizes. The bowl wall of the decanter is typically lined with ceramic tiles while the blade tip and the pressure face of the blade are protected by wear-resistant tungsten carbide.

Other mineral processing

When separating mineral slurries with a small amount of solids, disk centrifuges prove useful. Table 5.1 summarizes the application of disk and decanter centrifuges for mineral processing. In solvent extraction, leaching, and ion exchange, where the solids in the slurry stream are less than a few percent by weight, the intermittent-discharge disk centrifuges have been used for copper, uranium, cobalt, and so on. In processing yellow cake, where uranium is the product, both decanters and nozzle disks have been employed. The cake must be flowable under G-force for the nozzle disk to work.

TABLE 5.1 Mineral Applications Using Disks and Decanters

Mineral processing	Products	Centrifuges in separation step
Leaching (dissolve mineral valuable components with solutions)	Cobalt, copper, molybdenum, nickel, tungsten, uranium	Intermittent-discharge and nozzle disks, basket (clarify and polish leach solutions)
Solvent extraction (transfer valuables dissolved in aqueous solutions to water-immiscible organic solvent)	Same	Same
Ion exchange (concentrate after leaching, reagents used)	Copper, uranium	Intermittent-discharge and nozzle disks (recover expensive solvent)
Yellow cake	Uranium	Solid bowls, nozzle disks
Phosphoric acid	Phosphate fertilizer	Solid bowls, intermittent-discharge and nozzle disks
Classification (separate finer valuables from coarser sizes)	Kaolin, calcium carbonate	Solid bowls, nozzle disks
Dewatering (dry cake)	Calcium carbonate	Solid bowls
Degritting (remove foreign or oversize particles)	Calcium carbonate, kaolin	Solid bowls

Chemical applications

Plastics dewatering. Polyvinyl chloride (PVC) is the second most widely used thermoplastic polymer. The resin produced translates into over 1.4 million tonnes of compound worldwide. As an amorphous polymer, PVC resin is extensively formulated to produce an extremely large variety of compounds. Solid-bowl decanters have been widely applied to dewater PVC slurry from the reactors. Pipeline grade is the most common among the different resins for manufacturing PVC pipes. The lowest cake moisture as obtained from the decanters is typically between 15 and 25% w/w, depending on particle porosity, polymer property (molecular weight and inherent viscosity), and particle size. In operation with a decanter, the cake moisture curve exhibits a gradual increase at low solids throughput followed by a steep rise at high throughput, as shown in Fig. 5.5. All production decanters are generally operating near this transition or critical rate. The operating tempera-

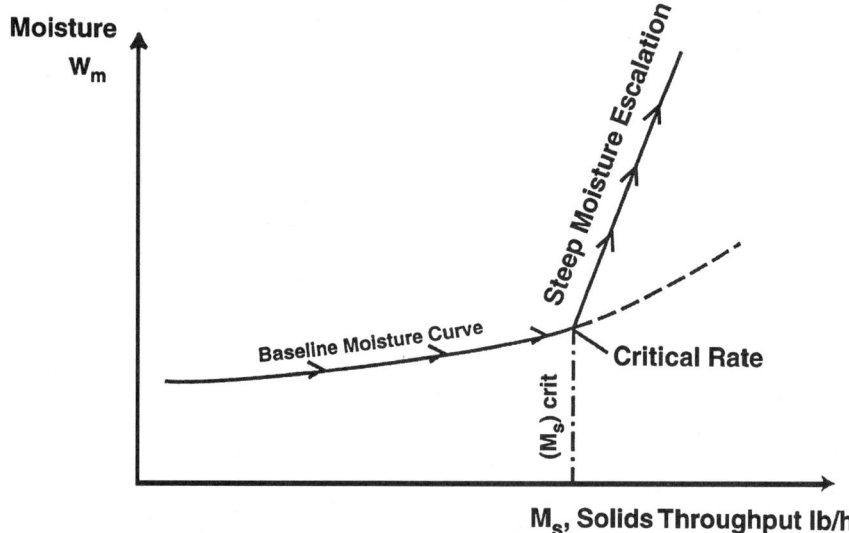

Figure 5.5 PVC dewatering characteristics.

ture ranges between 60 and 90°C (140°–190°F), depending on the process and the type of PVC. The high temperature helps reduce the liquid viscosity, which further facilitates dewatering. However, some processes cannot tolerate elevated temperatures due to decoloration of the PVC. The solids in the centrate are typically low, as the smallest particle size is fairly large, typically above 60 μm for most pipe-grade resins. Therefore the centrate solids are usually below 100 ppm. Some processes even require less than 50 ppm of solids.

Solid bowls are also used to dewater other plastics, such as high-density polyethylene (HDPE). With this application, given that the cake structure is more open, the moisture curve increases gradually with increasing feeding rate.

Sorting recycled plastics. The CENSOR®[2] can separate solids of different densities. A liquid medium is required as the suspension phase such that the lighter solids float on the liquid while the heavier solids sink to the bowl wall. Figure 5.6 shows an application where polyethylene (PE), polypropylene (PP), polystyrene (PS), polyvinyl chloride (PVC), polyamide (PA), and other residual plastics (RP) are sorted according to a two succeeding stages of separation.

The recycled plastics from films (plastic bags), bottles, automotive plastics, carpets, fibers, and other industrial plastics are shredded and milled typically to 0.5–10 mm. (The size to be shredded down depends on how well the different species of plastics can be separated out.)

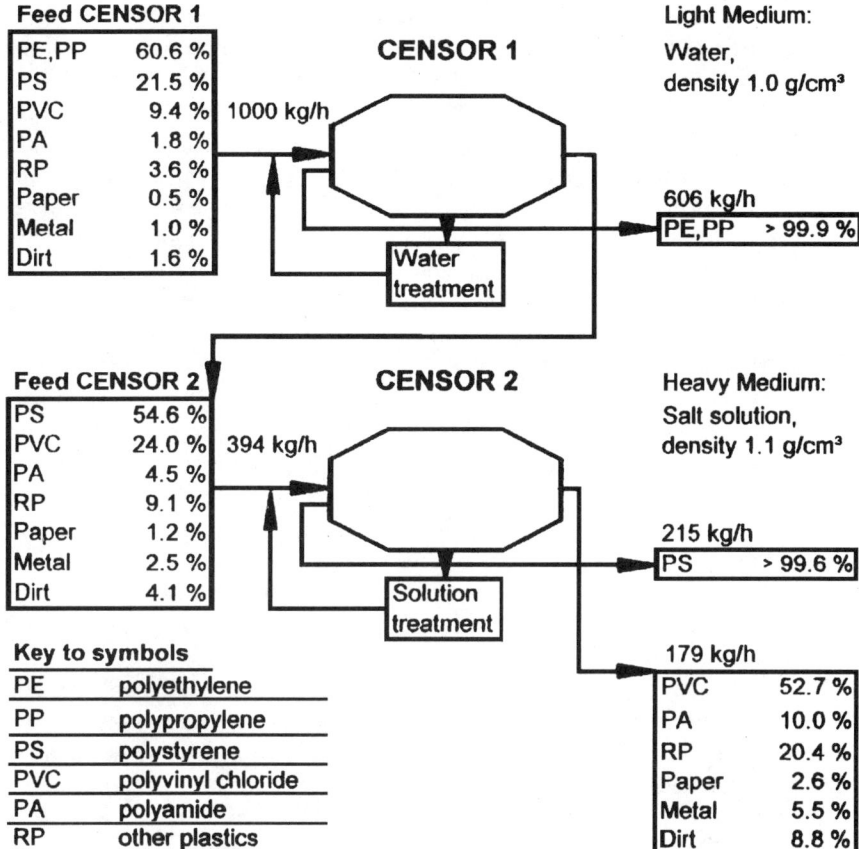

Figure 5.6 CENSOR™ used in sorting waste plastic. (*Courtesy of Bird Humboldt.*)

They are fed to a mixing tank where a liquid-water or a saline solution, of appropriate sg (specific gravity), is introduced as the suspending medium. The slurry is then fed to a two or more stage separation using the CENSOR® centrifuges which have two conical beaches, one for discharge of the light solids and the other for the heavy solids.

As shown in Fig. 5.6, in the first stage CENSOR® water is used as the suspending medium. The polyethylene and polypropylene, which have sg (specific gravity) between 0.91 and 0.94 float in water (sg = 1) and get removed on the first conical beach by the light-phase conveyor, while the remaining heavier (sg > 1) plastics sink to the bowl wall of the centrifuge under modest centrifugal force (generally 800–1200 g) and get removed at the second conical beach by the heavy-phase conveyor. They are further sent to a second-stage CENSOR® where typically a saline solution with sg = 1.07–1.1 is used to remove the polystyrene with sg = 1.05. The polystyrene gets separated as the light solids and the heavy solids reject contains polyamide, polyvinyl chloride, and other remaining plastics, all of which with sg > 1.1. To separate out the polyamide, a third stage can be implemented with a saline solution with sg = 1.7–1.2, whereby the polyamide (sg = 1.15) floats, while the polyvinyl chloride, latex (sg = 1.3–1.4) and the remaining heavier plastics sink. The number of stages as well as the appropriate suspending medium used depend critically on the objectives of the process separation. In all stages, the separated liquid phase from the CENSOR®, which is cleared of solids, are recycled back to the upstream mixing tanks conserving liquid use.

Nitration of aromatic chemicals. Disks are often employed for solid-liquid separation in the production of mono- and dinitroaromatic substances in a two-stage process. In the first step, nitric acid of 98% concentration is added to toluene in a reactor to produce mononitrotoluene. The spent acid is separated in a manual disk (with practically no solids residual). In the second step, nitric acid and sulfuric acid are both added to mononitrotoluene to yield dinitrotoluene. The spent acid again is separated out by a manual discharge disk. The dinitrotoluene is sent to pass a four-stage washing, first with water, then with caustic soda in two steps, and finally with water again to remove residual acids and byproducts containing hydroxyl. Subsequently the product is separated from the wash liquor. The spent acids from the first and second stages are further purified for reuse with another disk. The flow sheet of this application is similar to other applications using decanters.

Other chemicals. Disk centrifuges have been used to provide clarification and separation of phosphoric acid, separation of benzene, clarification of solvents, washing and thickening of terephthalic acid, and clarification of agricultural chemicals, separation of polycarbonate, and hydrogen peroxide production with catalyst recovery. When the solid particles in the feed slurry are above the submicrometer range, solid bowls can also be used for these applications.

Food processing

Nozzle disk centrifuges are used for the thickening and concentration of soft and fine solids such as vegetable and single-cell proteins, corn gluten, corn and wheat starch, the separation of corn gluten from cornstarch, and bakers' and brewers' yeast. For processing yeast, special yeast bowl designs allow the cake to be discharged from the nozzles at a smaller diameter so as to reduce power consumption.

Applications of Sedimenting Centrifuges

TABLE 5.2 Juice Applications Using Multibowl Centrifuges

Clarification*	Capacity, L/h (gal/h)†
Fruit juices	3000 (800)
Wine	5200 (1375)
Beerwort, cold to hot	9800–4170 (2600–1100)
Beer	6800 (1800)
Varnish and lacquer	13,265 (3500)
Pigment varnish	11,370 (3000)

*Disk centrifuges have also been used in these applications.
†Sludge holding capacity of largest-size multibowl is approximately 64 L (17 gal).

Both decanters and multibowl centrifuges have been used for the clarification of fruit juices such as grape, apple, pear, carrot, and cranberry. They have also been used to clarify wines and beer. Table 5.2 shows some typical rates for using the multibowl centrifuge. Disks and fin-screw decanters as described in Chap. 4 have also been used widely for juice clarification.

Pharmaceuticals and biotechnology

Sedimenting centrifuges are used in conjunction with the mixers for extracting valuable dissolved materials from the feed liquid into another solvent liquid (lighter phase). A simple schematic of a single-stage setup is shown in Fig. 5.7a. The transfer of solute from one liquid to another is effected by mixing and turbulence. However, the transfer of solute is still dictated by the concentration gradients in both liquid phases, with the transfer rate augmented by the optimal design and operation of the mixer. Therefore extraction from a single stage can be limited, and more frequently a multistage arrangement is used to enhance the result. Figure 5.7b shows a three-stage countercurrent extraction.

In the production of antibiotics such as penicillin, oxygen and nutrients are sent to a series of fermenters or reactors, where penicillin is produced after the metabolism of the mold under controlled conditions. Penicillin is subsequently separated from the by-product in the nutrient solution by two-stage countercurrent extraction using decanters. In the direct extraction process, the fermentation broth is set to the appropriate pH by means of sulfuric acid, and wetting agents are also added to the broth. The solvent (liquid effluent) from the second-stage decanter is combined with the fermentation broth feeding the first-stage decanter. The liquid solvent separated from the decanter and loaded with penicillin is further routed to a disk for polishing. In the disk, the residual amount of water and the mycelia are

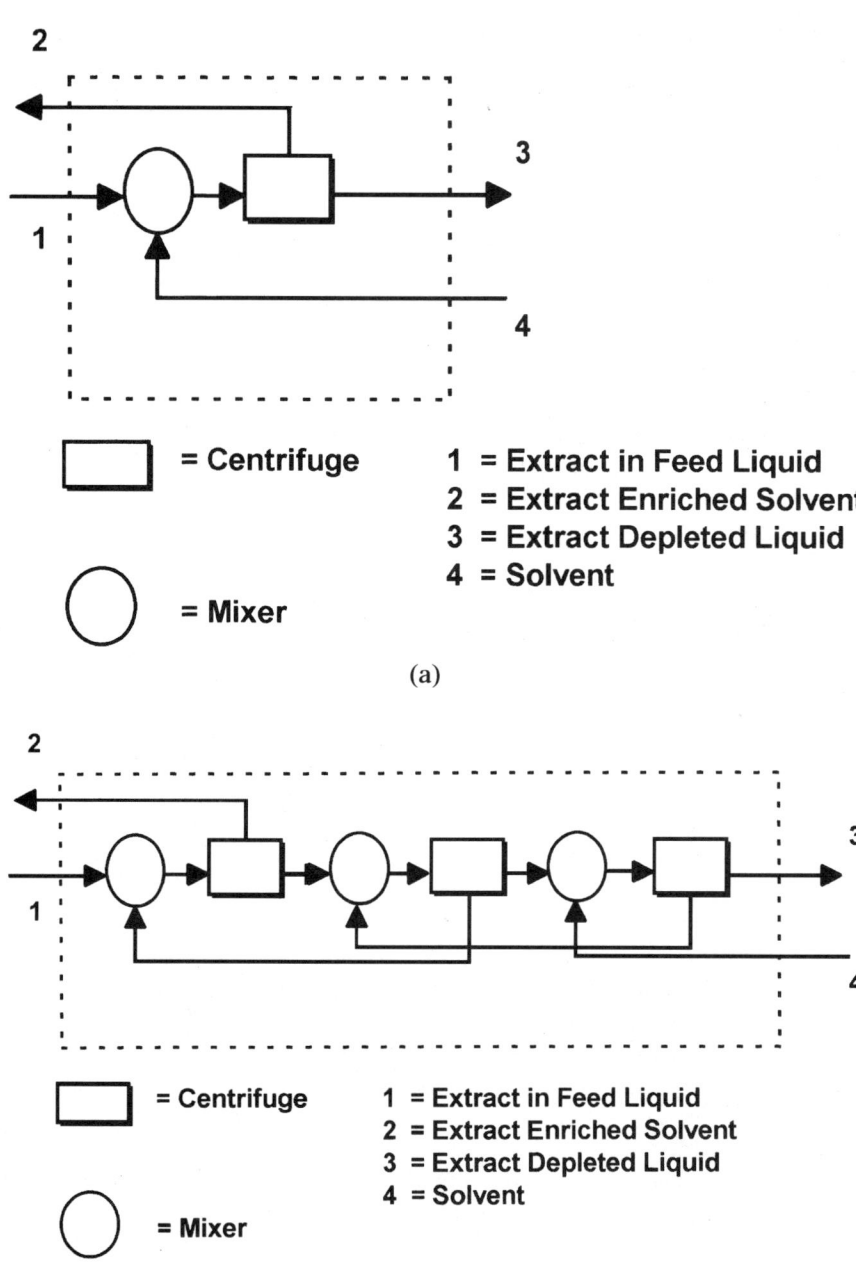

Figure 5.7 (a) Single-stage extraction. (b) Three-stage countercurrent extraction.

separated from the penicillin. The penicillin extract is subsequently precipitated as a raw salt, purified, and packed as a final product. The concentrated phase from the decanter in the first stage is reslurried and sent to the second-stage decanter, the concentrate of which contains a high level of mycelia. It is combined with the wastewater stream from the disk centrifuge for further treatment.

To strip small samples of bacteria from culture media manual and fully automatic high-speed basket centrifuges employing up to 20,000g are used for separation. This is particularly suitable for separating fine particles and where the density difference between solid and liquid is very small. Batch spin tubes are used when the sample quantity is limited. The high-speed version of the spin tubes, that is, the ultracentrifuges, can produce up to 500,000g.

Wastewater treatment

Industrial wastewater. In the wet processing of solids in industrial applications a large quantity of wastewater is generated. Centrifuges are employed to dewater wastewater of 0.2–2% feed solids concentration to a dry cake at very high solids content by mechanical means. As an example, a solid-bowl decanter takes a stream from a lagoon containing some catalyst fines with less than 0.1% solid by weight. It is used to thicken and dewater the stream to 30–60% w/w solid cake with a centrate containing less than 1% solid. Solid bowls 457 to 914 mm (18 to 36 in) in diameter are used to process the waste stream at flow rates of between 60 and 770 L/min (16–200 gal/min).

Municipal wastewater. Among many advantages such as compactness, odor containment, and drier cake, decanters are used to thicken and dewater municipal sewage sludge. Table 5.3 summarizes the applications of decanters in wastewater treatment. These include thickening and dewatering of waste-activated (secondary) sludge, aerobic and anaerobic digested sludge, raw primary, raw mixed primary and secondary sludge, trickling filter, lime sludge, heat-treated sludge, and extended aerated sludge. The percent suspended solid (%ss) in the feed is typically 0.5–2 for thickening, 1–6 for dewatering of nonheated sludge, and 10–12 for dewatering of heated sludge. As a general rule, the higher the feed solids concentration, the higher is also the dewatered cake as far as the concentration ratio is concerned, that is, the cake-to-feed solids ratio is between 4 and 10. Cake solids for dewatering depend on the type of sludge and the feed solids concentration. This can be seen in Table 5.3. The %ss in the cake can be as low as 10 for waste-activated sludge, 20+ for digested sludge, 30+ for raw mixed and 40+ for fibrous

TABLE 5.3 Solid-Bowl Decanters for Wastewater Treatment

	Application	Feed %ss	% Recovery	Cake solids,* % TS	Polymer,*† lb/ton
Waste-activated	Thickening	0.5–1	90–99	6–8	None
	Dewatering with primary	1.0–2.5	90–99	18–22 (28–34)	5–8 (8–15)
	Dewatering without primary	0.5–1.5	90–95	14–16 (16–20)	6–10 (15–20)
Digested	Thickening mixed sludge	2.0–3.5	87–97	5–9	0–5
	Dewatering mixed sludge	2.0–4.0	90–98	15–20 (18–24)	5–10 (11–15)
	Dewatering waste activated sludge	2.0–3.0	90–98	15–17 (18–20)	5–8 (15–20)
Raw primary	Dewatering	3.5–5.0	90–98	20–24 (30–40)	3–5 (8–12)
Trickling filter, rotary contactor	Thickening	1.0–2.5	90–97	6–8	None
	Dewatering	1.0–2.5	90–97	14–16 (17–20)	2–5 (10–15)
High lime sludge	Dewatering	3–6	90	50–55	None
Lime classification	Dewatering	3–6	75	40	None
Heat-treated sludge (without polymer)	Dewatering	10–12	85–90	30–50	None
Heat-treated sludge (with polymer)	Dewatering	10–12	92–99	30–50	3–6
Extended-aeration waste sludge	Thickening	0.4–1.5	70–90 85 90 95	5–13	None <5 5–10 10–15

*High solids dewatering in parentheses.
†Divide by 2 to convert to kg/t.

primary sludge, and up to 40–45 for heat-treated sludge. High-solids decanters can operate with 5–10+% higher cake solids than conventional decanters. Cake solids obtained by high-solids decanters for a given application are shown in parentheses in Table 5.3. The high-solids decanter technology is further discussed in Chap. 16.

The centrate clarity is measured by the solids recovery, which is generally above 90% and frequently exceeds 95%. Most dewatering requires polymer to agglomerate the fine particles so as to get the desired solids capture. The polymer dosage ranges between 2.5 and 5 kg/t (5 and 10 lb/ton) dry solid with conventional dewatering. A high-solids decanter requires typically 5–10+ kg/t (10–20+ lb/ton), depending on the process. On the other hand, polymer is used in small dosage rates in thickening to increase the volumetric throughput, by as much as 33%, to the decanter while maintaining a clear centrate and the desired underflow solids concentration.

Municipal wastewater thickening. High volumes of slurry flow are processed by solid-bowl centrifuges in the thickening of wastewater, mostly for municipalities, from an aqueous 0.5–1% feed to a typical concentrated 4–6% underflow, which is still a pumpable slurry, prior to downstream dewatering. The machine flow rate is generally between 1200 and 2400 L/min (320 and 640 gal/min). It can go up to 3300 L/min (880 gal/min) for the large 1120-mm (44-in)-diameter solid-bowl decanter. Given that centrifugation is energy-intensive and most importantly that the volumetric rate is significant, the power consumption in kilowatts, which is proportional to the product of these two variables, becomes a performance criterion that needs to be minimized. In the past decade, the "specific power consumption" (power consumed per unit volume of slurry feed treated) standard has been nominally at 1.0–1.2 kWh/m^3 (0.23–0.27 kW/gpm). This is achieved by operating with a very deep pool (see Fig. 5.8a), such that the power consumption can be reduced. This is because feed acceleration is the predominant power consumption (see Chap. 2) and is proportional to the second power of the pool radius when the liquid pool is in a solid-body rotation. Theoretically, as the pool radius approaches zero, there is virtually no power consumed in feed acceleration. However, as the centrate flow is brought to a small radius its tangential speed v_θ often increases inversely proportional to the radius, in accordance to a free vortex flow. To avoid the free vortex condition which does not reduce power consumption, the stream is brought into a separate chamber[3] attached to the bowl head with a set of radial or axial vanes corotating at the same angular speed Ω as the conveyor. This will enforce the solid-body rotation condition $v_\theta = \Omega R$ as the centrate liquid flows through the chamber, exiting the machine at a small radius. Further enhancement[3] is possible by turning the flow in the direction opposite to rotation to reduce power consumption. Another advancement has been made recently with a new pressurized centrifuge[4] which effects centerline discharge of both centrate and the thickened underflow. It has a straight cylindrical bowl and an arrangement of decelerating vanes and a center-line discharge back, pressure valve mechanism as shown in Fig. 5.8b and 5.8c. The power consumption is reduced significantly to 0.3–0.5 kWh/m^3 (0.09–0.15 HP/gpm).[4] This is half the power consumption used with conventional thickening technology, already reduced to 1 kWh/m^3 (0.23 kW/gpm). This represents a significant power saving in high-throughput thickening applications.

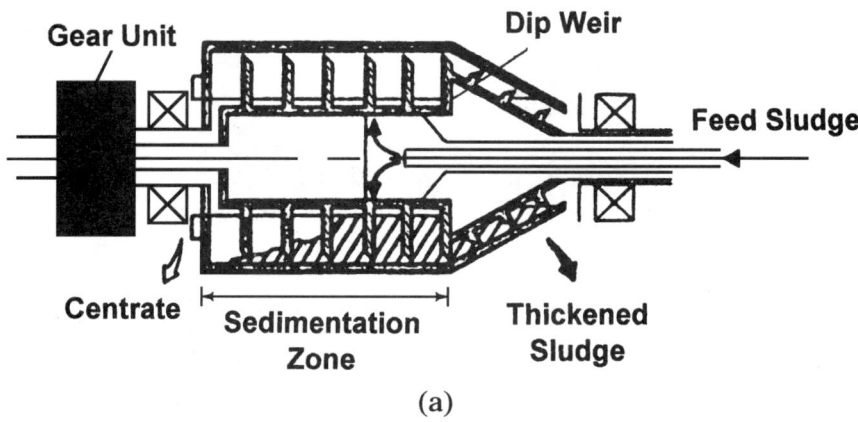

(a)

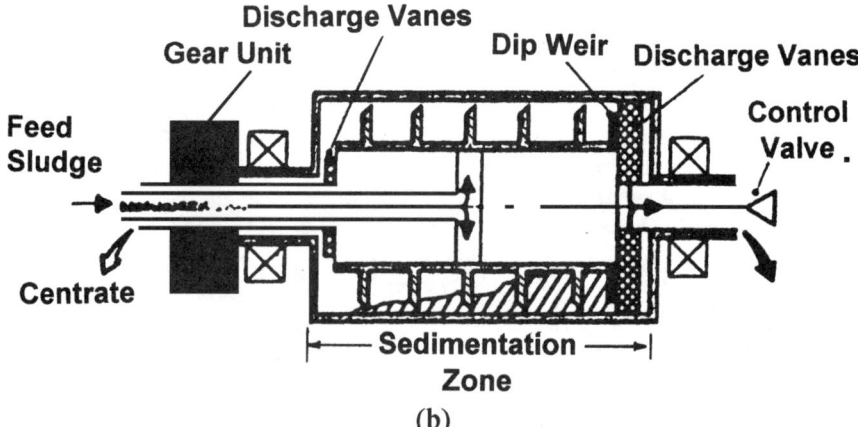

(b)

Figure 5.8 Thickening centrifuge. (a) Cross section of a conventional thickening centrifuge. (b) Cross section of new thickening centrifuge showing cylindrical bowl, discharge vanes, and back-pressure valve. (*After Miyano et al., reprinted with permission from the American Filtration and Separations Society.*)

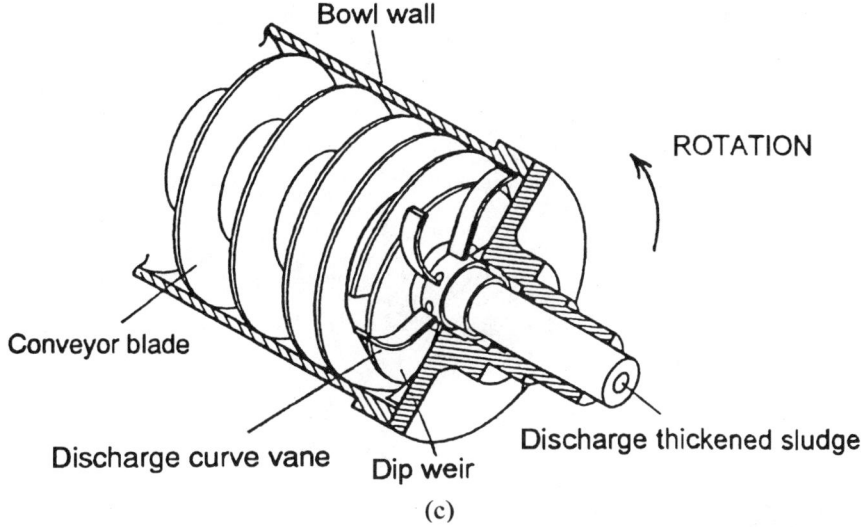

Figure 5.8 Thickening centrifuge. (c) Isometric view of exit discharge vanes in compartment. (*After Miyano et al., reprinted with permission from the American Filtration and Separations Society.*)

In this chapter only a handful of examples of solid-bowl decanters, disks, multibowls, tubulars, and spin tubes have been discussed. Table 5.4 gives a list of materials that have been processed successfully by solid-bowl decanters and Table 5.5 summarizes the typical applications of the disk centrifuges.

References

1. S. Suzuki," Three-Phase Separation Device Using Doubly-Canted Decanter," U.S. patent 4,729,830, 1987.
2. M. Wuensch, K. H. Unkelbach, and G. Arhelger, "CENSOR®—The Decanter with Two Cones and Two Flights—Washes, Separates, and Dewaters Plastics for Recycling," in *Proc. Am. Filt. Sep. Soc. Ann. Conf.* (Chicago, IL, May 3–6, 1993), vol. 7, W. W. F. Leung (ed.), pp. 260–270.
3. A. Shapiro, "Power-Efficient Liquid-Solid Separating Centrifuge," U.S. patent 5,147,277, Sept. 15, 1992.
4. K. Miyano, K. Nishida, and W. W. F. Leung, "Advanced Thickening Centrifuge Equipped with Discharge Vanes," in *Proc. Am. Filt. Sep. Soc. Ann. Conf.* (Valley Forge, PA, April 21–24, 1996), vol. 10, B. Scheiner (ed.), pp. 283–288.

TABLE 5.4 Typical Applications Using Solid-Bowl Decanters

Acetylene sludge	Hexachloroethane	Sodium sulfite
Adipic acid	Iron ore	Sodium thiosulfate
Alum	Iron oxide	Solvent extractions
Aluminum hydrate	Isophthalic acid	Spent grain
Alundum	Lead arsenate	Starch
Ammonium bicarbonate	Lead chromate	Synthetic resins
Ammonium chloride	Manganese dioxide	Tallow
Ammonium nitrate	Manganese ore	Terephthalic acid
Ammonium sulfate	Mica	Tin ore
Aniline sludge	Molybdenum concentrates	Titanium dioxide
Anthracene	Monosodium glutamate	Trisodium phosphate
Barium carbonate	Municipal and industrial	Water-softening sludge
Barium chloride	wastes	Zinc chromate
Barium sulfate	Nickel salts	Zinc glutamate
Beryllium hydroxide	Nylon chips	Zinc oxide
Beryllium sulfate	Packing house waste	
Borax	Paper mill waste	
Calcite	Paraxylene	
Calcium borate	Phenol	
Calcium carbonate	Phosphoric acid	
Carboxymethyl cellulose	Phthalic anhydride	
Carnallite	Pickle liquor	
Catalyst	Polyethylene	
Chromium oxide	Polyolefin	
Clay	Polypropylene	
Coal	Polystyrene	
Copper sulfate	Polyvinyl acetate	
Copper sulfide	Polyvinyl chloride	
Copperas	Potassium carbonate	
Corn germ	Potassium chlorate	
Corundum	Potassium chloride	
Cryolite	Potassium fluoride	
Dicalcium phosphate	Potassium hydroxide	
Dimethyl terephthalate	Potassium nitrate	
Disodium phosphate	Potassium sulfate	
Ethyl cellulose	Potassium tartrate	
Ferrous sulfate	Potato starch	
Fish meal	Refinery waste	
Flotation concentrates	Salt from caustic	
Flotation tails	Salt from glycerine	
Foundry sand	Sewage	
Foundry waste	Sodium bicarbonate	
Fruit extracts	Sodium fluoride	
Gland extracts	Sodium stannate	
Glutamic acid	Sodium sulfate	

TABLE 5.5 Typical Applications Using Disk Centrifuges

Application	Separation	Clarification	Classification	Thickening	Washing
Minerals					
Kaolin		×	×	×	×
Bentonite			×		
Pigments				×	×
Chemicals					
Solvents		×			
Phosphoric acid	×	×			
Agricultural fine chemicals		×			
Benzene	×				
Pharmaceuticals					
Mycelia broth	×				×
Citric acid		×			
Food					
Dairy	×				
Starch		×			×
Yeast		×		×	×
Gluten					×
Soy protein		×		×	×
Vegetable oil or syrup	×				
Juice		×			
Petrochemicals					
Slop and tar oil	×	×			
Catalysts		×			
Waste					
Biomass		×		×	
Magnesium hydroxide			×		×

Chapter

6

Separation in Continuous-Feed Sedimenting Centrifuges

In Chap. 2, Stokes' law modified for centrifugal sedimentation is presented. It has been used in Chap. 3 for transient batch sedimentation of a rotating bowl with zero inflow and outflow. We have yet to discuss the flow field established in the centrifuge bowl when inflow and outflow are both present, which even for the simplest case of a tubular centrifuge is quite complex. This is not to mention the flow pattern associated with the disk and decanter centrifuges discussed in Chap. 4, with the former involving the disk channels and the latter a conveyor screw rotating at a differential speed with respect to the bowl. In this chapter we examine more critically how one might apply Stokes' law to centrifugal sedimentation when there is a continuous flow of feed slurry into the centrifuge and simultaneously a stream of clarified liquid exiting the centrifuge. A new useful methodology on centrifugal clarification, classification, and degritting is introduced.

Plug-Flow Model

Stokes' settling velocity of a spherical particle in a centrifugal field is given by Eq. (2.9). Useful relationships can be established on continuous sedimentation by examining the kinematics of settling of a spherical particle with diameter d in an annular rotating pool,

$$v_s = \frac{dR}{dt} = cRd^2 \qquad (6.1a)$$

where the property constant $c = (\rho_s - \rho_L)\Omega^2/18\mu$. Therefore the time required for a particle to settle from an initial radius R in the pool to the bowl wall with radius R_b becomes

$$t = \frac{1}{cd^2} \ln\left(\frac{R_b}{R}\right) \tag{6.1b}$$

Assuming a plug flow with constant uniform axial velocity across the annular pool from the surface at radius R_p to the bowl wall at R_b, the axial velocity is taken as the mean velocity, which is

$$u = \frac{dx}{dt} = \frac{Q}{\pi(R_b^2 - R_p^2)} \tag{6.2a}$$

From Eq. (6.2a) the residence or retention time of a particle traversing an axial distance x along the bowl is then

$$t = \frac{\pi(R_b^2 - R_p^2)x}{Q} \tag{6.2b}$$

Equating the time the particle takes to settle along the radius [Eq. (6.1b)] to the time it takes to be carried along by the bulk flow in the axial direction [Eq. (6.2b)], we have

$$\frac{R}{R_b} = \exp\left[-\frac{\pi cx(R_b^2 - R_p^2)d^2}{Q}\right] \tag{6.3a}$$

As implied by Eq. (6.3a), a particle with size d initially located at $x = 0$ and radius R, where $R_p < R < R_b$, settles to the bowl wall as it travels along a distance x. Consider the special trajectory whereby a particle located initially at radius R^* settles to the bowl wall after it has traveled the entire bowl length L. This can be obtained from Eq. (6.3a) when $x = L$ and $R = R^*$. Thus

$$\frac{R^*}{R_b} = \exp\left[-\frac{\pi cL(R_b^2 - R_p^2)d^2}{Q}\right] \tag{6.3b}$$

Equation (6.3a) defines the general particle trajectory and Eq. (6.3b) the limiting (or critical) particle trajectory. Theoretically, particles introduced at $x = 0$ with radius $R < R^*$ would not be captured or settled against the bowl wall. Vice versa, particles with $R > R^*$ settle against the bowl wall and get captured. This is depicted in Fig. 6.1. Note that the particle trajectory depends on the size d.

Assuming all the particles of size d are uniformly distributed across the annular pool, the size recovery Rec_d of the particles [also known

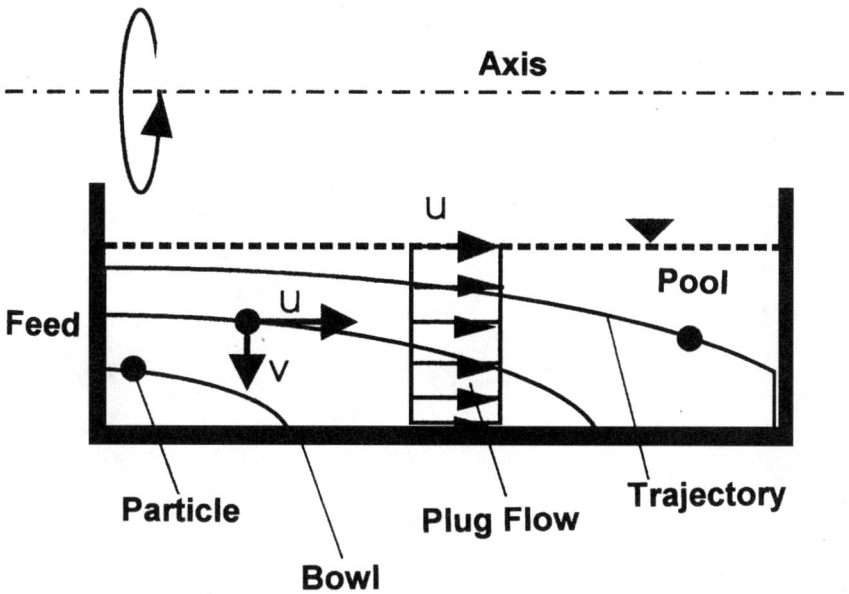

Figure 6.1 Plug-flow model.

as grade efficiency, which is the differential of the cumulative recovery $Z = 1-Y$, with Y given by Eq. (2.17a)] is thus equal to the ratio of the two annular areas,

$$\text{Rec}_d = \frac{\pi(R_b^2 - R^{*2})}{\pi(R_b^2 - R_p^2)} \quad (6.4)$$

Combining Eqs. (6.4) and (6.3b) and Stokes' law [Eq. (6.1a)], we obtain

$$Q = V_{ga} A_{\text{Recd}} = V_{gd} \left\{ \frac{2\pi\Omega^2 L}{g} \frac{R_p^2 - R_b^2}{\ln[1 - \text{Rec}_d(1 - R_p^2/R_b)]} \right\} \quad (6.5)$$

Equation (6.5) defines the maximum volumetric throughput to the centrifuge while it still meets the capture efficiency Rec_d as set for particles of size d. Note that in Eq. (6.5) V_{gd} is the Stokes' settling rate at $1g$. It is solely a function of particle size, particle density, and fluid properties. The group in curly brackets is defined as $A_{\text{Recd}} = Q/V_{gd}$. It is the equivalent settling area of a sedimentation tank at $1g$ to yield identical size recovery Rec_d. It is related to the operating speed, pool depth, and geometry of the centrifuge, as well as to the required size recovery. Note that the particle size d is taken out of this equivalent area definition as it is embedded in V_{gd}, and so is the liquid viscosity and the density difference. When the size recovery Rec_d is set to 50%, the general result [Eq. (6.5)] reduces to the special case corresponding to the well-known Ambler's sigma factor Σ, which for the case of a straight rotating bowl, applicable to tubular and decanter centrifuges, becomes

$$\Sigma_{50\%} = \frac{1}{2} A(\text{Rec}_d = 50\%) = \frac{\pi \Omega^2 L}{g} \frac{R_b^2 - R_p^2}{\ln\left[\frac{2 R_b^2}{(R_b^2 + R_p^2)}\right]} \quad (6.6a)$$

It can be approximated by the simpler form

$$\Sigma_{50\%} = \pi \Omega^2 L \frac{3 R_b^2 + R_p^2}{2g} \quad (6.6b)$$

Equation (6.6b) is adopted more frequently in place of Eq. (6.6a) because of its simple form. The error incurred with this approximation when compared to the exact Eq. (6.6a) is less than 4%.[1] Similarly, when Rec_d is set to 100%, Eq. (6.5) yields

$$\Sigma_{100\%} = \frac{1}{2} A(\text{Rec}_d = 100\%) = \frac{\pi \Omega^2 L}{g} \frac{R_b^2 - R_p^2}{2 \ln(R_b/R_p)} \quad (6.6c)$$

Note that Ambler's $\Sigma_{50\%}$ factor corresponds only to 50% size recovery, and it is well appreciated that $\Sigma_{100\%}$, the required area for 100% capture, is greater than $\Sigma_{50\%}$, the required area for 50% capture.[1] Also, by definition $\Sigma = \frac{1}{2} A_{\text{Recd}}$.

For disk centrifuges similar derivation results in

$$\Sigma_{50\%} = \frac{2\pi \Omega^2 (N-1)(R_2^3 - R_1^3)}{3g \tan \theta} \quad (6.7)$$

where N is the number of disks in the stack, R_1 and R_2 are the inner and outer radii of the disk stack, and θ is the conical half-angle. The Σ factor in units of 0.3×10^3 m² (10^4 ft²) units is in the range of 0.3–5 units for tubulars, 0.3–20 units for decanters, and 1–100 for disks. To scale up from one size to a different size of similar geometry, we use

$$\frac{\Sigma_2}{Q_2} = \frac{\Sigma_1}{Q_1} = \frac{1}{2V_{gd}} \quad (6.8a)$$

$$\frac{(A_{\text{Recd}})_2}{Q_2} = \frac{(A_{\text{Recd}})_1}{Q_1} = \frac{1}{V_{gd}} \quad (6.8b)$$

where the subscripts 1 and 2 correspond to two machines of different sizes, for example, a production machine and a smaller pilot test unit.

Separation in Continuous-Feed Sedimenting Centrifuges 151

Experience has demonstrated that often an adjustment factor is required with the Σ factor scaleup.[2] This is to account for discrepancies that can arise from the following:

1. Flow in the centrifuge is not a plug flow where axial velocity is uniform across the annular pool.
2. Feed introduced into the separation pool is underaccelerated and is often less than the tangential speed of the rotating pool in solid-body rotation, resulting temporarily in lower G force for separation turbulence and mixing (see Chap. 11).
3. Ambler's Σ factor corresponds to only 50% recovery and not 100%.
4. Feed may not have been distributed uniformly to all conical channels in the disk stack for a disk centrifuge.
5. Feed solids may not be distributed uniformly across the pool at the feed location in the clarifier and capture might not necessarily imply that the particle trajectories intersect the bowl wall of a decanter (that is, purely kinematic consideration without dynamics).
6. Secondary flow in a rotating flow disturbs the main flow pattern, which adversely affects sedimentation.
7. Particles settle with hydrodynamic or physical interference with each other.
8. Absence of complicated feed entrance and effluent exit effects.
9. Entrainment of already settled solids not accounted for.

Boundary-Layer Model

In this section we will address one of the more important difficulties ignored in the Σ model, namely, the flow pattern in a rotating bowl, especially for fluid flow with a free surface open to the atmosphere. For this case it has been known that a very thin boundary layer exists at the pool surface. The layer thickness is very small, typically on the order of millimeters, and is moving at a very high flow velocity. Injecting a small droplet of dye onto the surface of a rotating pool shows that the dye immediately shears off and spreads out into a thin film, moving at a much higher velocity as compared to that of an assumed plug flow across the annular pool. The incoming flow from the feed stream enters this boundary layer at one end of the clarifier and the effluent stream exits the boundary layer at the other end of the clarifier. Below this boundary layer (radially outward) is a "stagnant pool" as observed in the reference frame of the rotating bowl. As the particles from the fast-moving boundary layer settle out into the stag-

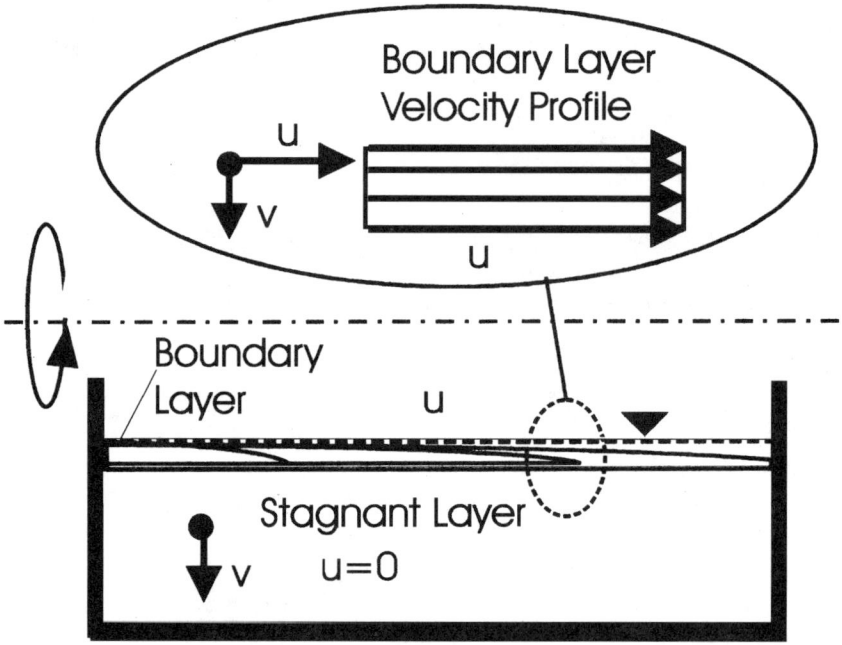

Figure 6.2 Boundary-layer model.

nant pool, essentially the particles settle unimpeded to the bowl wall. This is illustrated in Fig. 6.2.

As the particles settle from the boundary layer at the pool surface to the stagnant zone, an equivalent volume of clarified liquid is displaced. The rate of the "clarified overflow" can be determined from the product of Stokes' settling velocity [Eq. (2.19a)] and the surface area of the pool,

$$Q = V_{sG}(2\pi R_p L) = \frac{\pi L(\rho_s - \rho) d^2 (\Omega R_p \eta_a)^2}{9\mu} \quad (6.9)$$

where the acceleration efficiency of the feed $\eta_a = V_p/\Omega R_p$. Here V_p is the tangential pool speed, Ω the angular speed of rotation, and R_p the radius of the pool surface. From Eq. (6.9), for a given feed rate (which also equals the overflow rate Q), bowl length L, pool radius R_p, bowl speed Ω, density difference $(\rho_s - \rho)$, and liquid viscosity μ, the particles with size d_c that settle out (that is, the cut size) can be determined from

$$\frac{d_c}{d_0} = \frac{3}{\sqrt{\pi}} \mathscr{L}\!\varepsilon = 1.693 \mathscr{L}\!\varepsilon \quad (6.10)$$

where $\mathscr{L}\!\varepsilon$ is the dimensionless Leung number. It is defined as

$$\mathscr{L}\varepsilon = \frac{\sqrt{(Q/L)\mu/(\rho_s-\rho)}}{\Omega R_p \eta_a d_0} \tag{6.11}$$

In both Eqs. (6.10) and (6.11), d_0 represents a characteristic particle size in the slurry sample, say, the median size $d_{50\%}$ or the mean size. For our convenience here it is taken to be 1 μm. Note that $\mathscr{L}\varepsilon$ is a similarity parameter as used for clarification, classification, and degritting scaleup. Typically $\mathscr{L}\varepsilon$ is between 0.5 and 5 for high-speed small-throughput centrifuges with fine size cut of 1–10 μm, between 5 and 20 for medium-speed moderate rate centrifuges for medium-fine particles in the range of 10–45 μm, and above 20 for low-speed high-throughput centrifuges for relatively coarser particles above 45 μm.

Example: solid-bowl decanter. Given the following values:

30 in (762 mm) in diameter by 96 in (2438 mm) long

Q = 200 gal/min (757 L/min)

h_p (pool depth) = 3 in (750 mm)

L = 84.5 in (2146 mm) with 67.4-in (1712 mm) cylinder and 17.14-in (434 mm) beach for clarification

$R_p = R_b - h_p = 15 - 3 = 12$ in (305 mm)

$\eta_a = 0.9$

$\mu = 0.01$ g/cm · s

$\rho_s - \rho = 0.4$ g/cm^3

$d_0 = 1$ μm (10^{-4} cm)

Ω = 2700 rpm

Then

$$\mathscr{L}\varepsilon = \frac{\sqrt{\frac{200 \times 231}{84.5 \times 60}\left(\frac{0.01}{0.4}\right)}}{(2700\pi/30) \times 12 \times 0.9 \times 10^{-4}} = 1.56$$

$$\frac{d_c}{d_0} = 1.693 \times 1.56 = 2.65$$

or $d_c = 2.65$ μm.

Clarification Consideration

Consider the feed particle-size distribution (PSD) shown in Fig. 6.3a. Here $F_f(d)$ represents the cumulative size fraction under a given size

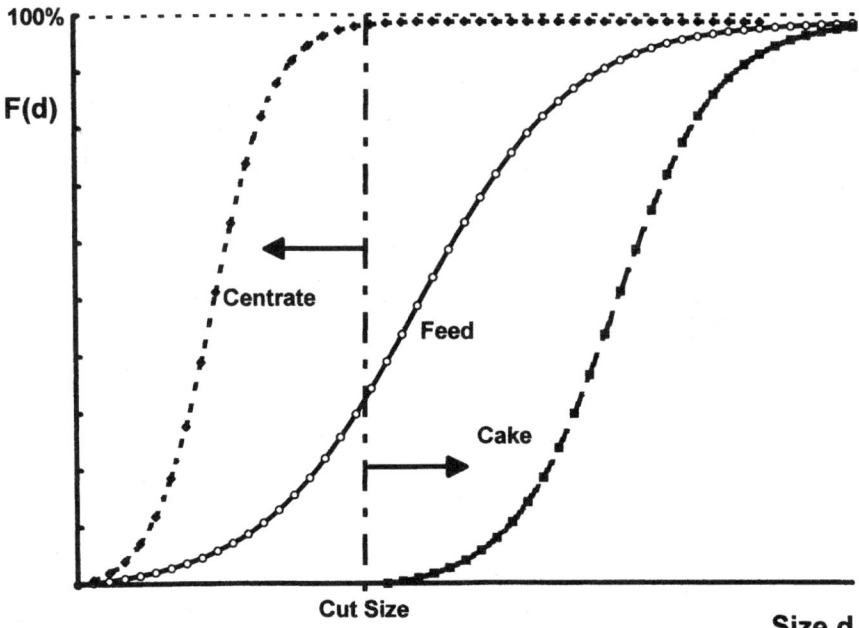

Figure 6.3 PSD and cut size. (a) Ideal situation.

d in the feed. If indeed there is a cut size d_c, predicted by Eq. (6.10), to which a centrifuge can be tuned, above which all the particles are captured by sedimentation and below which they escape uncaptured in the liquid centrate, then the PSD for the centrate and the cake can also be inferred from Fig. 6.3a. By definition of cut size, the particles in the centrate should have a size range $0 < d < d_c$ and those in the cake should have a size range $d_c < d < d_{max}$, where d_{max} is the maximum size of the distribution in the feed. Although Eq. (6.10) can predict d_c, in actuality the PSD for the centrate reveals that there are particles larger than d_c because of resuspension and entrainment of the sediment. Also, there are finer particles, smaller than d_c, which also settle in the cake. This is due to capture of the finer particles by the larger particles, forming an agglomerate during sedimentation. These nonideal deviations, shown in Fig. 6.3b, generally do not impact the results significantly.

Recovery Prediction

The total solids recovery can be predicted by the $\mathcal{L}e$ number. Under a given operating condition (that is, given $\mathcal{L}e$), the cut size d_c can be determined easily from Eq. (6.10). The total particles above d_c have a

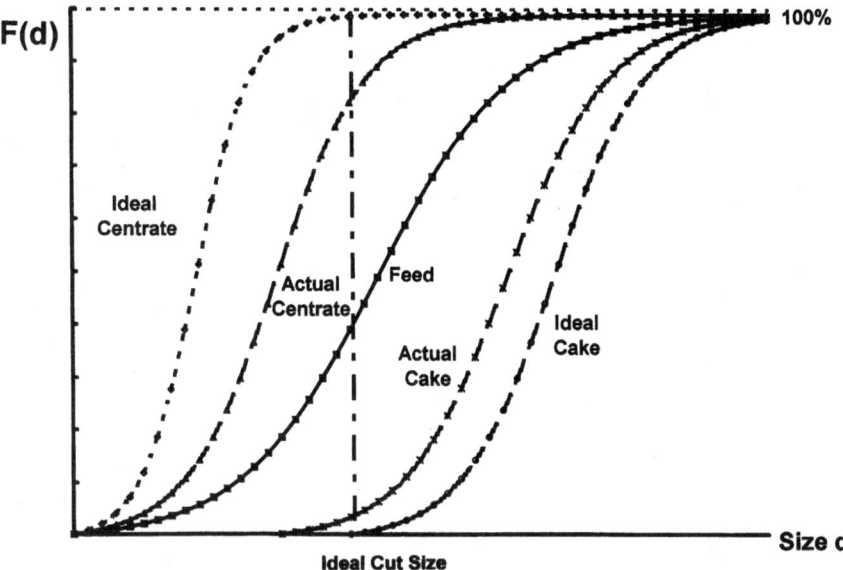

Figure 6.3 PSD and cut size. (b) Actual situation.

cake solids throughput of $[1 - F_f(d_c)]m$, where F_f is the feed PSD (% cumulative under a given size) shown in Fig. 6.3a based on a feed solids throughput m. Therefore the total recovery in cake becomes

$$\text{Rec}_s = 1 - F_f(d_c) = 1 - F_f(\mathscr{L}\varepsilon) \tag{6.12}$$

where d_c is determined by Eq. (6.10). Figure 6.4 shows a typical recovery curve for a polydispersed particle system. As can be seen, the solids recovery curve is a mirror image of the PSD curve. A gradual increase in $F_f(d)$ leads also to a gradual decline of Rec_s with an increase in $\mathscr{L}\varepsilon$. On the other hand, a sharp dropoff in size distribution also leads to a precipitous drop in recovery with $\mathscr{L}\varepsilon$ for a monodispersed particle size distribution (Fig. 6.5).

In classification and degritting, the product is in the fine-particle spectrum, where product recovery is in the liquid centrate and the reject is the coarser fraction which settles out as cake. The recovery in the centrate is thus

$$\text{Rec}_e = 1 - \text{Rec}_s = F_f(d_c) = F_f(\mathscr{L}\varepsilon) \tag{6.13}$$

In Fig. 6.6a $F_f(d)$ is shown for several different slurry applications (such as coal, tar sand, and silica), where the PSD and the fluid prop-

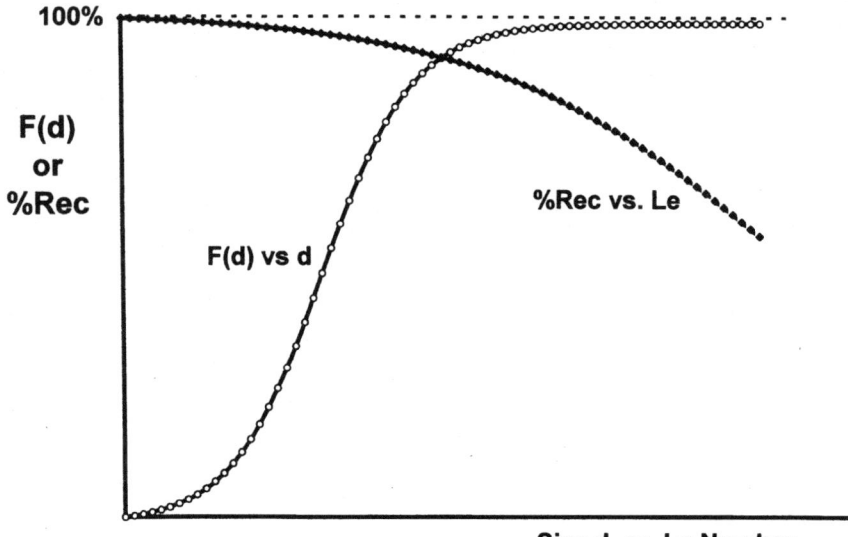

Figure 6.4 PSD of feed slurry for koalin classification.

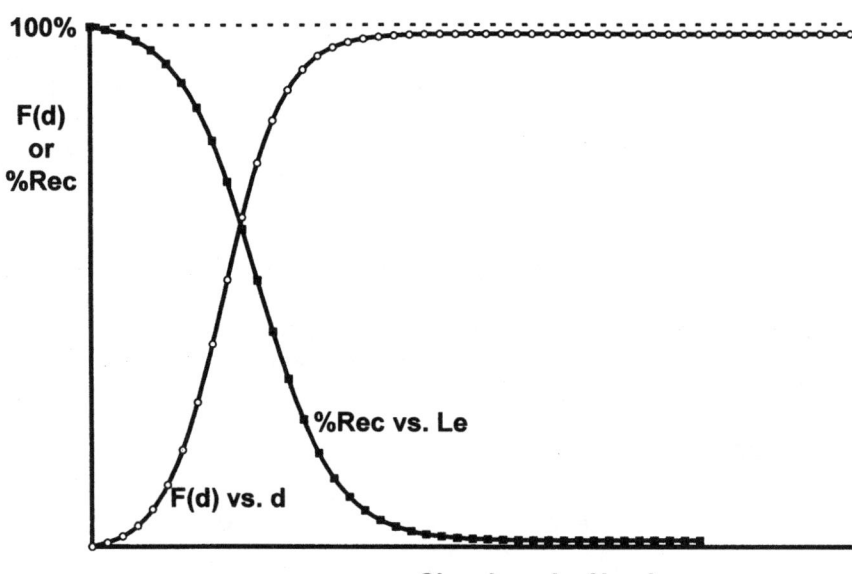

Figure 6.5 PSD and recovery for monodispersed distribution.

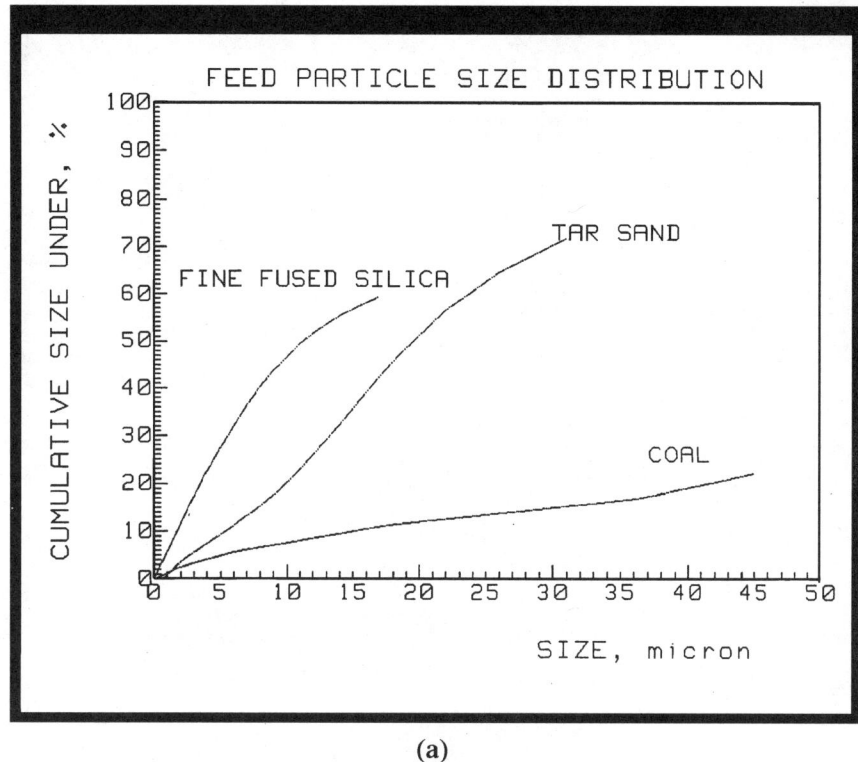

(a)

Figure 6.6 (a) Examples of PSD for mineral slurries.

erties are quite different. In addition, different centrifuge geometries operating at various Gs and pool depths are used to clarify the slurries. Figure 6.6b shows the test results in terms of Rec_s versus $\mathscr{L}\varepsilon$. The corresponding solids curves for each application material represent the theoretical predictions based on Eq. (6.12) and are compared with the test results. The agreement of both production and lab test results with theoretical predictions is very good. This further confirms the validity of the surface flow model and the $\mathscr{L}\varepsilon$ scaleup approach. Also, the mirror image between $F_f(d)$ and Rec_s is clearly demonstrated.

Classification

Sedimenting centrifuges have been used to sort particles by size and density. The PSD of the effluent $F_e(d)$ can be constructed theoretically based on the feed-slurry PSD $F_f(d)$ as follows. For a given centrifuge operating under a given condition this corresponds to a given $\mathscr{L}\varepsilon$. The cut size is determined by Eq. (6.10), or $d_c = 1.693 d_0 \mathscr{L}\varepsilon$. The total

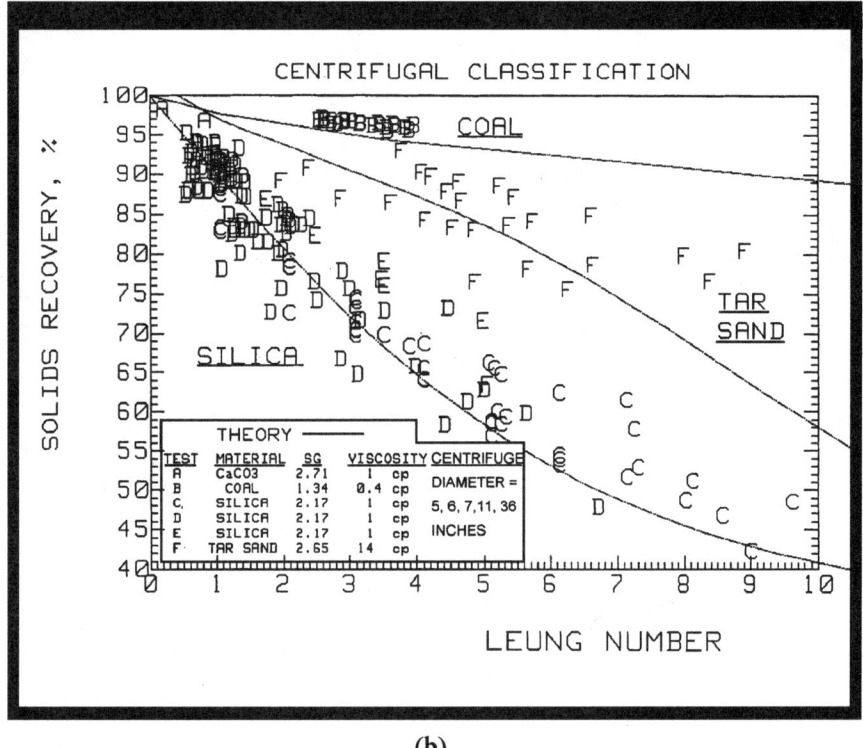

(b)

Figure 6.6 (b) Recovery versus Leung numbers for clarification for minerals in (a).

mass rate of solids leaving the centrate is thus $mF_f(d_c)$, where m is the total feed solids throughput on a dry basis. The cumulative fraction of the particles less than a given size d^* is also $F_f(d^*)$ and the mass rate associated with this fraction is $mF_f(d^*)$, which leaves with the centrate under the premise that $d^* < d_c$. Therefore the cumulative size fraction of particles under d^* in the centrate is then

$$F_e(d^*) = \frac{F_f(d^*)}{F_f(d_c)} \qquad (6.14)$$

In the preceding, $F_e(d^*)$ is a function of the particle size of interest d^* and the operating condition and geometry of the centrifuge through d_c, which is related to $\mathscr{L}e$ by Eq. (6.10). In fact, a plot of $F_e(d^*)$ versus $\mathscr{L}e$ gives an interesting chart (Fig. 6.7). At large values of $\mathscr{L}e$, $F_e(d^*)$ approaches the asymptote $F_f(d^*)$, where separation is absolutely ineffective, resulting in the feed PSD being the same as that of the effluent. The machine is being "flushed" by the feed slurry. As $\mathscr{L}e$ decreases, centrifugal separation takes effect with d_c also reducing, and $F_e(d^*)$ increases toward unity. Here d^* is the product size, presumably

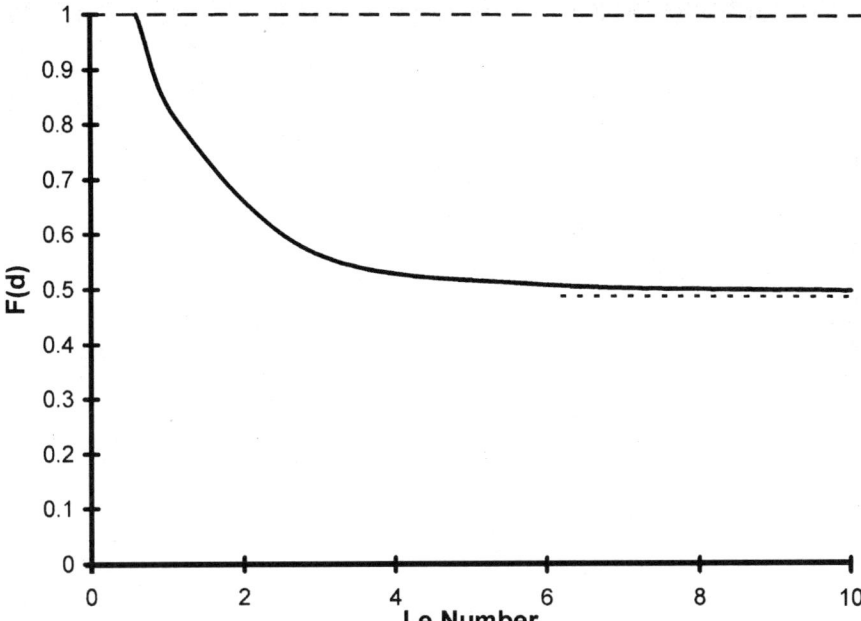

Figure 6.7 PSD of centrate versus Leung number for classification of a given particle size.

≤ 2 μm. At $\mathscr{L}\varepsilon = 0.591(d^*/d_0)$, from Eq. (6.10) the cut size equals the product size under consideration, $d_c = d^*$, in which

$$F_e(d^*) = \frac{F_f(d^*)}{F_f(d_c = d^*)} = 1$$

Based on the foregoing discussion, the behavior of $F_e(d^*)$ for any size d^* follows these two limits, (a) as $\mathscr{L}\varepsilon \gg 1$(or approach ∞), $F_e(d^*) \simeq F_f(d^*)$, and (b) as $\mathscr{L}\varepsilon = 0.591\,(d^*/d_0)$, $F_e(d^*) = 1$.

A family of curves can be generated for $F_e(d^*)$ versus $\mathscr{L}\varepsilon$ with d^* as a parameter. In kaolin classification it is useful to generate $d^* = 0.5, 1,$ and 2 μm, as these are the product sizes. Figure 6.8 shows the feed PSD and the results from a solid-bowl decanter in operation for kaolin classification. The solids curve in Fig. 6.8b is the prediction for the sizes of 0.5, 1, and 2 μm. (The 0.5 μm defines the minimum size of the product.) They compare well with the test data for the corresponding sizes from a large production decanter of 750-mm (30-in) diameter.

Degritting

In degritting, the feed contains a bimodal size distribution, as illustrated by the fine calcium carbonate slurry in Fig. 6.9. Here the fine-

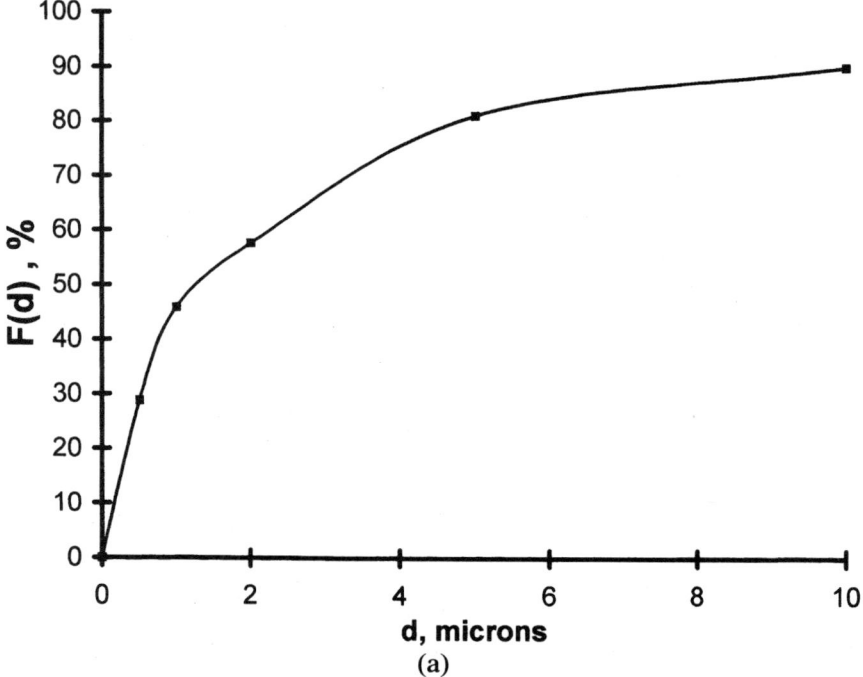

Figure 6.8 (a) PSD of slurry for kaolin classification.

size spectrum contains the valuable product with sizes less than the maximum size d_m (for example, $d_m = 2\text{--}3$ μm), mixed with the coarser grit of foreign materials (such as from grinding media) having minimum size d_g (d_g typically being between 25 and 45 μm), which needs to be removed. In sending the feed slurry through the disk or decanter centrifuge, it is desirable to keep the fine product in a suspension that eventually will leave the centrate stream as product while settling and removing the coarser grit in the cake. Consequently there are two process objectives—reduce the grit level in the centrate product to an acceptable level while keeping the size recovery of the product close to 100%.

Both the size recovery of the product and the grit level in the product are graphed, against $\mathcal{L}_{\varepsilon}$ in Fig. 6.10 for the feed slurry given by Fig. 6.9. The left curve represents the size recovery of the fine product in the centrate, which is simply

$$\text{Rec}_e(d_m) = \frac{F_f(d_c)}{F_f(d_m)} \tag{6.15}$$

where d_c is again related to the centrifuge parameters via $\mathcal{L}_{\varepsilon}$ and can be determined by Eq. (6.10). The right curve represents the fraction of

Separation in Continuous-Feed Sedimenting Centrifuges   161

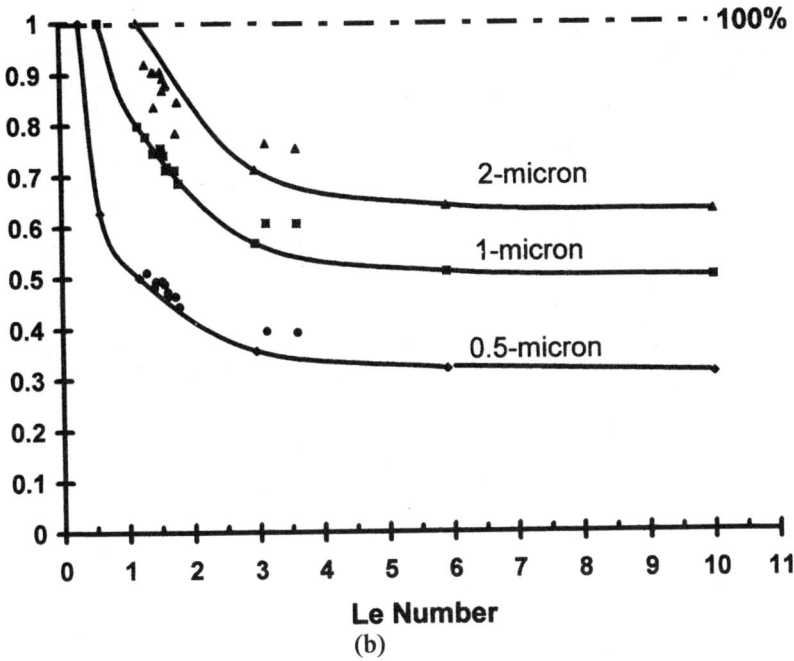

Figure 6.8 (b) PSD of concrete versus Leung number for kaolin classification.

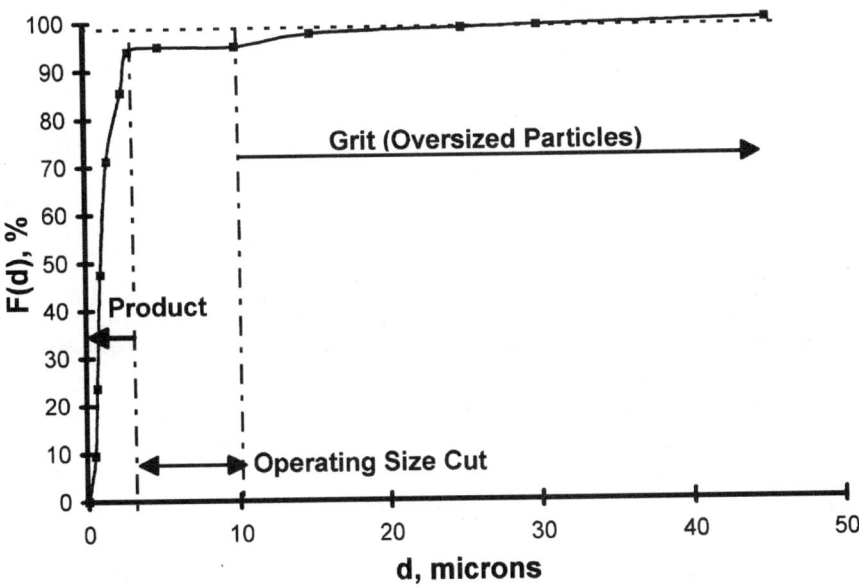

Figure 6.9 PSD of bimodal distribution for degritting.

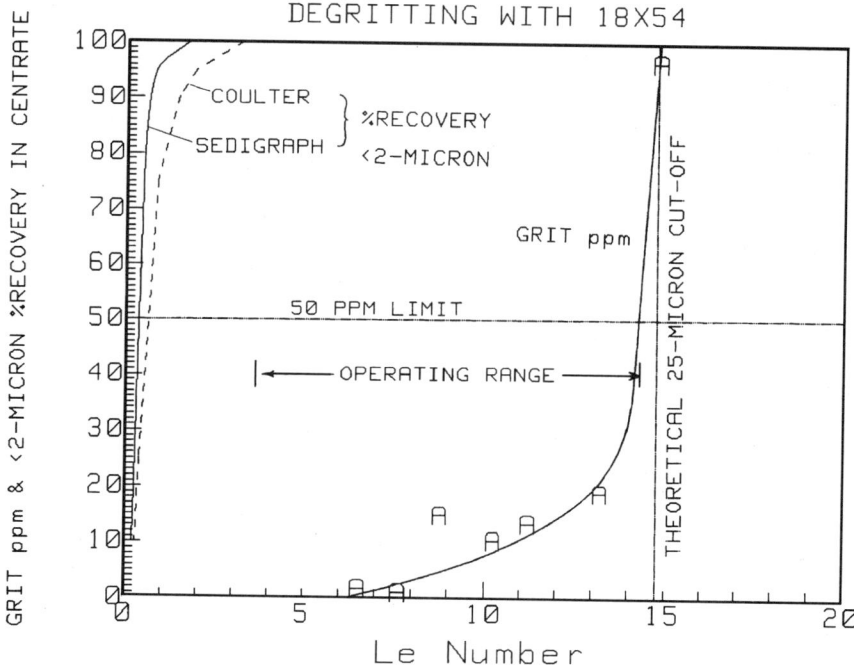

Figure 6.10 Product recovery and product grit concentration versus Leung number for degritting.

grit in the centrate solids, with d_g being the grit particle size. Thus in theory the grit level in the product is present only when $d_c > d_g$. Then the %grit in the product becomes

$$\%\text{grit} = \frac{F_f(d_c) - F_f(d_g)}{F_f(d_c)}, \qquad d_c > d_g \qquad (6.16)$$

$$= 0, \qquad d_c < d_g$$

From Fig. 6.9 it is evident that as $d_c > d_m$ or $\mathscr{L}\varepsilon > 0.591(d_m/d_0)$, the product recovery is 100% while the grit in the product is negligible, provided that $d_c < d_g$ or $\mathscr{L}\varepsilon < 0.591(d_g/d_0)$. This is possible when $d_m \ll d_g$ so that the cut size d_c is in between these two particle size limits.

In actuality, as shown in Fig. 6.10, there is entrainment of the oversize particles such that the grit fraction starts out at a lower $\mathscr{L}\varepsilon$ than $\mathscr{L}\varepsilon = 0.591(d_g/d_0)$. The measured grit should be graphed against $\mathscr{L}\varepsilon$ to determine this behavior, as shown in Fig. 6.10. Although not shown in Fig. 6.10, it is possible that the size recovery has not yet reached 100% when $\mathscr{L}\varepsilon = 0.591(d_m/d_0)$ because of the entrapment of fine particle product during sedimentation of the coarse grit particles. Despite this possible complication, it is still best when $d_m \ll d_g$ so that the centrifuge can be tuned to the appropriate size cut, as facilitated by the use of the scaleup factor $\mathcal{L}e$. This would result in a much better balance between high product recovery and low grit level in the product.

In this chapter we reviewed the conventional Σ approach and introduced the new $\mathcal{L}e$-number approach. This new approach is simple, and it has the correct mechanics of separation associated with free-surface rotating flow. The methodology has been proven extensively with sedimenting applications related to centrifugal clarification, classification, and degritting.

References

1. W. W. F. Leung, "Centrifugation" in P. Schweitzer (ed.), *Handbook of Separation Techniques for Chemical Engineers,* 3d ed., McGraw-Hill, New York, 1997, pp. 4-78 to 4-86.
2. W. W. F. Leung, "Centrifuges" in Perry and Green (eds.), *Chemical Engineers' Handbook,* 7th ed., McGraw-Hill, New York, 1997.

Chapter

7

General Principles of Filtering Centrifuges

In filtering centrifuges, under the centrifugal field, both the liquid and the solids move toward the filtration medium, which is supported inside a perforate basket. The solids are withheld by the filter medium building a cake while the liquid filters through the deposited cake and subsequently the filter medium. The filter medium can be a wedge-wire screen or a thin perforated plate secured inside a basket or a cage; or a piece of fabric or filtering paper supported inside a perforated bowl. Because the solids in the slurry are coarser than those processed by sedimenting centrifuges, settling under centrifugal gravity is relatively fast, leaving filtration generally as a possible process-limiting step. Washing and subsequent dewatering of the cake are common practice using filtering centrifuges, and each of these steps may also become limiting. Filtering centrifuges are also known as centrifugal filters, wringers, extractors, or dryers. There is a difference between the various types of machines as to the condition of the feed—batch, intermittent, or continuous feed—and in the manner with which the cake solids are removed from the basket. The various types of filtering centrifuges will be discussed in the next two chapters. In this chapter our discussion centers on the principles of bulk filtration, desaturation, and washing of centrifuge cake.

Bulk Filtration

In the case of a fast settling solid, the cake builds up rapidly, leaving the overlying pool of clarified liquid to filter through the cake and

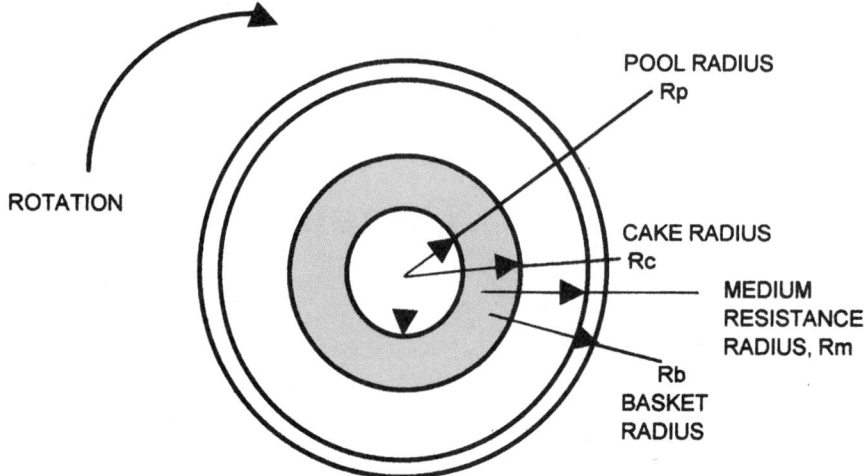

Figure 7.1 Cross section of filtering basket.

subsequently the filter medium. For finer solids or for particles with a density close to or even lighter than that of the liquid, sedimentation of the solids is slow—sometimes the solid particles even float in the latter—and sedimentation and filtration occur simultaneously. However, even in this case the slurry pool still has to filter through the cake that continues to grow and the filter medium. Some useful relationships can be derived from a close examination of the problem.

The bulk filtration rate Q for a basket with an axial width b can be determined by Darcy's law. The resistance to flow in the three regions—cake, pool, and filter medium—is shown in Fig. 7.1.

Cake. For cylindrical coordinates in the rotating reference frame, the filtration velocity $u(R)$ is according to Darcy's law,

$$u = -\frac{K}{\mu}\left(\frac{dp}{dR} - \rho\Omega^2 R\right) \qquad (7.1)$$

where μ and ρ are the viscosity and the density of the liquid, Ω is the angular speed of the centrifuge, and K is the permeability of the cake (m^2), which is related to the specific cake resistance α_s (m/kg) by the relationship $\alpha_s K \rho_s = 1$, with ρ_s being the solid density. R_p, R_c, and R_m the radii of the liquid pool surface, the cake surface, and the filter medium adjacent to the perforated basket wall, respectively. These quantities are labeled in Fig. 7.1. Note that as $dp/dR < 0$, u is positive along the increasing radius, which is indicative of the fact that fluid flows in the direction of the decreasing pressure gradient.

Another driving force with regard to filtration velocity u along the increasing radius is the centrifugal acceleration $\rho\Omega^2 R$, as evident from Eq. (7.1). By invoking mass conservation,

$$\tilde{Q} = \frac{Q}{b} = 2\pi R u = \text{constant} \tag{7.2a}$$

or

$$\tilde{Q} = -2\pi R \frac{K}{\mu}\left(\frac{dp}{dR} - \rho\Omega^2 R\right) = \text{constant} \tag{7.2b}$$

where $\tilde{Q}$ is the filtration rate per unit width of the basket. For incompactible cake, the cake permeability K is constant, and Eq. (7.2b) can be integrated between the radius of the cake surface R_c and the radius of the filter medium R_m. Otherwise an average cake permeability across the cake thickness can also be used. Thus

$$p_m - p_c = -\frac{\mu \tilde{Q}}{2\pi K}\ln\left(\frac{R_m}{R_c}\right) + \frac{1}{2}\rho\Omega^2(R_m^2 - R_c^2) \tag{7.3}$$

where p_m and p_c are the pressures at the medium and at the cake surface.

Pool. Integrating the pressure equation $dp/dR = 1/2\rho\Omega^2 R$ between the pool surface R_p and the cake surface R_c,

$$p_c - p_p = \frac{1}{2}\rho\Omega^2(R_c^2 - R_p^2) \tag{7.4}$$

where p_p is the pressure at the pool surface, which also corresponds to the ambient. It is evident that the pressure distribution in the liquid pool above the cake is quadratic.

Filter medium. It is reasonable to assume, similar to Darcy's law, that the pressure drop across the filter medium is proportional to the flow rate. Hence

$$p_m - p_b = \mu r_m \frac{\tilde{Q}}{2\pi R_b} \tag{7.5}$$

where p_b is the pressure at the basket wall, which is also at ambient. Note that the pressure drop across the filter medium also includes the pressure drop across the cake heel. This is taken into account by defining r_m (m^{-1}) as the combined resistance due to both the cake heel and the filter medium.

The total pressure drop is then the sum of all three contributions,

$$dp_{total} = p_b - p_p = (p_b - p_m) + (p_m - p_c) + (p_c - p_p) = 0$$

The net change in pressure should be zero given that the pool and the basket wall are both at ambient. Substituting Eqs. (7.3) to (7.5) into the preceding equation, and integrating and rearranging,

$$Q = \tilde{Q}b = \frac{\pi b K \rho \Omega^2 (R_b^2 - R_p^2)}{\mu [\ln(R_b/R_c) + Kr_m/R_b]} \tag{7.6}$$

This is the well-known centrifugal filtration equation. The driving force is the differential pressure established between the basket wall and the pool surface, which is used to overcome the two resistances in series—the cake and the filter medium, including the cake heel. The logarithmic term is atypical for resistance configured in a cylindrical coordinate system, similar to those problems related to conduction of heat, mass, or electricity.

For incompactible cake, the pressure distribution and the filtration rate depend on the resistance of the filter medium and the permeability of the cake. Figure 7.2 shows an example where $R_p/R_b = 0.6$ and

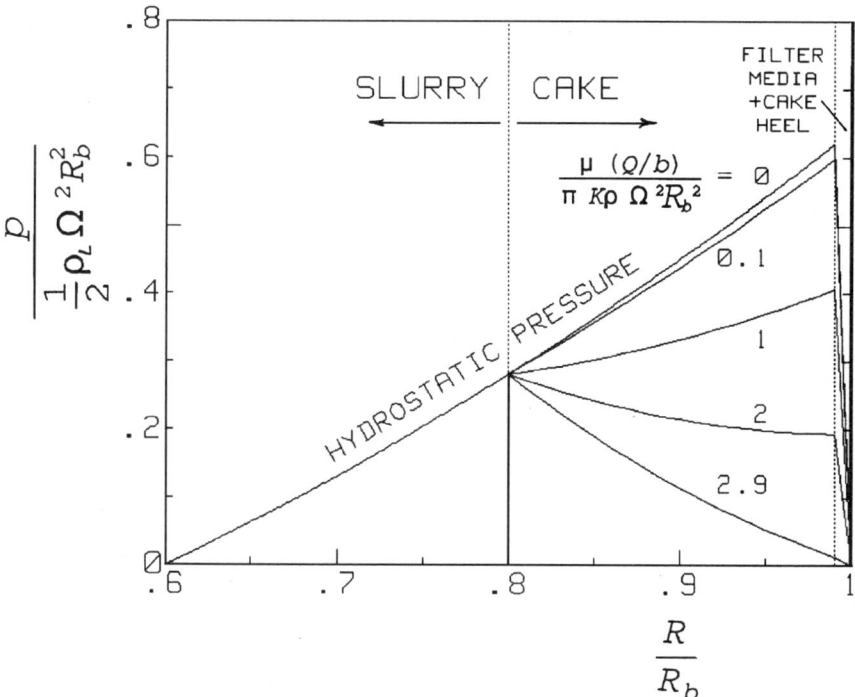

Figure 7.2 Pressure distribution in filtering centrifuge under bulk filtration.

$R_c/R_b = 0.8$ for increasing values of filter medium resistance. The pressure at $R = R_b$ corresponds to the pressure drop across the filter medium Δp_m with the fluid pressure at the basket wall equal to the ambient, which has a gage pressure $p = 0$. As can be seen in Fig. 7.2, the filtration rate as well as the pressure distribution depend on the resistance of the medium and the cake. High medium resistance is due to either medium blinding or high cake-heel resistance as a result of glazing of the cake from mechanical pressure exerted by repeated contact with the scraper knife as in a basket, or both. This additional resistance will result in a greater penalty on the filtration rate. In the extreme case, when the medium resistance becomes too high, to the extent that the filtration rate stalls, the pressure distribution at equilibrium is once more reduced to the quadratic form, as discussed in Chap. 2.

Optimal condition

For a continuously filtering centrifuge undergoing bulk filtration, the machine is optimally operated such that there should be no free-standing liquid column in the feed zone. An excess amount of liquid might cause flooding in form of either liquid flowing over the cake or liquid streams cutting through the cake. The condition that there be no free liquid column above the cake requires $R_p = R_c$ in Eq. (7.6), which leads to the following result:

$$\frac{Q/b}{\pi K \rho \Omega^2 R_b^2} = \frac{\pi[1 - (R_c/R_b)^2]}{\mu[\ln(R_b/R_c) + Kr_m/R_b]} \qquad (7.7)$$

Equation (7.7) is plotted in Fig. 7.3 for various values of filter medium resistance. For a given filter medium the bulk filtration rate is low at both large cake thickness, where R_c/R_b is small, as well as small cake thickness, where R_c/R_b is large. In the former case, the driving force is too little despite the resistance also being small, whereas in the latter case the resistance is too large despite the increased driving force. It follows that between these two extremes there must be a maximum filtration rate independent of the cake radius, but which depends primarily on the resistance of the filter medium. For example, with $Kr_m/R_b = 0.05$, as illustrated in Fig. 7.3, the maximum rate occurs at $R_c/R_b = 0.85$, which corresponds to a cake thickness of $h/R_b = 0.15$. For optimal operation, the medium resistance to cake resistance should be small, with $Kr_m/R_b < 5\%$ (see Fig. 7.3). Otherwise the filtration rate can be limiting. Based on these considerations, the cake thickness can be reduced somewhat. However, it should not be too small so that the throughput to the centrifuge is compromised. For negligible medium resistance the filtration flux decreases monotonically as the cake thickness increases with decreasing R_c.

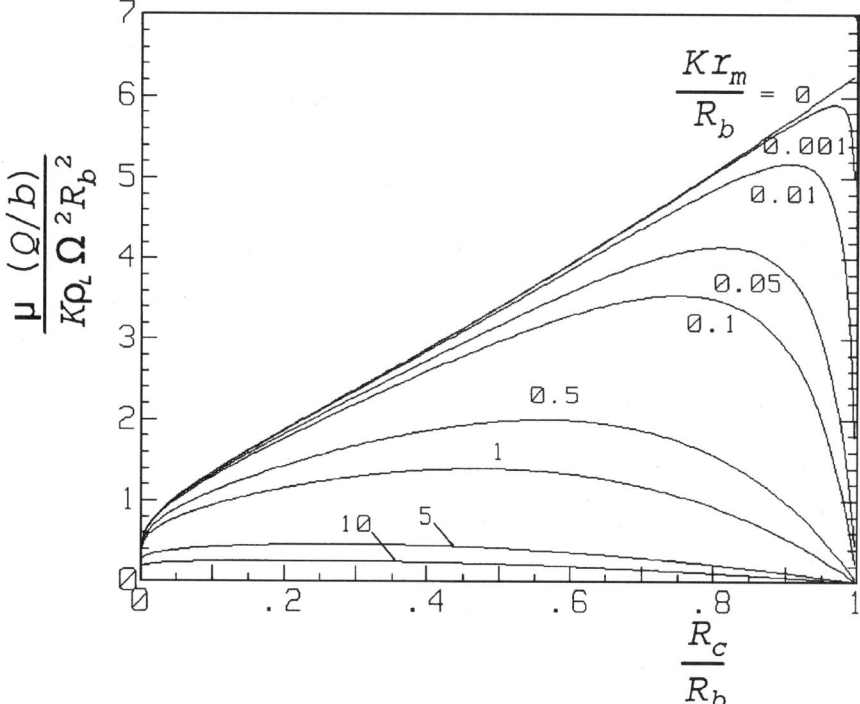

Figure 7.3 Centrifugal filtration rate as a function of both cake and medium resistance.

Thin cake with negligible medium resistance

It is known that the specific resistance for centrifuge cake, especially for compactible cake, is greater than that obtained from either pressure or vacuum filtration. Therefore the specific resistance has to be measured directly from centrifuge tests (and not taken from pressure or vacuum filtration data) for different cake thicknesses so as to scale up accurately to reliable centrifuge performance. In the case of negligible resistance of the filter medium Eq. (7.6) can be written as

$$\frac{\tilde{Q}/2\pi R_b}{u_0(h_p/R_b)(R_b + R_p/2R_b)} = \frac{1}{\ln\left(\dfrac{R_b}{R_b - h}\right)}$$

$$\approx \frac{1}{h/R_b + 1/2(h/R_b)^2 + 1/3(h/R_b)^3 + \cdots} \quad (7.8)$$

where the characteristic superficial filtration velocity, or more correctly, the filtration flux is

General Principles of Filtering Centrifuges 171

TABLE 7.1 Typical Operating Range of Filtering Centrifuges

Type of centrifuge	G/g	Minimum feed solids, %w/w	Minimum particle size, μm	u_0, m³/m²·s (gal/min·ft²)
Vibrating	30–120	50	300	7.0×10^{-3} (10)
Tumbler	50–300	50	200	3.5×10^{-3} (5)
Screen scroll	500–1500	35	75	2.1×10^{-3} (3)
Pusher	300–1200	40	75	1.4×10^{-3} (2)
Screen bowl	500–2000	20	20	7.0×10^{-4} (1)
Peeler/advanced vertical basket	300–1600	10	10	$3.5 \times 10^{-5} – 3.5 \times 10^{-4}$ (0.05–0.5)
Pendulum	200–1200	10	5–10	$3.5 \times 10^{-5} – 7.0 \times 10^{-5}$ (0.05–0.1)

$$u_0 = \frac{GK}{\nu} \qquad (7.9)$$

with $G = \Omega^2 R$, and where $h_p = R_c - R_p$ is the pool depth and $h = R_b - R_c$ is the cake height. It is of interest that u_0 depends only on the filtrate liquid kinematic viscosity ν, the cake permeability K, and the centrifugal gravity G. In Table 7.1 u_0 has been estimated for various types of filtering centrifuges based on their typical applications and operating conditions. It can range from 7×10^{-3} m³/m²·s (10 gal/min·ft²) for a vibrating centrifuge processing coarse particles with 300 μm minimum size to as little as 5×10^{-5} m³/m²·s (0.1 gal/min·ft²) for a vertical centrifuge processing fine particles with particle sizes of 5–10 μm. In the former case, the throughput is high and the product cake has much lower selling value measured on a dollars per tonne basis, whereas in the latter case the cycle time is long, corresponding to much lower average throughput, and the high-value product cake is measured on a dollars per gram basis, from which the lower throughput can be justified.

It has been shown in Eq. (7.8) that the logarithmic term can be approximated by an infinite series expansion in h/R_b, which in most cases is less than 0.2 and rarely exceeds 0.3. It is likely that for all practical purposes the leading terms in this infinite series can be used to approximate the exact form, especially when $h/R_b \ll 1$. Figure 7.4 compares the logarithmic term (that is, the exact solution) and the first and second terms in the series as leading approximations. At $h/R_b = 0.3$ the exact computation yields a value of 1.40, whereas the first-term approximation gives only 1.67 (a 19% error) and the first two terms combined give 1.45 (a 3.5% error), which approaches the exact value much more closely. Likewise at $h/R_b = 0.15$, the exact value gives 3.08, whereas the approximation using the first term gives 3.33 (8% error) and the first two terms give 3.1 (0.007% error). It is evident from Fig.

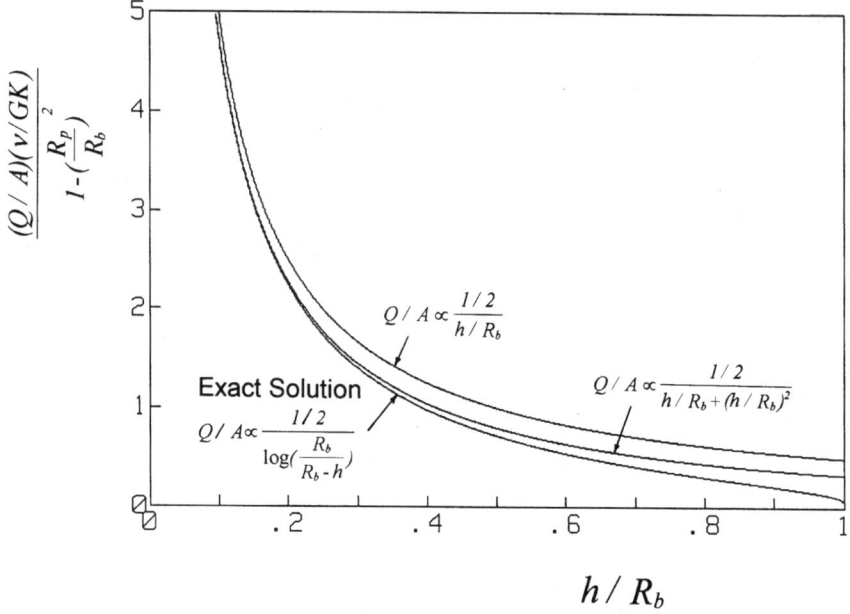

Figure 7.4 Centrifugal filtration rate as a function of cake thickness.

7.4 that the two-term approximation yields fairly accurate results for the range of cake height encountered, and the first term gives a reasonable estimate only when $h/R_b < 0.15$.

For the special case when $h/R_b \ll 0.1$ the leading approximation gives a simple linear relationship,

$$\frac{q}{2\pi R_b} \approx u_0 \frac{h_p}{h} \tag{7.10}$$

Note that in this approximation the filtrate flux is proportional to the characteristic filtration velocity u_0 and the driving liquid head h_p, and inversely proportional to the cake height h. This "linearized" version of the centrifugal filtration equation is applicable when the cake thickness is small compared to the bowl radius, to the extent that the cylindrical geometry is essentially replaced by a linear planar geometry.

Desaturation

Desaturation is the process whereby the liquid from the pores of the cake drains out under external body or surface force. Here desaturation is by the centrifugal body force exerted on the liquid. In Chap. 2 saturation S is defined as the fraction of the void space in the cake occupied by the liquor (mother liquor or wash liquid). Figure 7.5a shows a simplified drainage model, where the void space of the cake, which was previously filled with liquid, starts to drain under centrifugal gravity. A typical desaturation curve that changes with time for a fixed cake mass is shown in Fig. 7.5b. Here S_{total} denotes the total liquid saturation, which has both the transient component S_T and the equilibrium component S_∞. The transient component S_T decreases with time under centrifugal gravity, say G_1. At large time, S_{total} eventually reaches an equilibrium saturation $S_{\infty 1}$ which does not change with t. At a higher gravity G_2, the desaturation curve falls below that for G_1, as illustrated in Fig. 7.5b. Again, an equilibrium $S_{\infty 2}$ is reached at large t. In fact, this equilibrium value may be the same as, or perhaps lower than, that at G_1. This depends on the capillary number, as will be discussed later in this chapter.

The equilibrium saturation is determined by a balance under equilibrium of the capillary forces, which hold the liquid in the cake, and the centrifugal force, which acts to drive the liquid out of the cake. An analogy is the leakage of a liquid out of an originally filled container that has a hole punched in the bottom. A tube is inserted through this hole with the opening of the tube inside the container, positioned at a small distance above the container bottom. The liquid in the container drains through the tube until the liquid level drops below the tube opening, after which the drainage stops.

Liquid saturation

Before examining the various desaturation mechanisms, it is useful to review the sources of the liquid moisture in a centrifuge cake. With reference to Fig. 7.6, liquid is trapped as (1) bound/inherent liquid inside particle pores or at the particle surfaces; (2) pendular liquid at particle-particle contact points; (3) capillary liquid column in fine pores forming continuous passageways; and (4) free liquid, the bulk of which can be drained out relatively fast except for the moisture coated on the surface of the particles, which takes longer to drain. Liquids (2) to (4) can be removed to various degrees by centrifugation, but (1) cannot be removed since the surface forces (physical force, such as capillary or electrical, or chemical force) are stronger than the applied centrifugal force. Among the three sources of moisture that can be removed by centrifugation, only desaturation of the free liquid (4) and, to a much lesser extent, the liquid at contact points (2) are a function of time. Depending on the dewatering time and the surface morphology and roughness of the particles, the wet cake desaturates from a state of full saturation with $S = 1$ and drops to a condition where $S < 1$. At increasing time, S approaches equilibrium saturation S_∞, which is a function of G, the capillary force,

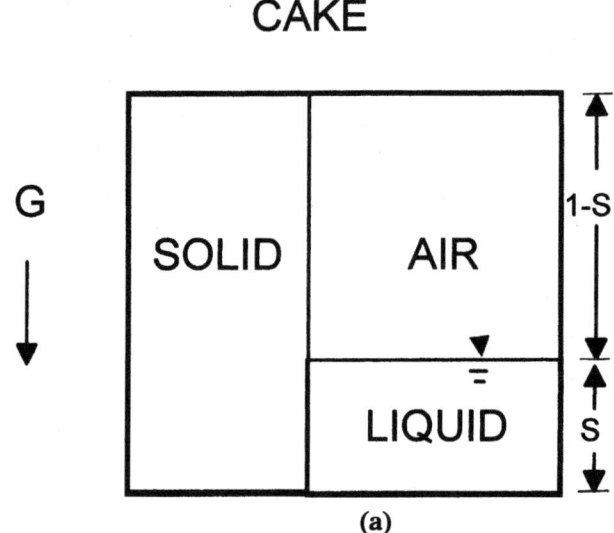

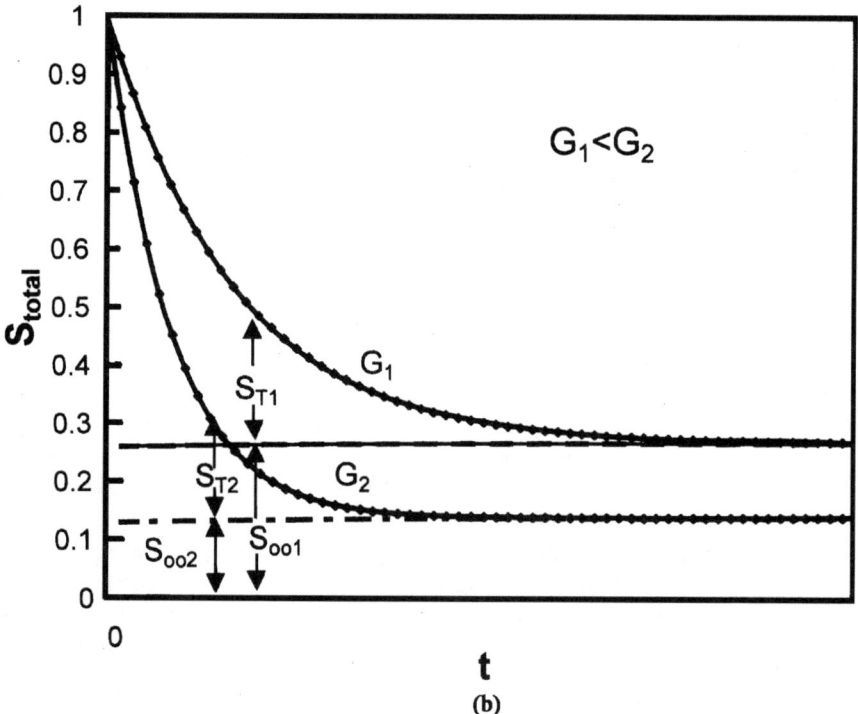

Figure 7.5 (a) Schematic of centrifuged cake showing void and water saturation. (b) Total liquid saturation of centrifuged cake as function of time.

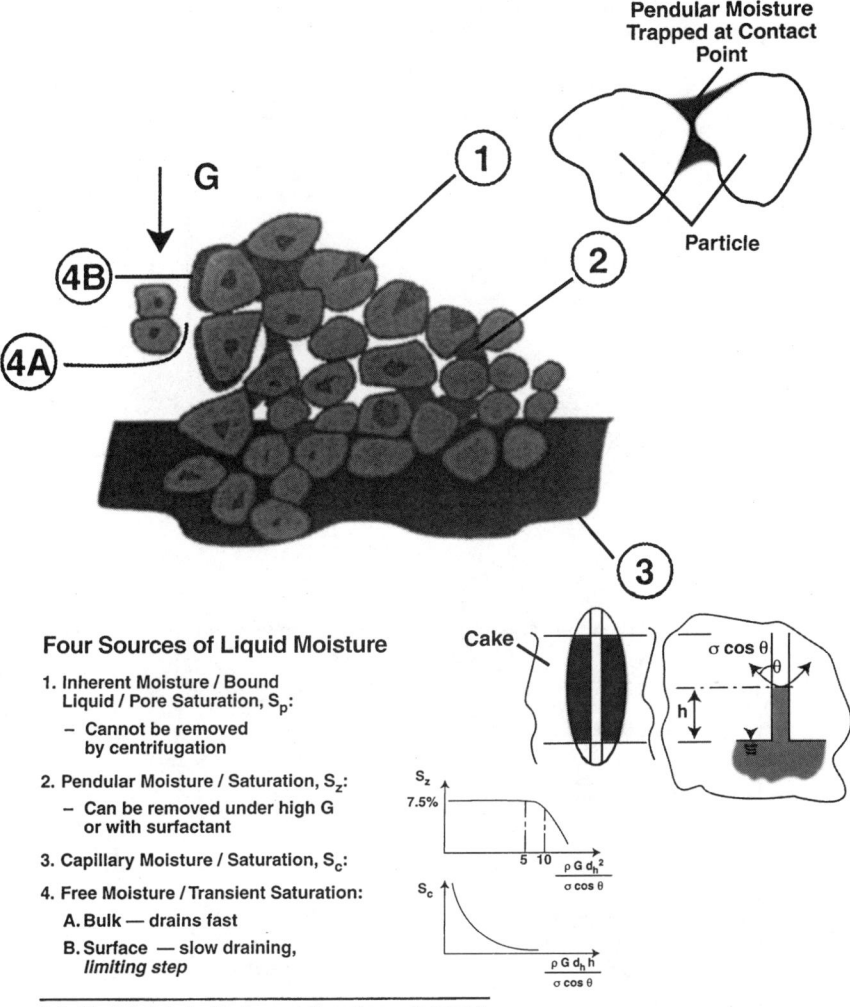

Figure 7.6 Sources of liquid saturation in cake.

and the amount of bound liquid in the particles trapped internally or externally.

Mechanics of desaturation

Desaturation of the liquid cake ($S < 1$) begins promptly after bulk filtration is completed, at which time the liquid level starts to recede below the cake surface. The liquid in the coarse capillaries with higher permeability drains out first, followed by that in the finer capillaries with lower permeability. Figure 7.7 shows scenarios A, B, C, and D of cake desaturation, representing a time sequence in the case of a batch

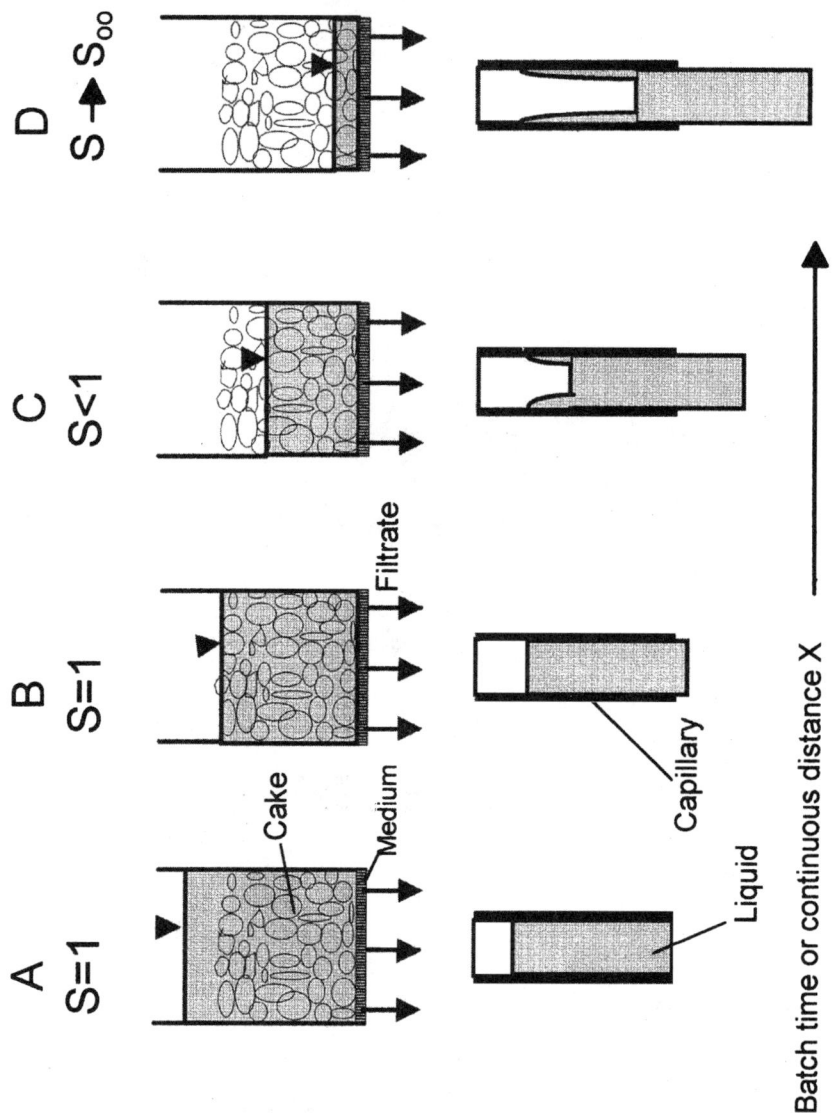

Figure 7.7 Time or spatial sequence of filtration and desaturation for batch and continuously filtering centrifuges.

basket, or spatial "snapshots" along the basket in the case of a continuously filtering centrifuge. The upper diagrams in Fig. 7.7 show schematic cross sections through the cake. The lower diagrams correspond to the drainage of a typical capillary, which initially starts with bulk drainage similar to the pipe flow with macroscopic drop in liquid level inside the capillary, to be followed by a film flow of the residual

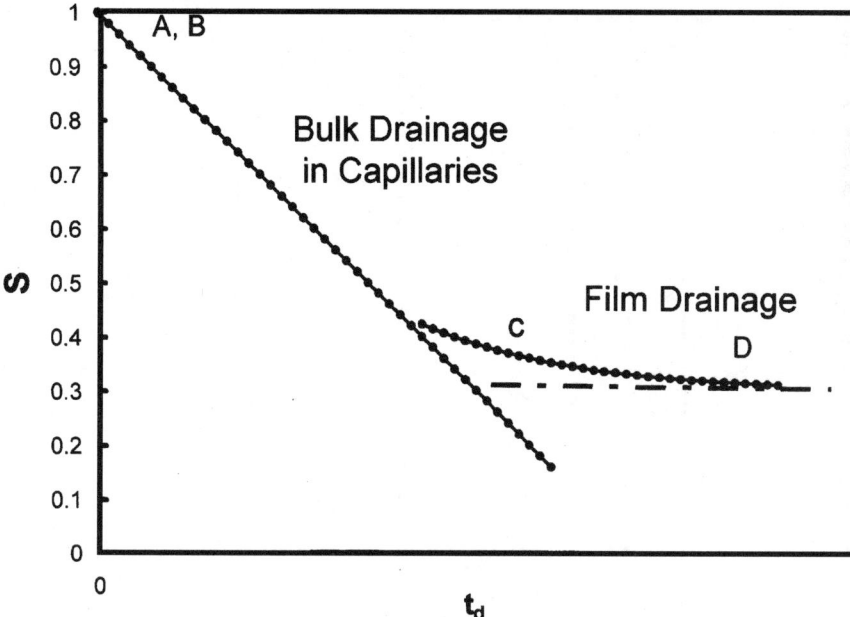

Figure 7.8 Bulk drainage and film drainage of capillary in cake as a function of dimensionless time.

liquid over the inner surface of the tube. This scenario takes place over time t in a batch basket or in a continuously filtering centrifuge as the cake travels over the basket under constant G. The location of the time sequence A, B, C, D marked along the S versus t curve is seen in Fig. 7.8. It can also be interpreted as a spatial sequence in a Langrangian description for a continuously filtering centrifuge where $X = Vt$, with X being the cake travel distance, t the time, and V the average cake velocity. In this figure points A and B correspond to conditions where $S = 1$, point C corresponds to film flow, and finally point D to equilibrium, where $S = S_\infty$.

Liquid level in capillary. Figure 7.9 shows an element of fluid in the capillary tube with radius r_0 subject to the centrifugal body force. Balancing the fluid shear stress and the normal stress acting on a fluid element, the governing equation for the shear stress τ can be written as

$$\tau = -\mu\left(\frac{dV}{dr}\right) = \frac{1}{2}\rho G r$$

Integrating the right-hand side of this equation and using the no-slip condition at the capillary wall, where $V = 0$ at $r = r_0$,

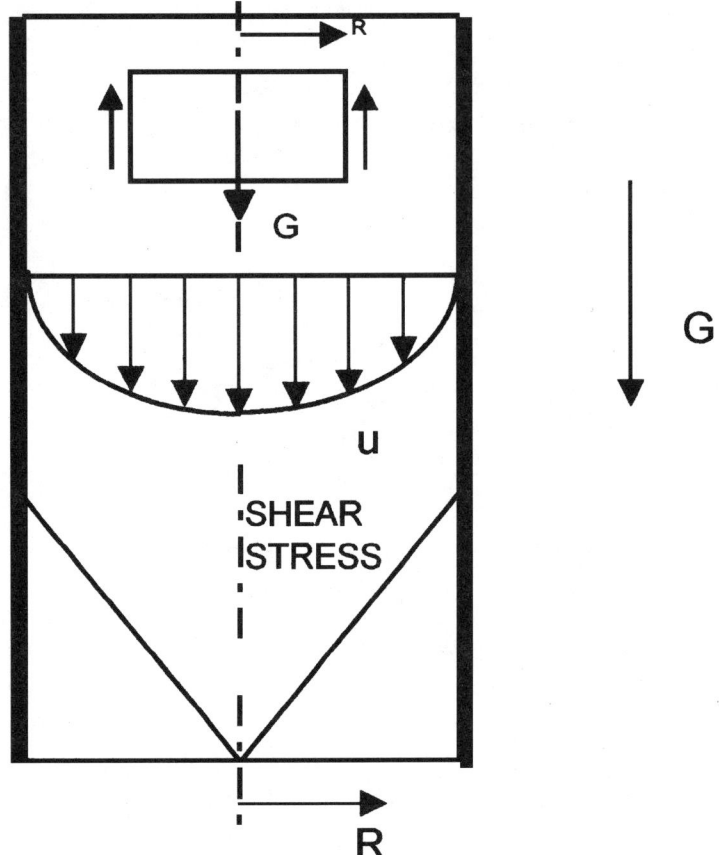

Figure 7.9 Flow in a capillary in cake.

$$V = \frac{G}{4\nu}(r_0^2 - r^2)$$

from which the flow rate through the capillary becomes

$$Q = 2\pi \int_0^{r_0} Vr\, dr = \frac{\pi G r_0^4}{8\nu}$$

From conservation of mass, the change in volume of the fluid in the capillary is equal to the net flow of fluid out of the capillary,

$$Q = -\pi r_0^2 \frac{dy}{dt}$$

where y is the level of the air-liquid interface in the cake as measured from the cake bottom and h is the total cake thickness or height. By definition, $S = y/h$, and after combining the results,

$$S = 1 - \left(\frac{Gr_0^2}{8\nu h}\right)t \qquad (7.11)$$

Thus liquid saturation S decreases linearly with t, in which the rate of decrease dS/dt depends on G, the radius r_0 of the capillary, the liquid kinematic viscosity ν, and finally the cake height h or, equivalently, the length of the capillary. This is shown by the solid line in Fig. 7.8. This description is valid provided that the fluid in the capillary tubes of the cake flows as a bulk liquid, as opposed to the film flow to be discussed in the next section.

Film drainage and residual moisture content. Subsequent to bulk drainage, the liquid adhered to the wall of the capillaries also drains out in thin films under G, where the film thickness decreases over time. This results in lower saturation until no further liquid drains out and an equilibrium is reached (see points C and D on the desaturation curve in Fig. 7.8). Useful results have been developed and tested[1] using a model glass-bead slurry.

Total saturation. As discussed, the total saturation can be considered as the sum of the equilibrium component S_∞, which is reached ultimately at steady state, and a transient component $S_T(t)$. The latter can also be considered as the difference between the total saturation at any given time t and that at equilibrium. In other words,

$$S_{\text{total}} = S_\infty + S_T(t) \qquad (7.12)$$

As can be seen in Fig. 7.5b, the transient component decreases with time. Time for the batch centrifuge refers to the dewatering or drying (as it is being referred to) time of the cake in a batch basket, or in a continuously sedimenting (in the dry beach) or filtering centrifuge. The dewatering time is well defined for the batch basket, but not so with some of the continuous centrifuges due to the complicated cake motion associated with the cake conveyance motion (such as with a scroll conveyor), which often results in complicated recirculatory flow.

Equilibrium component. A detailed examination of the equilibrium component S_∞ reveals that it is made up of contributions from the capillary saturation S_c, the cake pore saturation S_p, and the pendular saturation S_z, which are all related by the equation

$$S_\infty = S_c + (1 - S_c)(S_p + S_z) \tag{7.13}$$

Transient component. Likewise, the transient component is made up of the same contributors and in addition a film of residual liquid coated on the particle surface with saturation S_F which takes time to drain, especially when the particle surface is rough or "nonsmooth,"

$$S_T(t) = (1 - S_c)(1 - S_p - S_z)S_F(t) \tag{7.14}$$

The results from a mathematical model of each of the four components are given in the next section.

Drainage of free liquid in thin film. At large t, the thin-film saturation S_F is given by

$$S_F(t) = \frac{4/3}{t_d^n} \tag{7.15a}$$

with the dimensionless time t_d defined by

$$t_d = \frac{\rho G d_h^2 t}{\mu h} \tag{7.15b}$$

where for smooth-surface particles n is ideally 0.5, and for particles with rough surfaces, or for fine particles where strong surface forces are adhering to the moisture which it takes time to drain, it can be as low as 0.15. This implies that smooth particles can reach the equilibrium cake moisture much faster than particles with nonsmooth surface texture. This is analogous to observing the rate of drainage of liquid coated on a bath towel after removing it from a tub of water and comparing this same process for a panel of glass with identical surface area. Certainly, the smooth-surface glass panel drains much faster than the nonsmooth surface of the bath towel.

Bound liquid saturation. The bound liquid refers to liquid trapped by pores within the particles, or dead nonflowable space within the cake structure, or where chemical or physical forces are so predominant that the liquid would not release under a centrifugal field. It follows that the bound liquid saturation S_p depends on the particle characteristics.

Pendular saturation. The pendular liquid saturation is the liquid trapped at particle-particle contact points by surface tension or capillary forces, which can be overcome by very high centrifugal forces. From analysis supported by experiments on model slurry,[1] the pendular saturation is given by the following experimental correlation, depending on the value of the capillary number N_c,

$$S_z = 0.075, \quad N_c \leq 5 \qquad (7.16a)$$

$$S_z = \frac{5}{40 + 6N_c}, \quad 5 \leq N_c \leq 10 \qquad (7.16b)$$

$$S_z = \frac{0.5}{N_c}, \quad N_c \geq 10 \qquad (7.16c)$$

where the capillary number,

$$N_c = \frac{\rho G d_h^2}{\sigma \cos \theta} \qquad (7.17)$$

The capillary number N_c is the ratio of the centrifugal force to the capillary force. Frequently when $N_c < 5$, S_p and S_z are combined for convenience, with $S_z = 0.075$. Only when the centrifugal force overcomes the capillary force, corresponding to $N_c > 5$, can pendular saturation be reduced in accordance with Eqs. (7.16b) and (7.16c).

Saturation due to capillary rise. The capillary rise y can be determined from a balance of the force acting on the air/liquid/solid interface, $\pi d_h \sigma \cos\theta$, and the weight of the liquid column in the capillary acting under the influence of centrifugal gravity, $\rho G(\pi d_h^2 y)/4$. Simplifying $S_c = y/h$ becomes

$$S_c = \frac{4}{B_o} \qquad (7.18)$$

where the dimensionless Bond number B_o is defined as

$$B_o = \frac{\rho G h d_h}{\sigma \cos \theta} \qquad (7.19)$$

Here ρ and μ are the density and the viscosity of the liquid, θ is the wetting angle of the liquid on the solid particles, σ the interfacial tension, h the cake height, d the mean particle size, and t the dewatering time. The hydraulic diameter of the particles can be approximated[1] by the equations $d_h = 0.667\varepsilon d/(1-\varepsilon)$ or $d_h = 7.2(1-\varepsilon)K^{1/2}/\varepsilon^{3/2}$, where ε is the cake void fraction or porosity and K is the cake permeability.

The transient saturation component depends not only on G, the cake height, and the cake properties, but also on the dewatering time, which corresponds to the solids throughput for a continuous centrifuge, and the cycle time for batch centrifuges. If the throughput is too high or the dewatering cycle is too short, the liquid saturation can be unacceptably high and becomes the limiting factor. If this is not the case, dewatering of the liquid lens at particle contact points requires a much higher G force as discussed, because the residual saturation depends on N_c, a ratio of the G force to the capillary force. The maximum value is 7.5%, which is still quite significant. For the case when the cake is not disturbed (scrolled and tumbled) during conveyance and dewatering such as with a batch basket or a single-stage pusher, additional liquid can be further trapped in fine capillaries due to liquid rise, the amount of which is a function of B_o, which depends on a balance of the G force and the capillary force. This amount of liquid saturation can be small for a thick cake, and it can be large for a relatively thin cake. Lastly, liquid can be trapped by chemical, electrical, or surface forces at the particle surface and capillary or interfacial forces in the pores within the particles. Because the required desaturating force is extremely high, this portion of moisture cannot be removed by mechanical centrifugation. Fortunately, for most applications if it exists it is only a small percentage. The following worked example shows the relative magnitudes based on the saturation equations presented.

Example 1. Given:
$\rho = 1000$ kg/m³, $\rho_s = 1200$ kg/m³, $\mu = 0.004$ N · s/m², $\sigma \cos\theta = 0.068$ N/m, $h = 0.0254$ m (1 in), $d = 0.0001$ m (100 μm), $\varepsilon = 0.4$, $G/g = 2000$, $t = 2$ s, $S_p = 0.03$, and for the two separate cases where $n = 0.25$ (for rough-surface particles) and $n = 0.5$ (for smooth-surface particles).
Calculate:

$$d_h = 0.67 \left(\frac{0.4}{0.6}\right) \times 10^{-4} = 4.4 \times 10^{-5} \text{ m}$$

$$N_c = \frac{(1000 \text{ kg/m}^3)(2000 \times 9.8 \text{ m/s})(0.449 \times 10^{-4} \text{ m})^2}{0.068 \text{ kg/s}^2} = 0.58$$

Because $N_c = 0.58 < 5$ and $S_z = 0.075$,

$$B_o = \frac{(1000 \text{ kg/m}^3)(2000 \times 9.8 \text{ m/s}^2)(0.449 \times 10^{-4} \text{ m})(0.0254 \text{ m})}{0.068 \text{ kg/s}^2} = 329$$

$$S_c = \frac{4}{B_o} = 0.012$$

$$S_p + S_z = 0.03 + 0.075 = 0.105$$

$$S_\infty = S_c + (1 - S_c)(S_p + S_z) = 0.0122 + (1 - 0.0122)(0.105) = 0.1159$$

$$S_T = (1 - S_c)(1 - S_p - S_z)S_F(t) = (1 - 0.0122)(1 - 0.105)S_F(t) = 0.8841 S_F(t)$$

TABLE 7.2 Rough-Surface Particle, $n = 0.25$

t, s	t_d	S_F	S_T	S_∞	S_{total}	W_s
2	779	0.2518	0.2226	0.1159	0.3385	0.8417
10	3893	0.1684	0.1489	0.1159	0.2648	0.8718
20	7786	0.1416	0.1252	0.1159	0.2411	0.8819
60	23,359	0.107	0.0951	0.1159	0.2110	0.8951
120	58,814	0.0854	0.0755	0.1159	0.1914	0.9039
240	117,628	0.0718	0.0635	0.1159	0.1794	0.9094

TABLE 7.3 Smooth-Surface Particle, $n = 0.50$

t, s	t_d	S_F	S_T	S_∞	S_{total}	W_s
2	779	0.0477	0.0421	0.1159	0.1580	0.9193
10	3893	0.0213	0.0188	0.1159	0.1347	0.9304
20	7786	0.0151	0.0133	0.1159	0.1292	0.9330
60	23,359	0.0087	0.0077	0.1159	0.1236	0.9357
120	58,814	0.0055	0.0048	0.1159	0.1207	0.9371
240	117,628	0.0039	0.0034	0.1159	0.1193	0.9378

$$t_d = \frac{(1000 \text{ kg/m}^3)(2000 \times 9.81 \text{ m/s}^2)(0.449 \times 10^{-4} \text{ m})(t)}{(0.004 \text{ kg/m} \cdot \text{s})(0.0254 \text{ m})} = 389t \text{ (seconds)}$$

$$S_F(t) = \frac{1.33}{t_d^{0.25}}$$

$$W_s = \frac{1}{1 + (1/1.2)(1/0.6 - 1)S_{total}} = \frac{1}{1 + 0.5556 S_{total}}$$

The results are detailed in Tables 7.2 and 7.3 for rough- and smooth-surface particles, respectively. Figure 7.10 compares the saturation curves of the rough- and smooth-surface particles. The smooth particles reach the equilibrium saturation of 0.116 at about 50 s, whereas it takes the rough-surface particle at least 600 s to attain the same end result.

Example 2.

For some applications, such as dewatering of pharmaceutical solids, the capillary saturation can be quite significant. This can be seen with the following example. Take the particle size of the solid to be 10 μm instead of the 100 μm in the previous example, and reduce the G to 1000g. The capillary number N_c then becomes

$$N_c = \frac{(1000 \text{ kg/m}^3)(1000 \times 9.8 \text{ m/s})(0.449 \times 10^{-5} \text{ m})^2}{0.068 \text{ kg/s}^2} = 0.0029$$

Because $N_c = 0.0029 < 5$ and $S_z = 0.075$ (unchanged)

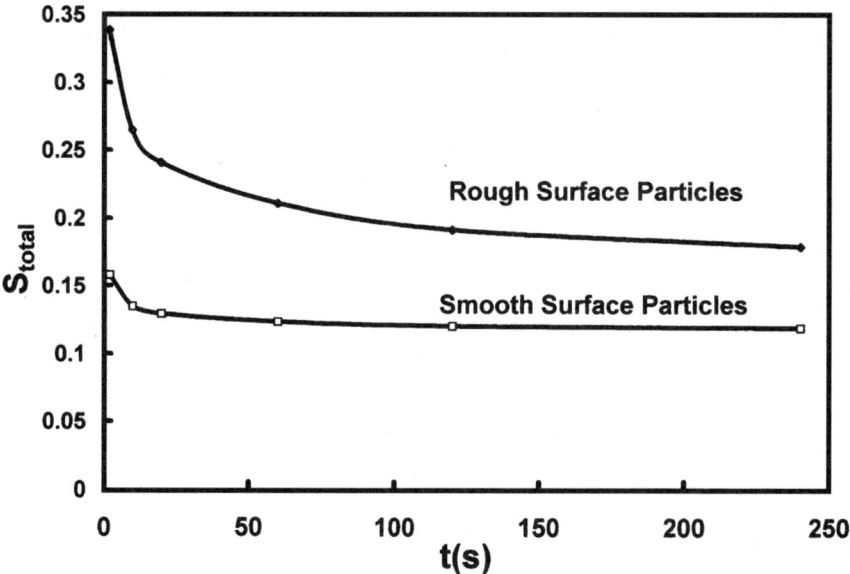

Figure 7.10 Comparison of desaturation of rough-surface and smooth-surface particles.

$$B_o = \frac{(1000 \text{ kg/m}^3)(1000 \times 9.8 \text{ m/s}^2)(0.449 \times 10^{-5} \text{ m})(0.0254 \text{ m})}{0.068 \text{ kg/s}^2} = 16.45$$

$$S_c = \frac{4}{16.45} = 0.243$$

The pendular saturation remains at 7.5% as before, whereas the capillary saturation for the finer solids jumps up from 1.2% to 24%. This yields $S_\infty = S_c + S_z = 0.315$. (See operating point A on the upper curve in Fig. 7.11. Even if the G force increases to 2000g in the previous example, the capillary saturation $S_c = 12.2\%$ is still quite high.) One means of compensating for this shortcoming is to process the material with a larger-diameter basket, such as a 1200–1500-mm-diameter basket, so as to accommodate a thicker cake, say, 125-mm (5-in) thick, to reduce the capillary rise effect. Repeating the same calculations at 1000g with the 125-mm (5-in)-thick cake shows that S_c diminishes to 4.9%. Because $t_d \propto 1/h$, this also results in a lower t_d. Despite the lower t_d, the resulting saturation actually drops to 19% from the previous 31.5%. This is illustrated in Fig. 7.11 as the new operating point B on a lower curve. This is due to the lower equilibrium saturation of the thicker cake at 12.5% ($S_z + S_c = 7.5\% + 4.9\%$), which is significantly reduced from the previous 31.5%. Also, as the kinetics are slower for a thick cake, an in-

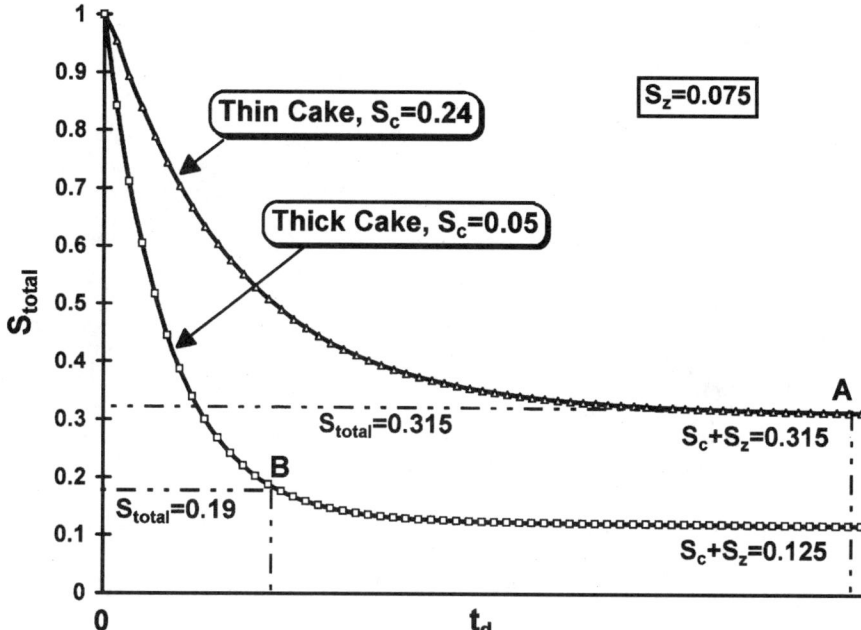

Figure 7.11 Dewatering mechanics due to capillary saturation difference for thick and thin cakes.

crease in the dewatering time which results in a longer cycle time and thus a lower throughput would further reduce the saturation, approaching the lower equilibrium saturation of 12.5% (see Fig. 7.11).

Superimposed pressure

Another means of reducing the capillary saturation is to apply both centrifugal gravity G and pressure to the cake. This can be understood from the fact that the capillary rise h_{cap} is resisted by both the centrifugal force on the liquid column and the imposed pressure force Δp. At equilibrium, a balance of forces on a capillary with radius r_0 in the cake yields

$$2\pi r_0 \sigma \cos \theta = \pi r_0^2 (\rho G h_{cap} + \Delta p) \qquad (7.20)$$

from which

$$S_c = \frac{h_{cap}}{h} = \frac{4}{B_o} - p_d \qquad (7.21)$$

where the dimensionless p_d is a measurement of the pressure imposed versus the hydrostatic pressure of a liquid column of height h (same as the cake thickness) under G,

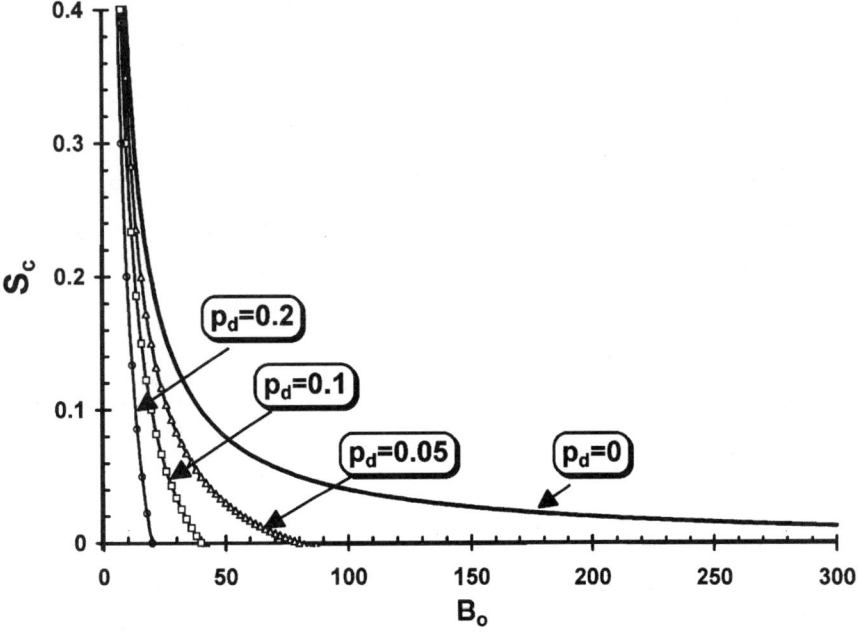

Figure 7.12 Effect of superposition of pressure on centrifugation.

$$p_d = \frac{\Delta p}{\rho G h} \qquad (7.22)$$

The behavior of Eq. (7.21) is best understood in Fig. 7.12, where S_c is plotted against B_o with p_d as a parameter. The effect of the superimposed pressure p_d on further desaturation is evident. With $\Delta p = 2$ bar, $G = 1000g$, and $h = 0.0254$ m (1 in),

$$p_d = \frac{2 \times 10^5 \text{ Pa}}{(1000 \times 9.8 \text{ m/s}^2)(1000 \text{ kg/m}^3)(0.0254 \text{ m})} = 0.8$$

Therefore,

$$S_c = \frac{h_{cap}}{h} = \frac{4}{B_o} - p_d = 0.24 - 0.8 \rightarrow 0$$

Note that the negative value is just an artifice. The foregoing exercise shows that with a superimposed pressure of 2 bar it is possible to reduce the capillary saturation to zero. In reality, for a distribution of pore sizes, especially for sizes less than those considered, the superimposed pressure might not be able to reduce capillary saturation completely. Therefore, other than considering the mean pore size, the pore-size distribution is also an important factor. For finer particles, forming capillaries with mean diameters less than 5 μm, the application of filtering basket centrifuges would be restricted due to the desaturation limitation, as discussed. In addition the bulk filtration rate is slow because of the significant reduction in cake permeability. Commercial basket centrifuges are available with superimposed pressures of up to 6 bar. While the foregoing discussion has focused on steady-state desaturation, the transient pressure superimposed centrifugation has also been studied in some details.[2]

Cake Washing

Filtering centrifuges have been used for washing cake to remove impurities dissolved in the mother liquor. There are three different mechanisms:

1. Displacement
2. Dissolution
3. Diffusion

Washing in filtering centrifuges is accomplished predominantly by displacing the mother liquor by the introduced wash liquid since the kinetics of diffusion and dissolution are too slow. Typical cake washing characteristics are shown in Fig. 7.13, where the concentration of impurities is plotted against the wash ratio, which is defined as the ratio of the amount of wash liquid to that of the residual mother liquor. As expected, the concentration of impurities drops with increases in G force and cake washing time and with improved wash nozzle design and screen design.

With the nozzle design it is important to provide wash liquid to displace the mother liquor. The wash liquid needs to attain the same tangential speed and, hence, the same gravity as the mother liquor in the cake in order to displace the latter. Consequently, improved multi-spray wash nozzles[3,4] have been designed to provide uniform coverage of cake with wash liquid and to accelerate the wash liquid to tangential speed in order to obtain some moderate radial penetration into the cake.

If the impurities of the filtrate are monitored concurrently with those of the cake, the level of impurities in the cake after the washing is always higher than that of the spent wash liquid, and the two measured variables approach each other as the wash ratio increases.

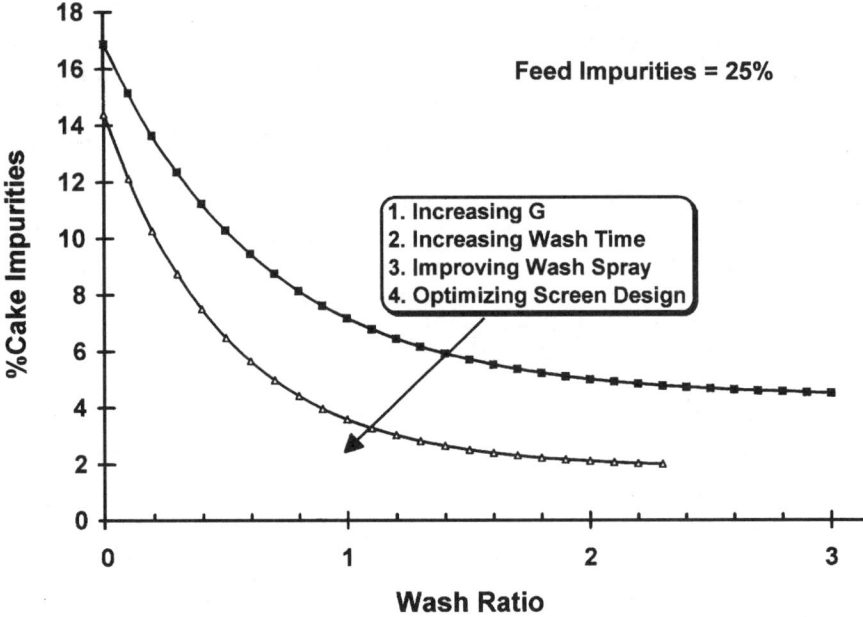

Figure 7.13 Cake purity as a function of wash ratio and G force.

Wash nozzle sizing

For filtering centrifuges which require cake washing, wash liquid is introduced into the machine either through a stationary feed pipe at a high velocity such as used in batch baskets and pushers. Alternatively, it is introduced into the process area of the machine from a rotating compartment in the conveyor through a set of wash nozzles as used in screen bowls and screen scrolls.

Stationary wash pipe

For the case where wash liquid is introduced via a stationary wash pipe and nozzle, it is important that the opening of the nozzles are oriented in the direction of rotation. More importantly, the velocity of the wash liquid needs to match closely the rim speed $V_\theta = \Omega R$ of the cake for a given wash rate Q_n. This is accomplished by crimping the nozzle at the end of the wash pipe to change its cross-sectional area. Also, the cross-sectional profile of the nozzle can be shaped to provide a wide-angle spray of wash liquid onto the cake surface. In addition, some radial momentum further helps the wash liquid to penetrate into the cake interior. Industrial spray nozzles have also been used for this purpose.

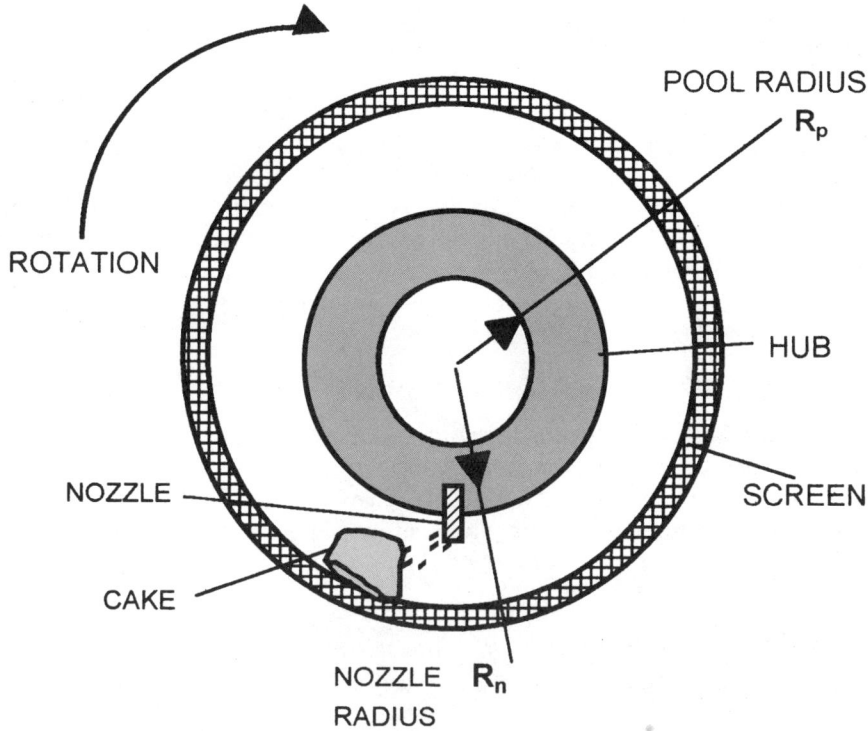

Figure 7.14 Cross section of a wash compartment of a centrifuge showing a wash liquid pool and the wash nozzles delivering wash liquid to the cake on the screen at a larger diameter.

Wash compartment

Figure 7.14 shows a cross section of the rotating wash compartment with nozzles extending from the outer diameter of the compartment into the space between adjacent conveyor blades to deliver wash liquid to the cake resting on the screen. For a given wash rate Q_n as determined from the wash ratio WR, measured by kg wash liquid/kg of cake, and cake rate m_s, the wash rate per nozzle q_n from N_{noz} wash nozzles uniformly distributed along the wash area of the screen becomes

$$q_n = Q_n/N_{noz} = \frac{m_s WR}{\rho N_{noz}} \qquad (7.23)$$

where ρ is the density of the wash liquid.

On the other hand, the rate of the delivering the wash liquid depends on the pressure difference Δp applied across the wash nozzle. The exit of the nozzle is at atmospheric pressure, whereas the intake to the nozzle is immersed in a pool of wash liquid at a pressure p_n given by

$1/2\rho\Omega^2(R_n^2 - R_p^2)$, where R_n is the nozzle intake radius and R_p the wash liquid pool radius, and Ω the rotation speed. Therefore, it follows that $\Delta p = 1/2\rho\Omega^2(R_n^2 - R_p^2)$. The maximum Δp corresponds to the minimum R_p where the liquid pool level reaches the inner radius of the baffle of the wash compartment. Further increase in pool depth leads to spillage of the liquid over the baffle.

Given the wash nozzle is a flow resistance element, which for a given geometry has a flow-pressure drop characteristic, where $\Delta p = f(q_n)$. Therefore, by matching the required pressure drop for a given rate through a nozzle of a given design to the driving pressure available from the head of the liquid pool,

$$\Delta p = f(q_n) = 1/2\rho\Omega^2(R_n^2 - R_p^2) \qquad (7.24)$$

The nozzle geometry and arrangement are selected such that the nozzles can deliver a given total wash rate to the filtering centrifuge and operate at the optimal condition of the nozzles with the wash liquid pool well contained in the wash compartment.

Example

Figure 7.15 shows pressure drop characteristcs of an industrial spray nozzle of a given design and for various nozzle openings. Suppose a noz-

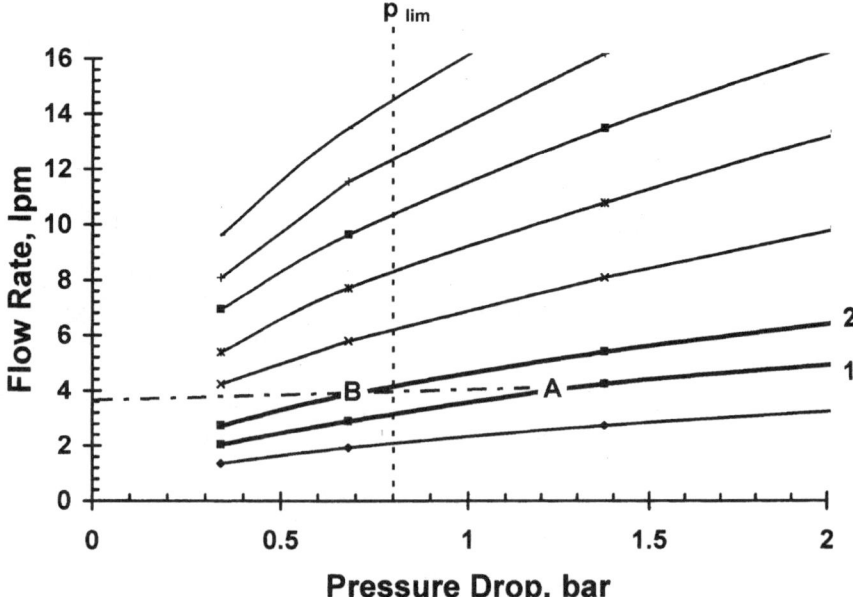

Figure 7.15 Flow rate-pressure drop characteristics of wash nozzles of different sizes with higher flow rate curves corresponding to nozzle diameters.

zle geometry (design and size) is selected such that the operating characteristic is given by the solid curve 1 in Fig. 7.15. Given the wash requirement is that each nozzle has to deliver q_n = 3.8 L/min (1 gal/min), this requires to operate at p_n = 1.22 bar, point A on curve 1 of Fig. 7.15. Suppose the maximum p_n is limited to 0.8 bar without the liquid spilling over the baffles in the wash compartment at a given operating rotation speed, therefore the required pressure head is greater than the available and the flow resistance of the nozzle needs to be reduced at 3.8 L/min (1 gal/min). A larger wash nozzle is selected with the flow characteristics given by curve 2. The operating point at the same flow rate for this larger nozzle yields p_n = 0.7 bar which is within the operating pressure range of 0.8 bar. Another alternative is to increase the number of the smaller wash nozzles with the operating curve 1 (providing the spacing of the nozzles is not limiting) such that the flow rate per nozzle is reduced below 2.5 L/min.

On the other hand, it is equally undesirable to use a large wash nozzle to deliver a small flow rate such that the wash liquid does not run full in the nozzle resulting in dripping of the liquid onto the cake surface. This is especially pertinent for the fan spray nozzles where the fan spray characteristic requires the nozzles to run full and has a minimum Δp or liquid head to operate these nozzles at optimal performance. In general, for a given selected q_n the operating point should lie somewhere at the midpoint on the operating curve of a given nozzle geometry, between p_n = 0 and the maximum p_n referred in Fig. 7.15 as $p_{\lim}$, corresponding to the maximum liquid pool in the wash compartment at a given rotational speed of the machine. The number and positioning of the nozzles can be tailored to the specific application.

In this chapter bulk filtration, desaturation, and cake washing have been discussed. The first two subjects have been studied quite extensively both theoretically and experimentally.

References

1. G. Mayer and W. Stahl, "Model for Mechanical Separation of Liquid in a Field of Centrifugal Force," *Aufbereitungs-Technik*, no. 11, 1988.
2. C. Stadager and W. Stahl, "The Superposition of Centrifugal and Gas Pressure Forces for Cake Filtration," in *Proc. Am. Filt. Sep. Soc. Ann. Conf.* (Nashville, TN, Apr. 23–26, 1995), vol. 9, K. -J. Choi (ed.), pp. 551–559.
3. W. W. Leung, "Method of Accelerating a Liquid in a Centrifuge," U.S. patent 5,403,486, Apr. 4, 1995.
4. W. W. Leung, "Method of Accelerating a Liquid in a Centrifuge," U.S. patent 5,527,474, June 18, 1996.

Chapter

8

Batch Filtering Centrifuges

Despite the tendency of the industry to favor continuous filtering centrifuges in order to reduce down time and increase productivity, batch filtering centrifuges still find significant use as improvements in modern control allow certain operational flexibilities unique to these batch operations. At the low end, applications include small machine shops for the recovery of excessive cutting oil from metal chips and the high-tonnage sugar industry. At the high end they are used for pharmaceuticals and specialty chemicals, where product solids are measured on a dollar per gram basis. For the pharmaceuticals and specialty chemicals their feeding and filtration, washing, and dewatering characteristics may often change from batch to batch, even on supposedly similar materials and certainly on different feed-slurry materials. This necessitates adjustments in the operation of the centrifuges, such as feeding rate and duration, cake height, wash rate and duration, dewatering time, G force, or unloading speed, in order to regulate the quality of these high-value products. The cycle time can vary from 10 min to as long as several hours. These flexibilities cannot be accommodated by continuous filtering centrifuges. Furthermore, in most installations surge tanks with controlled inlet and bypass loops permit the integration of batch centrifuges into continuous processes.

Basket Cycle

As shown in Fig. 8.1, a typical cycle consists of initial acceleration, feeding, final acceleration, washing, dewatering, deceleration, cake discharge, and basket cleaning.

Feeding. The basket accelerates to an intermediate speed after which feed slurry is introduced, typically in a few charging steps with intermittent breaks so as to prevent the feed slurry from overflowing the

194 Chapter Eight

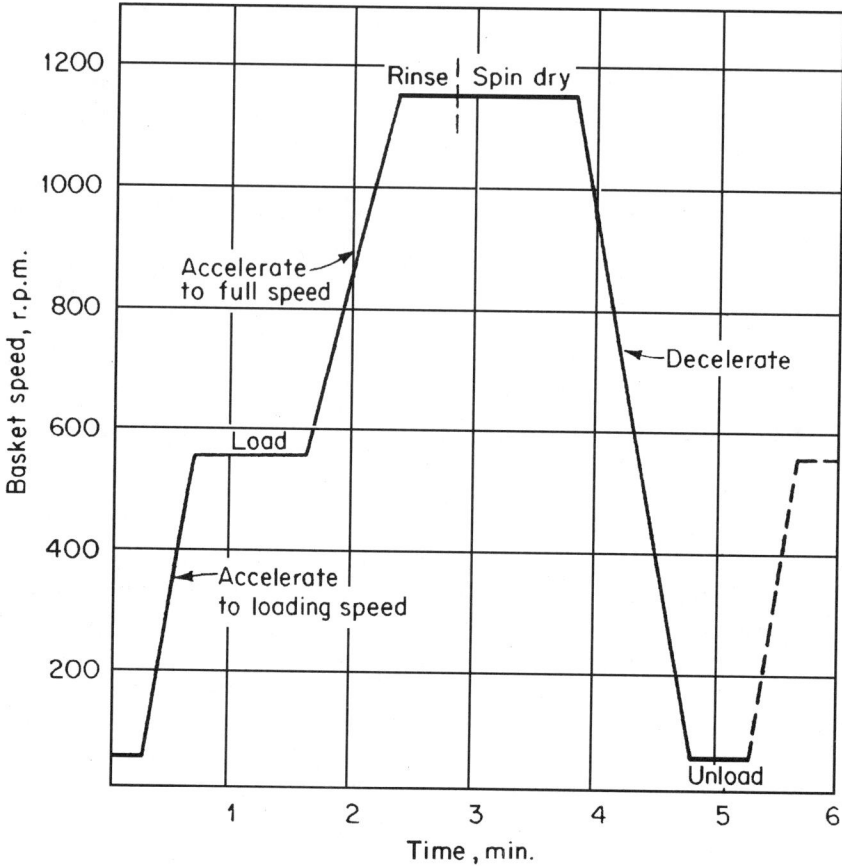

Figure 8.1 Typical cycle of batch basket centrifuge.

ring weir of the basket while filtration catches up and cake solids build up in thickness ultimately to reach about 75–85% of the overflow limit as set by the weir. Typically a load detector, either a mechanical contact type, a radio-frequency transmitter and receiver, or a capacitance measurement sensor such as an ultrasonic sensor [a programmable logic controller (PLC), a distributor control system (DCS), or others] is used to sense and control the cake buildup during feeding of the basket. The sensor ensures that the slurry pool is always below the weir height, and a control will shut off the feed once the pool level reaches a prescribed limit, allowing filtration to catch up with the feeding. The same sequence repeats until the cake reaches the desired level.

Washing. Subsequently, both the slurry and the basket are further accelerated to the operating speed. After reaching the full speed, the cake is washed. For fast-filtering solids it is also common practice to introduce the wash liquid to full operating speed immediately after feeding before the basket accelerates to full operating speed in an attempt to shorten the cycle time. A more important reason is that if wash liquid is introduced too late, when the liquid saturation S in the cake falls below 100%, the void spaces in the cake are exposed to air entrapment. This results in undesirable air blocking the effective displacement of the mother liquor by the displacing wash liquid and therefore leads to higher impurities in the cake after wash. On the other hand, for slow-draining cake it is equally undesirable to introduce wash liquid to the cake when there is a standing liquid pool above the cake.

Dewatering. After washing, the basket is spun for an extended period to dewater the cake to the desired residual moisture level under the maximum G force. Finally the basket decelerates to an unloading speed. Depending on the agility of the cake solids, it is typically below 50 rpm that the cake is unloaded by a peeling knife. The reasons for the low-speed discharge are (1) to reduce the G force, which holds the cake to the basket wall, and (2) to reduce the differential speed between the stationary knife and the rotating cake, which for certain solids may lead to particle attrition. Most materials, especially shear-sensitive materials, cannot tolerate unloading speeds greater than 50 rpm. The unloading can be either single-acting or double-acting. In single-acting unloading the knife spans the entire length of the basket and is brought radially inward to plow the cake in its entirety. (The knife falls slightly short of the filter medium to avoid ripping of the medium.) In double-acting unloading the unloading knife spans only a fraction of the basket length. In the rest position it is at the top of the basket; during cake unloading it is brought into the cake and plows the cake at the top of the basket. Subsequently it travels axially along the basket, continuously plowing the cake at the middle and lower portions of the basket. Upon finishing, it retracts axially to the top of the basket and subsequently radially inward back to its original position.

The rate of acceleration depends on the inertia of the basket and its content, as well as on the available torque from the drive, whereas the rate of deceleration (that is, coast down) depends on the inertia of the rotating mass (basket and cake) and the available braking force. The unloading time depends on the torque available for turning the basket at unloading speed, the rheological properties of the cake to be unloaded, and the design of the unloading knife. The remainder of the cycle depends on the processing characteristics of the slurry to be centrifuged—the filtration rate of the mother liquor, the wash and drain rate of the wash liquor, and the final drying of the cake to a residual moisture level. The operating cycle may be either manual or fully automatic through a sequence of programmed steps using DCS or PLC interfacing with reset timers, speed sensors, limit switches, load cells, and radio-frequency capacitance switches.

Vertical baskets can be classified into two general categories—solid bottom and open bottom. The solid-bottom type is used typically for small quantities and fragile solids which cannot tolerate mechanical handling. After the machine is filled, manual discharge of the cake solids is required from the top after the machine has been brought to rest. With the open-bottom type, where solids can be discharged through a series of openings at the bottom of the basket, the operation is automatic. In both cases a ring weir is used to cap the top of the basket in order to contain the contents and prevent overflowing. With the open-bottom type the inner diameter of the openings has to be smaller than that of the ring weir. Also, an unloading and peeling knife or simply a peeler is used to scrape the cake out, typically at a rotational speed of less than 50 rpm to prevent breakage of cake solids, especially with crystalline materials. The basket can be driven from either the bottom or the top. Various suspension mechanisms are available. The filter medium can be a screen or a polypropylene filter cloth with a support backing. Typically, it is rated by its air permeability or by the micrometer-size openings.

Solid-Bottom Basket

Smaller-scale solid-bottom batch basket centrifuges are available. These are for small test samples, when the sample cannot tolerate mechanical handling or when the traces of solids remaining in a more automated centrifuge would be subject to decomposition or spoilage. Two common supports are used—base-bearing and link-suspended arrangements.

Base-bearing design

As shown in Fig. 8.2a, the drive shaft of the centrifuge is supported from below on a thrust bearing, which often is held and pivoted in a ball joint. The shaft is centered by radial springs or rubber in compression. This provides damped freedom of motion of the axis of rotation to compensate for an out-of-balance basket load. This type of centrifuge is used extensively in chip wringers, which recover excess oil from metal chips and turnings. The basket wall is usually solid and tapered radially outward toward the top, ending in an annular lid from where the oil is freely discharged while the chip is withheld. A typical unit is 660 mm (26 in) in diameter at the top and 584 mm (23 in) at the bottom, holds up to 0.15 m^3 (5 ft^3) or about 225 kg (500 lb) of crushed steel chips, and is driven at 1025 rpm (370g) by a 7.5-kW (10-hp) motor.

Batch Filtering Centrifuges 197

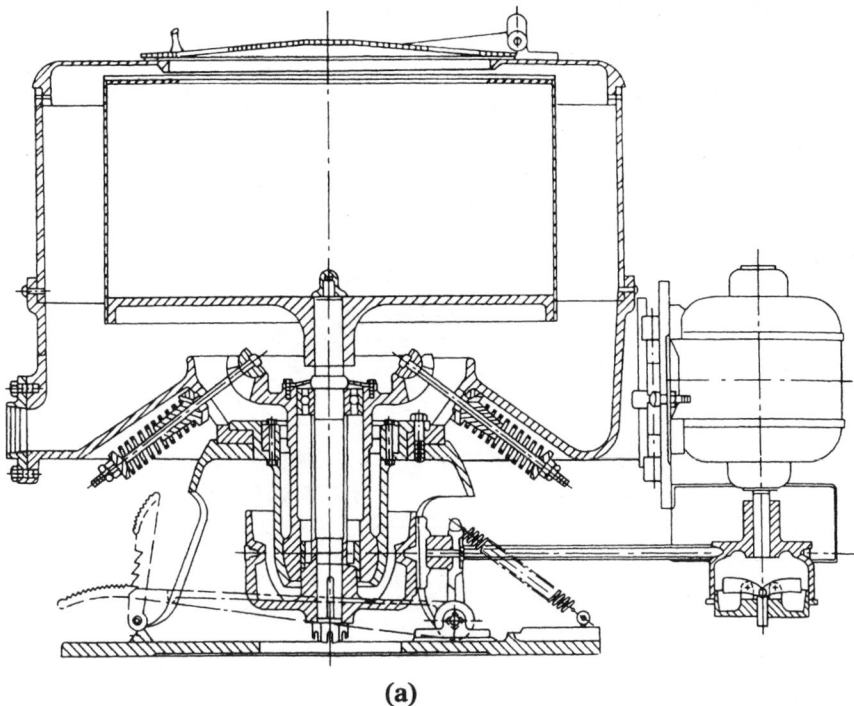

Figure 8.2 (a) Base-bearing solid-bottom basket. (*Reproduced with permission from Perry and Green.*[1])

Link-suspended design

In centrifuges with diameters larger than 762 mm (30 in), the basket, curb, curb cover, and drive form a rigid assembly, which is suspended flexibly from three fixed posts spaced out evenly around the circumference (also known as three-column centrifuge). The three suspension members may be either chain links or stiff rods in ball-and-socket joints and are spring loaded. The suspended assembly has restrained freedom to oscillate to compensate for normal out of balance. The drive is vertical and has a more efficient power transmission than the base-bearing type.

This type of centrifuge is usually loaded at zero speed. Therefore loading should be uniform inside the basket. Subsequently the machine is brought to operating speed and maintained there until the free liquid has drained off through an opening in the bottom of the curb. Then the basket is brought to rest and the cover lid opened for unloading. The latter may be facilitated if the filter medium is in the form of a bag contoured to fit the inside of the basket. Link-suspended

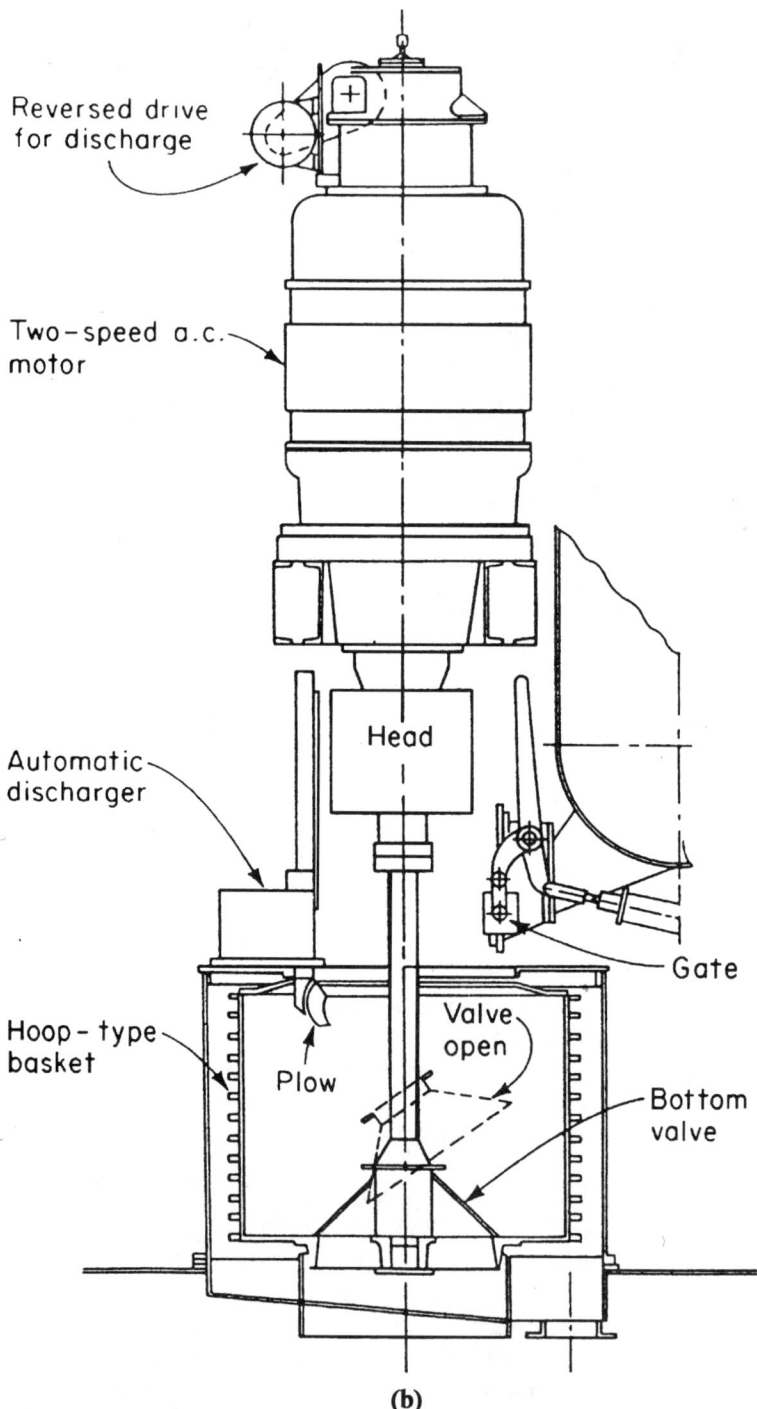

Figure 8.2 (*b*) Top-suspended pendulum centrifuge. (*Reproduced with permission from Perry and Green.*[1])

centrifuges with solid bottoms are available with inside diameters ranging from 305 to 2743 mm (12 to 108 in).

Open-Bottom Basket

Used for heavier duties, these centrifuges are either top-driven or bottom-driven. In both cases the bottom bowl head consists of three functional components contained in a single fabrication: (1) the central nave by which the basket is attached to the drive shaft, (2) an outer ring to which the cylindrical shell is attached and whose inside diameter is less than that of the liquid ring weir on the opposite end of the basket, and (3) spokes connecting the nave to the outer ring. A typical cycle follows that shown in Fig. 8.1. The control of the cycle may be manual, semiautomatic, or fully automatic. The drive may be a variable-speed electric motor, either direct or through V-belts; a high-pressure, fixed-volume hydraulic motor receiving its energy from a constant-speed variable-volume pump, or, rarely in modern practice, a steam or water turbine.

The unloading may be accomplished manually with a hand-driven plow or knife mounted on the curb cover or automatically with a double- or single-acting knife with hydraulic or pneumatic piston actuators. In some designs the opening through the basket is covered with a plate that is lifted during unloading.

Top-suspended, top-driven design

The top-suspended centrifuge shown in Fig. 8.2b, also known as pendulum centrifuge, is used widely for purging molasses from crystallized sugar as well as for many other applications. Conventionally the drive is suspended from a horizontal bar supported at both ends from two A-frames. The drive head, which is connected to the motor or a driven pulley through a flexible coupling, carries the thrust and radial bearings that support the shaft and its load. The rotating cylindrical perforate bowl has a ring weir at the top. The outer stationary casing is attached to the A-frame. The filtrate is collected and diverted to an outlet in the casing. The entire weight of the motor is carried on the frame with no component of force other than its rotation reacting on the centrifuge proper. This permits the use of very large special motors for rapid cycling. On white-sugar service, up to 24 cycles/h with 364 kg (800 lb) of sugar per load is provided by wound-rotor variable-speed ac motors with a power rating of 112 kW (150 hp).

Typical pilot-plant top-suspended baskets are 305 mm (12 in) in diameter by 127 mm (5 in) deep. Commercial machines are available in sizes from 510 mm (20 in) in diameter by 305 mm (12 in) deep to 1520 mm (60 in) in diameter by 1020 mm (40 in) deep and develop up to $1800g$ in the smaller and intermediate sizes. Except for sugar service, operation with a two-speed motor (half-speed for loading and full speed for purging) is typical. Hydraulic and electric (ac VFD) drives with variable-speed capability are commonly used in the chemical industry. To maximize the number of cycles per hour, a combination of electrical and mechanical braking is employed to minimize the period of deceleration, which is a transition period of no value to the process.

Link-suspended, top-driven versus bottom-driven designs

With the previous top-suspended design, the only limit to the size and weight of the drive motor is the strength of the supporting structure, which can be made as strong as possible. With the link-suspended design, the weight of the side-mounted motor constitutes an overturning moment, and therefore this weight is limited to a proportion of the remaining suspending components. For a basket with a diameter of 1219 mm (48 in) and a depth of 762 mm (30 in), only a relatively lightweight 45-kW (60-hp) motor is employed. Typically it is a two-speed operation, as with other basket centrifuges, half-speed for loading and full speed for deliquoring. Greater torque transmission and more flexibility in speed control are possible with hydraulic drives.

Recent design with top-mounted motors together with a more rugged link suspension eliminates this concern. Figure 8.3 shows a top-driven link-suspended centrifuge for bulk material dewatering.

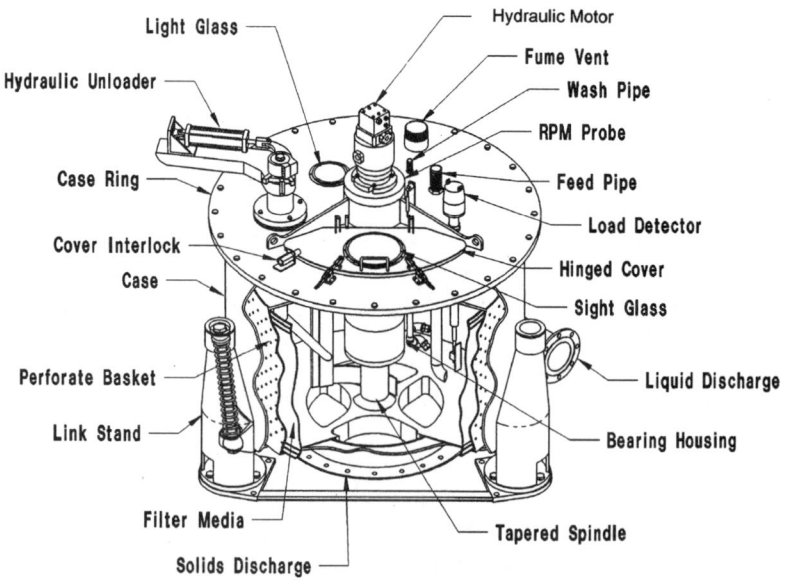

Figure 8.3 Top-driven, open-bottom basket. (*Courtesy of Bird / Ketema.*)

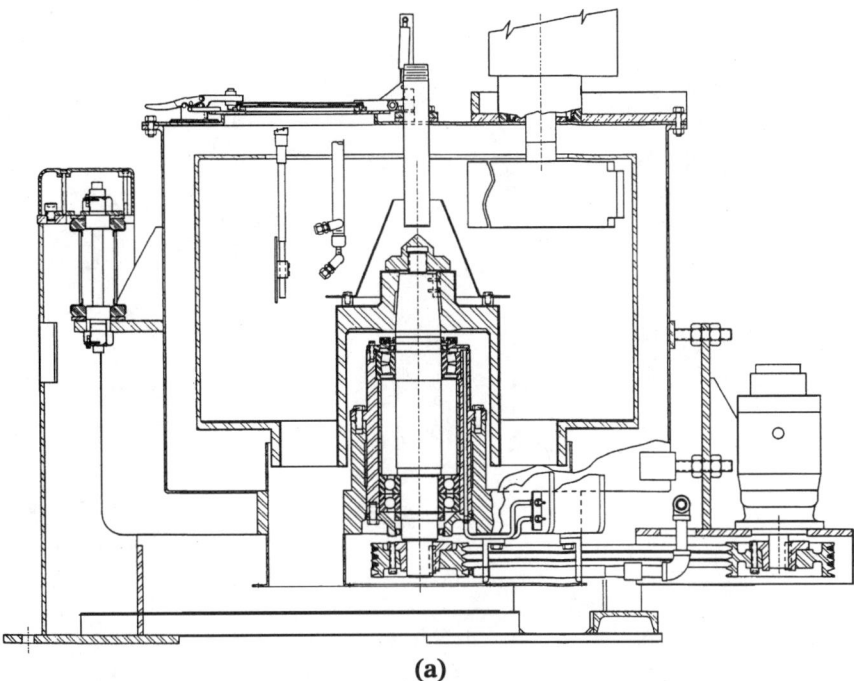

Figure 8.4 (*a*) Bottom-driven basket with cone-disk feeding mechanism of 1000 by 600 mm (40 by 24 in). (*Courtesy of Bird / Ketema.*)

This unit is equipped with updated features such as fumetight design and vents to purge accumulated toxic fumes, a load detector to gage a preset amount of feed load to the basket, a hydraulic unloading knife, and an rpm probe (tachometer). Unlike the top-driven design, the drive in the design shown in Fig. 8.4*a* is from the bottom. This allows the basket to be more accessible from the top. Figure 8.4*b* further reveals the clean-in-place procedure, where the basket is washed thoroughly inside as well as outside to remove any process residues.

Semiautomatic and Fully Automatic Baskets

Some batch centrifuges operated automatically on a preprogrammed cycle can be semiautomatic or fully automatic centrifugal filters. Examples are the advanced vertical peeler centrifuge, the horizontal peeler, the rotary siphon centrifuge, the pressurized siphon, and the inverting filtering centrifuge.

Vertical peeler

A vertical peeler centrifuge has been developed recently[2] to address the needs of pharmaceutical and specialty chemical applications (Fig. 8.5).

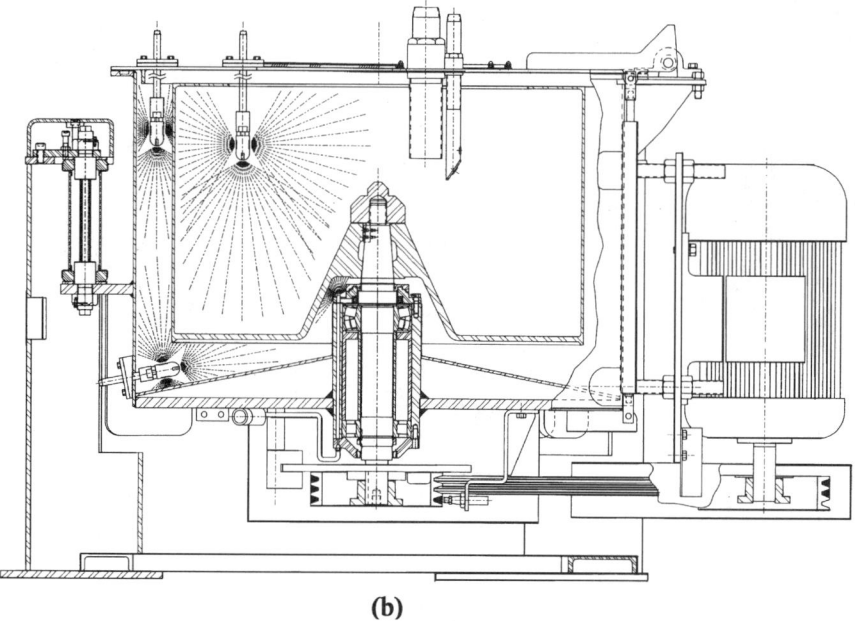

(b)

Figure 8.4 (*b*) Bottom-driven basket showing clean-in-place nozzles of 760-mm (30 in)flip top. (*Courtesy of Bird / Ketema.*)

The basket is bottom-driven with the bearing housing outside the basket for easy access. This allows more effective clean-in-place operation and prevents solids from getting into the bearing area as with conventional baskets. The entire basket and accessories sit on a platform, which in turn rests on four "visco dampers." This allows added stability and at the same time isolates and dampens any out-of-balance vibration passed to the foundation. The blow back the filter cloth from the outer diameter of the basket in conjunction with air jets from the scraper blade at both corners further reduces any residue cake heel left on the cloth. In addition, the low-profile cloth-retaining top and bottom rings ensure that the scraper blade will get close to the cloth, removing the last millimeter of cake remaining on the cloth. All these features are provided in an effort to reduce the cake heel, which otherwise reduces filtration since the cake heel glazes as a result of compaction by the scraper knife over several cycles. The variable-speed ac drive and the regenerative power feature during deceleration of the basket allows flexibility in the G force to meet various process requirements and reduce power consumption. A higher G force, as compared to conventional vertical baskets, of up to 1000–1500g is implemented for 970–1520-mm (38–60-in)-diameter baskets. Smaller baskets of

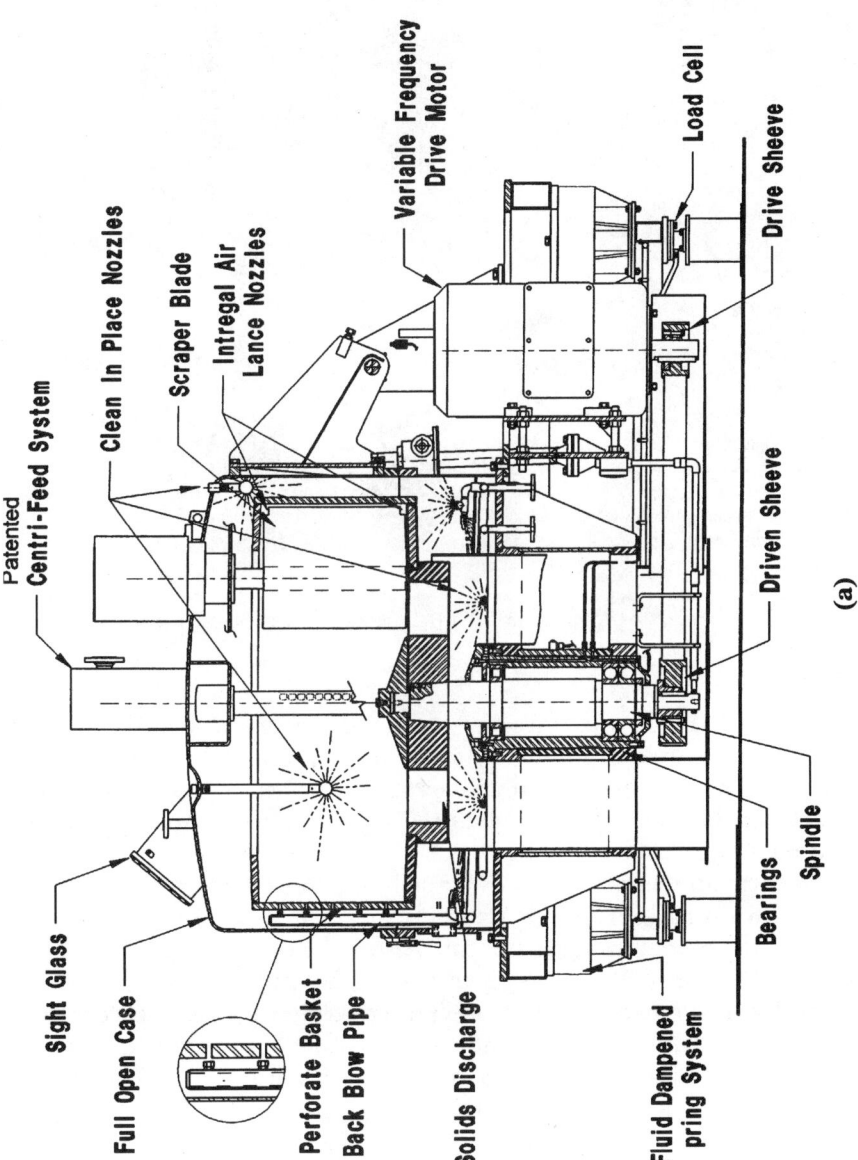

Figure 8.5 (a) Vertical peeler. (*Courtesy of Bird / Ketema.*)

(b)

Figure 8.5 (*b*) Process 2000®, 1524 mm (60 in) diameter vertical peeler basket. (*Courtesy of Bird/Ketema.*)

610–710-mm (24–28-in) diameter can attain up to 2000g. This takes full advantage of the higher G for liquid drainage of incompactible cake. Further, an operator touch-screen interface and sophisticated controls further incorporate advanced information technology into a traditional piece of equipment.

Conventional baskets use a J-shaped pipe for feeding (see Fig. 8.3). For fast-draining slurry this leads to thicker cake pile at the basket bottom and thinner cake at the top. Also a cone-disk feed distributor has been used (see Fig. 8.4*a*). The cake is better distributed circumferentially. The most important advantage is that it prevents feed liquid from hanging onto the underside of the J-pipe and dripping, which would contaminate the cake solids. Despite these advantages, the cake profile is such that its thickness is maximum at the point of feed, especially with fast-draining slurry (see Fig. 8.4*a*). This limitation is not as serious for slow-filtering slurry when there is sufficient time for the slurry to spread out under the G force along the basket. Both J-pipes and cone distributors typically result in nonuniform cake distribution along the basket length. In addition, for the J-pipe design, it does not guarantee uniformity in the circumferential direction, which further leads to vibration from any unbalance in solids loading. For a cone-disk feed accelerator, because of the cone feeding at 360°, the geometry requires a double-acting unloading arrangement (see Fig. 8.4b).

The vertical peeler basket utilizes a rotary distribution pipe assembly,[3] where the feed is distributed uniformly along the entire length of the basket (see Fig. 8.5a, b). The design has two concentric tubes. The outer tube has a set of holes along the length at one angular position, whereas the inner tube has another set of holes distributed lengthwise along a helix. As the inner tube rotates slowly within the stationary outer tube, feed comes out periodically in different axial directions as the holes from the inner and outer tubes align, in turn forming a continuous-flow passage. This design overcomes the nonuniform cake distribution associated with use of the J-pipe and feed cone, especially for fast-filtering slurries. The centrifuge is also adapted for steam sterilization.

A versatile PLC is used with customer-defined programs on operating machine parameters such as the duration and speed of feeding, washing, dewatering, and unloading segments of the cycle. The basket weight is metered on-line by load cells installed at the four pedestal supports. The measured weight of the basket contents and the rate of change during feeding provide information on the buildup of mass (solid and liquid) in the basket during feeding and the rate of filtration during the breaks (feed temporarily stops) in the feeding segment of the cycle. Thus an optimal sequence can be obtained during feeding for a given feed slurry. Compensation can be made in the operational sequence based on the variability in filtration characteristics of the feed materials. On the other hand, the rate of change in the basket contents during dewatering provides information on the loss of moisture from the cake. The dewatering cycle can be stopped when the rate of reduction of mass falls below a set limit. The measured weight of the basket contents versus time trend follow the desaturation-time curve discussed in Chap. 7. Thus the machine can advance to the next cycle, reducing idle time and loss in throughput capacity. The advanced sensors and controls have provided a great amount of flexibility and versatility to this machine.

The clean-in-place procedure ensures a contaminant-free product changeover, which can involve parts-per-million detection criteria. This is speed up by automated closed-system protocols of the vertical peeler, which adheres to the U.S. Food and Drug Administration's (FDA) current Good Manufacturing Practices (GMP) standards. The FDA approval requires that the light-sensitive test chemical introduced into the centrifuge be completely removed from the centrifuge during the clean-up cycle, which is facilitated by adequate wash from nozzles placed both inside and outside the basket. Inspection of the wet surfaces under ultraviolet light after the cycle is completed is used to ascertain the completeness of contaminant removal.

One minor drawback is that unlike with the horizontal peeler, which will be discussed next, cake unloading of the vertical peeler cannot be carried out at full speed of operation. Instead, it takes place at a much lower speed, similar to the conventional baskets. This applies to materials that could be unloaded at high speed and not to delicate crystalline solids, which require low-speed unloading regardless of the equipment.

Horizontal peeler

As shown in Fig. 8.6, the conventional horizontal peeler centrifuges operate at a constant speed (that is, near maximum basket speed) during the entire sequence of feeding, dewatering, and cake discharging. Since the cake can be discharged at speed, no process time is lost in acceleration and deceleration. All horizontal peelers operate with the drive shaft supported by fixed bearings on a horizontal axis. The basket may be cantilevered at one end of the drive shaft, with the driven pulley at the other end. In another design the drive shaft extends through the basket and is also supported by an outboard bearing. The through-shaft design may carry two baskets with a common hub to allow feeding one basket while the other is spinning to smooth out the power demand.

A cake distributor may be provided to level the load during feeding,

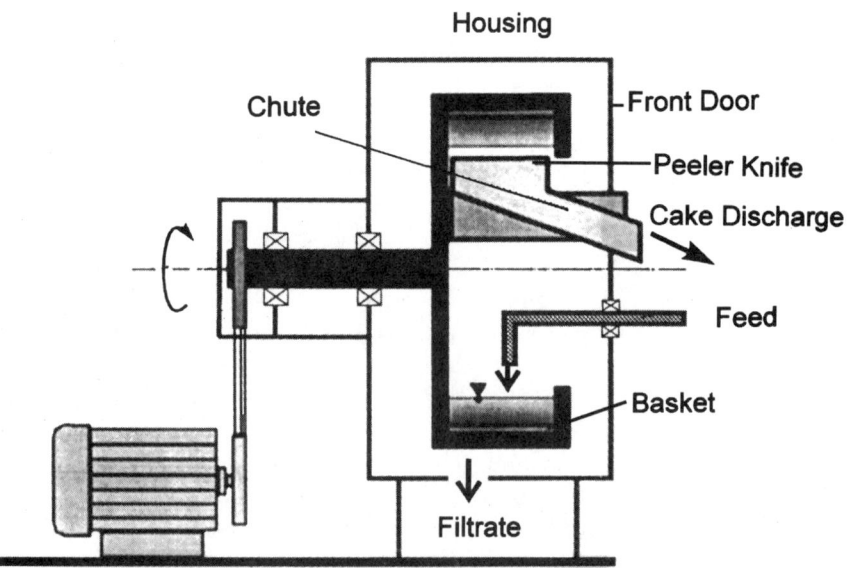

Figure 8.6 Horizontal peeler.

to indicate cake depth, and to activate the feed-valve closure when the desired cake depth has been reached. In other designs the unloader knife is smaller and double-acting, and in very large sizes the discharge of the cake through the front of the basket is facilitated with a horizontal screw conveyor. Because the unloader knife cannot get close to contact with the filter medium, a heel of product remains in the basket after each unloading. This serves as a precoat to prevent loss of fines through the screen during the next cycle. The disadvantage is that it also adds resistance to filtration, similar to the filter medium. The heel may become glazed and impervious from the rubbing action of the knife, and a rinse or backwash of the screen is frequently required during a production stop to restore the permeability. (The heel problem is especially serious with the horizontal peeler where unloading is carried out at higher speed, as compared to the vertical peeler, which requires reduced unloading speed.) Other possible options are pneumatic blow-off, hydraulic dilution and dissolution, and resuspension by flooding. Some horizontal peeler designs are designed such that the basket, shaft, and bearings form a module which can be removed backward on rollers to the drive end. This conveniently allows changing the basket outside the process area, which is especially suitable for pharmaceutical applications.

The horizontal peeler centrifuges are used to dewater insoluble solids such as starch, citric acid, and food stuff. Their most common application is for dewatering and washing solids such as specialty chemicals, pharmaceuticals, and thermo plastics having medium to fast drain rates, typically 48–200 mesh (75–300 μm). The drive is predominantly an electric motor and should be sized so that it can bring the loaded basket to operating speed. For optimum performance the feed rate should match the cake drain rate so that a minimum of free mother liquor is left on the surface of the cake when the feed valve closes. Some modern versions have a gas-tight design for fume control.

Diameters of the horizontal peelers range from 630 to 1250 mm (25 to 49 in) with widths of 406 to 1422 mm (16 to 56 in). Small units can achieve $1700g$ whereas large units operate at $1000g$.

Table 8.1 compares the Gs of vertical and horizontal peelers. As can be seen, the vertical peeler has a higher G than its horizontal counterpart for the same size basket.

Siphon horizontal peeler

For siphon horizontal peelers a partial vacuum is drawn on the outer diameter of the filter such that the filtrate flows through the cake under centrifugal force as well as under a pressure difference of about 1 atm (14.7 lb/in^2). Thus a higher filtration rate results from the increased driving force.

In the siphon horizontal peeler shown in Fig. 8.7 the filtrate leaving

TABLE 8.1 Comparison of G Forces for Horizontal and Vertical Peelers

Diameter, mm (in)	Horizontal peeler, G/g	Vertical peeler, G/g
600 (24)	1750	2000
800 (32)	1450	NA
1000 (40)	1200	1450
1250 (50)	1000	NA
1500 (60)	NA	1000

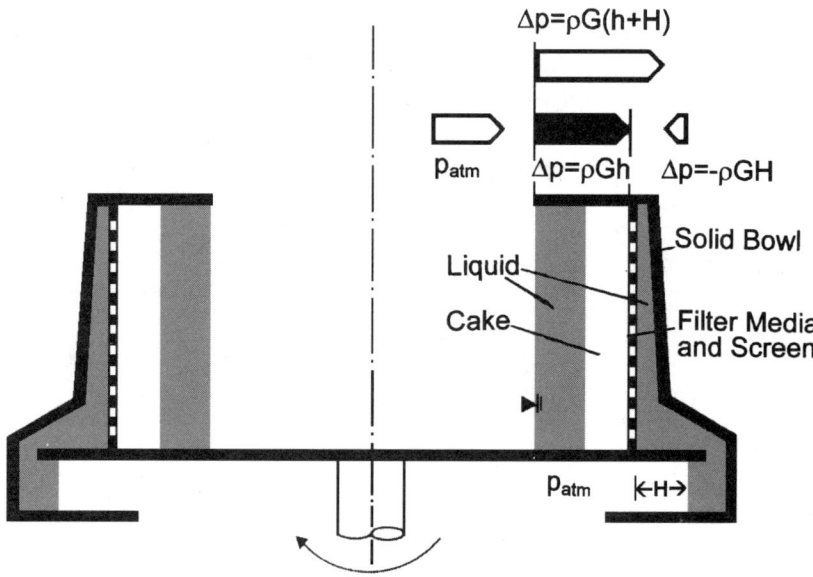

Figure 8.7 Rotary siphon peeler.

the rotating basket is collected at the outer rotating solid bowl, which has a small tapered section. Under G force the liquid flows axially along the bowl toward the larger-diameter end of the taper, through an annulus to a chamber with a larger diameter. In this chamber an annular baffle is used to maintain a differential liquid head across the baffle. Thus the liquid head is higher (smaller diameter) toward the basket, while it is maintained lower (larger diameter) downstream by a stationary skimmer tube which opens tangentially along the pool surface. Similar to the centripetal pump, the kinetic energy of the filtrate from rotation sustains the flow through the skimmer tube and subsequent connecting pipings so that the liquid is ultimately discharged against back pressure, in the form of a liquid column in a storage tank. A portion of the liquid from the storage tank is returned back to the rotating chamber of the centrifuge, whereas the remainder overflows the tank under a constant liquid head and exits the system.

The suction pressure generated is equal to the liquid head difference across the annular baffle as modulated under centrifugal gravity. The maximum suction that can be realized assuming negligible vapor pressure is 1 atm. For example, with a differential liquid level of 2 mm (0.79 in) of water across the annulus under 500g, the suction generated by the siphon is 0.98 atm (1.44 lb/in^2).

Pressurized inverting filter

In Fig. 8.8a an additional pressure difference of as much as 6–7 atm is imposed across the centrifugal filter to further enhance dewatering.[4] This is primarily used to reduce liquid saturation due to the capillaries in the cake. The principle has been considered in the previous chapter. Generally prior to applying the high pressure, the centrifuge speed is reduced so that the cake formed is less compact and the positive pressure gradient provides better filtration on an otherwise low-permeability centrifuged cake. The basket is pressurized and so is the feed pipe. In one design as shown in Fig. 8.8b the filter is inverted during cake discharge to remove the cake heel left on the filter. There is axial travel to facilitate the inversion of the filter bag.

A nonpressurized version of the inverting filter centrifuge has been used in the past. Of course, in inverting the bag to discharge the cake, in both cases there is always the possibility that the filter bag materi-

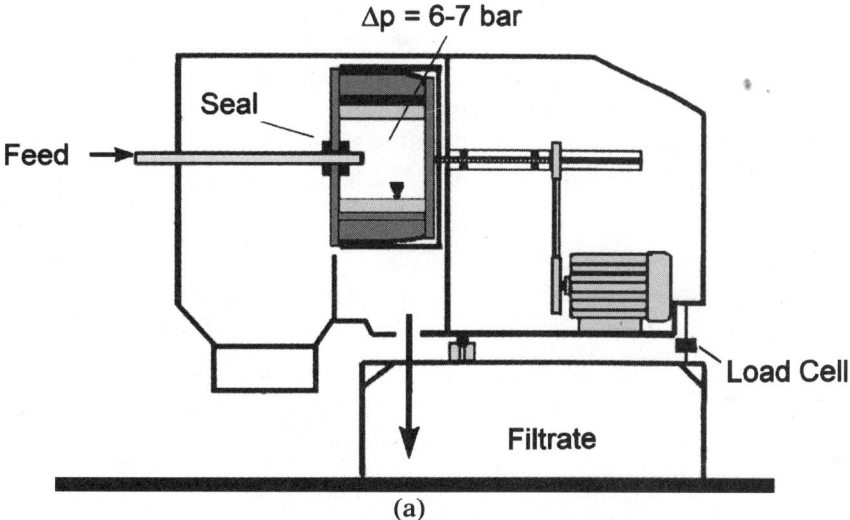

Figure 8.8 Pressurized inverting filtering centrifuge. (*a*) Filtration under imposed pressure.

Figure 8.8 Pressurized inverting filtering centrifuge. (b) Cake discharge with inverted bag.

al will be discharged together with the cake, especially when it is aged and worn.

In another design dewatering of the centrifuged cake is further enhanced by thermal drying, where hot inert gas, such as nitrogen, is blown into the sealed centrifuge basket.

Basket centrifuges have gained wide acceptance in pharmaceutical and special chemicals applications. In addition to matching individual needs to the special features available with the basket centrifuges, other operational advantages of these machines[5] are available and need to be considered in making the selection.

References

1. W. W. F. Leung, "Centrifuges" in Perry and Green, *Chemical Engineers' Handbook*, 7th ed., McGraw-Hill, New York, 1997.
2. W. Wilkie and T. Patnaik, "Advances in Vertical Basket Centrifuge Design—It Is Not Just a Filter Any More," in *Proc. Am. Filt. Sep. Soc. Ann. Conf.* (Valley Forge, PA, April 1996), vol. 10, B. Scheiner (ed.), pp. 274–282.
3. W. Wilkie et al., "Rotary Distribution Pipe Assembly," U.S. patent 5,582,742, Dec. 10, 1996.
4. G. Mayer, "Hypercentrifugation," in *Proc. Am. Filt. Sep. Soc. Ann. Conf.* (Chicago, IL, May 1993), vol. 7, W. W. F. Leung (ed.), pp. 271–277.
5. J. Wright, "Practical Guide to the Selection and Operation of Batch-Type Filtering Basket Centrifuges," *Filtration and Separation,* Nov. 1993, pp. 647–653.

Chapter

9

Continuous Filtering Centrifuges

The trend toward continuous processing and reduction in labor has resulted in more stringent requirements on continuous processing equipment, and without exception this also applies to centrifuges. Despite the elimination of downtime, which translates to lower solids throughput, the continuous centrifuge is less flexible than its batch counterpart because there is no separate control for duration and G force during feeding and filtering, washing, and dewatering, for which the optimal condition of operation may differ depending on the feed material. Frequently with continuous centrifuges, when a given function is emphasized through design or operation, the other functions are somewhat compromised. There are several types of continuous filtering centrifuges, the principal distinction being the solids conveyance mechanism and the associated retention time of the cake in the machine.

Conical-Screen Centrifuges

When a conical screen rotates about its axis, the component of the centrifugal force normal to the screen surface drives the filtration of liquid through the cake and the screen, whereas the component parallel to the screen in the longitudinal direction conveys the cake along the screen toward the larger discharge diameter. The sliding of the solids on the cone is favored by smooth perforated plates or wedge-wire sections with slots parallel to the direction of cake motion. For unassisted conveyance, this cake motion is along the longitudinal direction of the screen. As a corollary, woven wire with the intermesh arrangement provides undesirable resistance to cake movement.

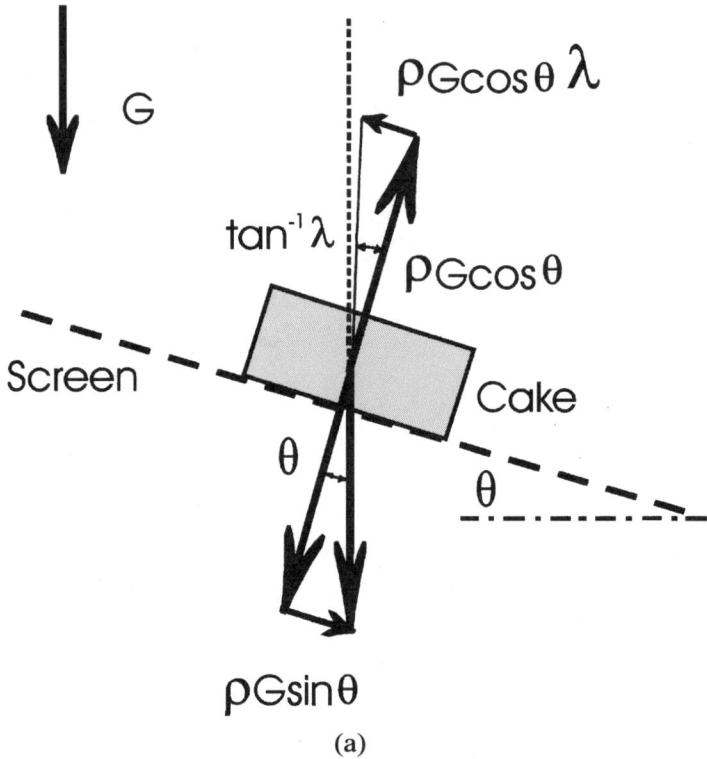

Figure 9.1 Frictional-angle and cone-angle relationship. (a) Frictional angle is less than half cone angle.

Steep-angle conical screen

If the half-angle of the conical screen θ is greater than the frictional angle $\tan^{-1}\lambda$ of the cake solids, where λ is the Coulomb typed frictional coefficient between the cake and the screen, as shown in Fig. 9.1a, the solids slide across the screen at a velocity that depends on the frictional properties of the cake but not on the feed rate. On the other hand the frictional property of the cake depends on the solids property, such as shape and size, as well as on the moisture content. If θ greatly exceeds $\tan^{-1}\lambda$, the cake slides across the screen at a high velocity, thereby reducing the retention time for dewatering. Typically θ is between 30 and 45° with respect to the axis of the machine. The angle selected is therefore highly critical with respect to performance on a specific application. Both steep-angle and compound-angle centrifuges are used to dewater coarse coal and rubber crumb and to dewater and wash crude sugar and vegetable fibers such as from corn and potatoes.

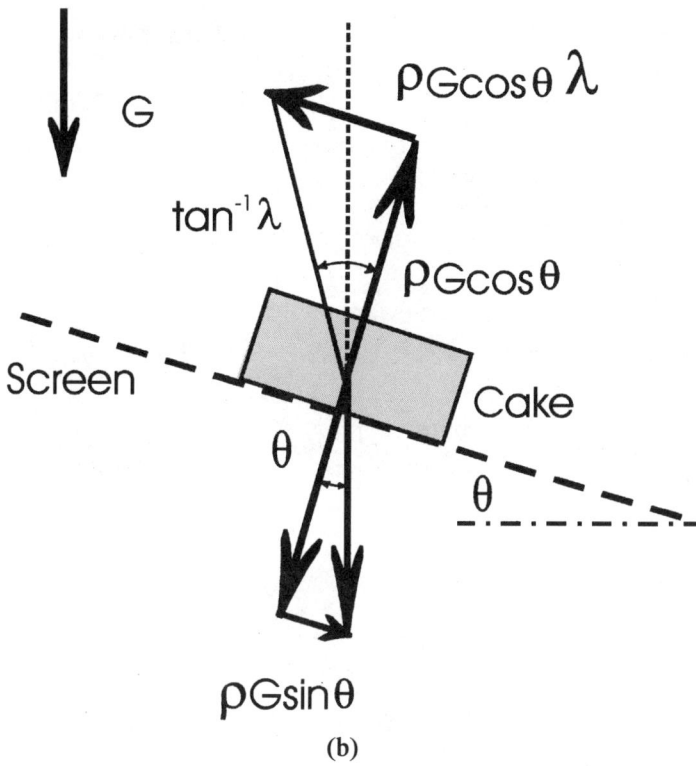

Figure 9.1 Frictional-angle and cone-angle relationship. (b) Frictional angle exceeds half cone angle.

Shallow-angle conical screen

By having a conical-screen half-angle less than the frictional angle of the cake (see Fig. 9.1b) and providing a cake conveyance mechanism, a longer and better controlled retention time is available for the shallow-angle conical-screen centrifuge for cake dewatering. The half-angle θ is typically between 18 and 25°. Three methods are in common use for cake conveyance.

Vibrating conveyance. A relatively high-frequency excitation force is superimposed on the rotating assembly for a vibrating centrifuge. This produces an axial or a torsional vibration. In either case, the cake under inertial force from the vibration is partially "fluidized" and propelled down the screen under a quasi-steady pace toward the large diameter of the basket. Figure 9.2a shows the rotating eccentric weights which translate to an axial vibration on the screen.

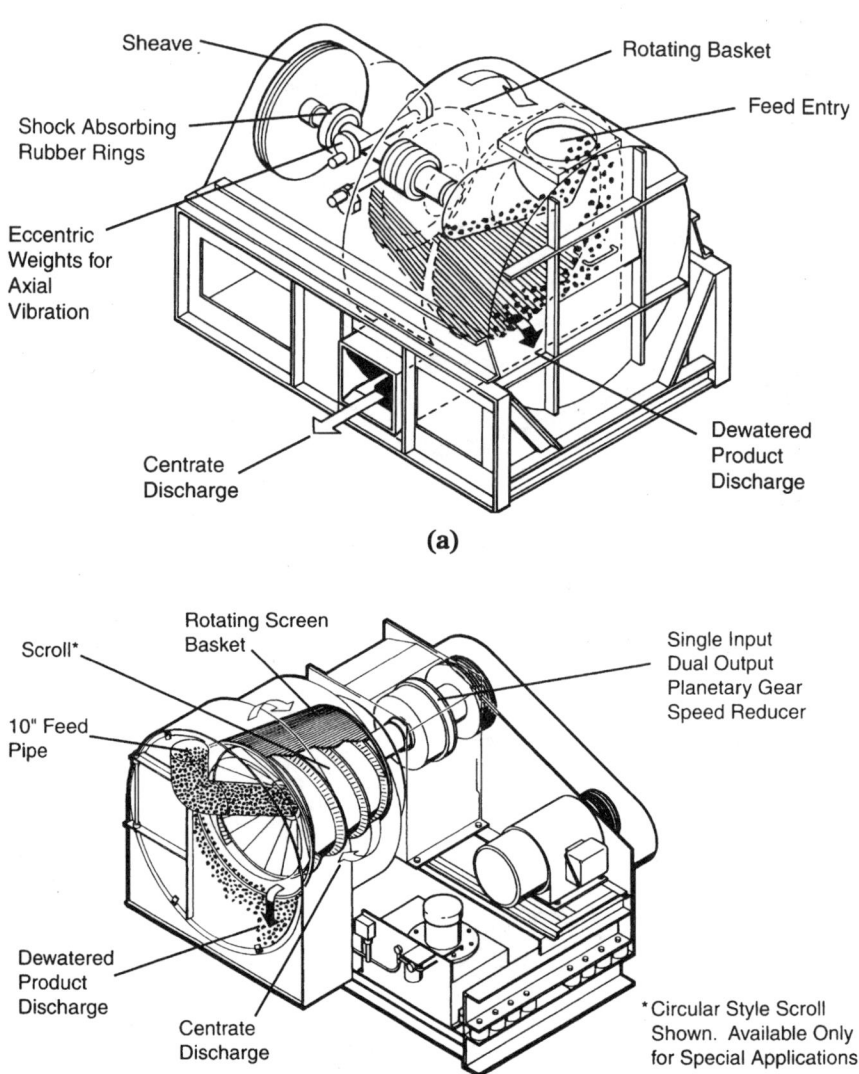

Figure 9.2 (a) Vibrating conical-screen centrifuge using eccentric weights. (b) Conical-screen-scroll centrifuge. (*Courtesy of Bird Machine Company.*)

Oscillating conveyance. This oscillating or tumbling conveyance centrifuge is commonly known as the tumbling centrifuge. The drive shaft is supported at its lower end on a pivot point. A supplementary power source causes the shaft and the rotating bracket it carries to gyrate about the pivot at a controlled amplitude and at a frequency lower than the rate of rotation of the basket. The inertia force generated also provides partial fluidization of the bed of solids in the basket, causing the cake to convey toward the large end, as in the vibrating conveyance.

Scroll conveyance. The screen-scroll centrifuge (Fig. 9.2b) is the most popular shallow-angle conical-screen centrifuge. The cake conveyance is controlled either by a continuous helical-screw conveyor, or with a discrete set of helical scraper blades, usually in sets of four or eight blades. The conveyor guides the cake solids down the conical screen by the differential rotation between the conveyor and the screen, which is maintained through a cyclo gear (Europe) or a gear box (United States).

Conical-screen centrifuges are designed with either a vertical or a horizontal axis of rotation for various applications and installation requirements. While the vertical unit utilizes gravity to facilitate cake discharge, one of the key disadvantages is that it is inconvenient for mechanical repair, such as changing the screen or repairing the bearing or the gear, as all the components are stacked vertically one on top of each other. The energy requirement for the conical screens is low. For example, as little as 792 J/kg (0.2 kWh/ton) is needed to dewater a feed size range between 6 and 19 mm ($\frac{1}{4}$ and $\frac{3}{4}$ in) stoker coal at the rate of 38 kg/s (150 ton/h) to 6% surface moisture on a large oscillatory centrifuge. The screen-scroll centrifuge can handle finer-size coal, with the lower end at 0.5 mm (28 mesh). Its liquid-handling capacity is however limited, and high feed-solids consistency, such as 40–50% w/w, is preferred to obtain best performance and solid capacity. For this reason the amount of cake wash that can be applied is restricted, and therefore the wash efficiency is relatively low as compared to a screen bowl or a pusher. As with any filtering centrifuge, performance is also optimized with operation on large and uniformly sized solid particles. Screen thickness, and hence screen life, is a function of the openings that will support the solids of the size being centrifuged, the conveyance mechanism, and finally whether the feed slurry is accelerated to speed before it is introduced to the small-diameter section of the conical screen, with a much lower tangential speed as compared to the rim speed of the screen. Under accelerated feed slurry, induces wear as the feed slips and rolls on the screen while it is accelerated by the screen surface.

Recently the performance of screen-scroll centrifuges has benefited from improved feed accelerator technology.[1] Not only is the screen wear at the small end of the screen where the thickened feed is introduced being reduced significantly due to bringing the feed tangential velocity closer to that of the rotating basket, in addition the liquid-handling capacity is also increased because the feed acquires promptly the necessary G force for bulk filtration. Furthermore the coal is more evenly distributed onto the basket, utilizing the full basket area for filtration and dewatering. Because the cake is more spread out on the screen, a much thinner cake is formed from which filtration gets further benefit. For fine-coal application, using a better feed accelerator, a 900-mm (36-in)-diameter centrifuge operating at a speed of 700 rpm that used to process 10–15 kg/s (40–60 ton/h) can now handle 20–25 kg/s (80–100 ton/h), yielding comparable cake moisture of 5%.

Among the conical screens, the screen-scroll centrifuge is most versatile in terms of the ability to handle lower feed-solids concentration or higher liquid filtration rates, and relatively finer-size solids. This is partly due to the higher G force and the longer cake retention time on the screen. The screen scroll can provide fair to good washing efficiency. This type is often used together with the steep-angle conical screen for the dewatering of cellulose fibers. The screen-scroll centrifuge has also been applied to process the crystal form of edible lactose having size 150–65 mesh (100–230 μm). With a 450-mm (18-in) outer-diameter screen, the machine can process 5 ton/h at 50% solids. The wash rate is 3.5×10^{-8} to 7×10^{-8} m^3/s · kg (0.5–1 gal/min · ton). The lactose solids after being processed through the centrifuge are further dried to 4–8% moisture.

Pusher Centrifuges

Single-stage pusher

The pusher centrifuge consists of a rotating perforated cylindrical basket lined with wedge-wire screen, with the slots parallel to the axis of rotation. The cylinder is open toward the solids discharge end and cantilevered at the opposite end through its hub to a hollow drive shaft. Fitting close to the cylindrical screen inside the basket is an annular pusher plate. This is mounted on its own shaft concentric with the drive shaft. It rotates at the same speed as the basket and concurrently reciprocates axially via a hydraulic mechanism. The slurry is fed into a feed chamber through a stationary feed pipe located at the centerline of the machine. The slurry is distributed and accelerated in the rotating feed chamber, which usually takes the form of a cone or a disk accelerator. As the pusher plate retracts, a clean surface of screen is exposed for bulk drainage of the new feed material. As the pusher plate advances, the incremental annulus of the cake thus formed transmits its "pressure" to the annular cake already formed from the previous feed cycles in the basket, causing an equivalent amount of dry cake to discharge out of the basket at the open end. Cake wash is applied as a spray through which the cake advances stepwise. Pusher stroke, usually of between 30 and 75 mm (1.2 and 3 in) with a stroke rate of under 100 cycles/min—typically about 40–60 cycles/min, depending on the size of the unit and the load requirement—is generally controllable from the exterior of the machine.

The axially oriented wedge-bar screen construction minimizes friction between the screen and the advancing cake to permit the use of a relatively long screen, and a correspondingly long retention time for the cake without buckling. The slot opening varies typically from 100 to 300 μm, with some finer openings down to 20–40 μm made by laser cutting.

In one modification the pusher plate consists of a conical screen with an angle slightly greater than the angle of repose of the cake solids. This accelerates the feed slurry and provides extra area for bulk drainage, permitting operation over a wider range of feed concentrations.

Multistage pusher

In the multistage variation shown in Fig. 9.3, the basket consists of a series of concentric stages of baskets that increase in diameter in the direction of the solids discharge. The first (smallest-diameter) stage and each alternate stage, which are fixed to the inner shaft, rotate and reciprocate. The final stage (largest diameter) and the alternate stages, which are fixed to the outer shaft, rotate but do not reciprocate. Each screen section is relatively short, with its own pusher action from the preceding stage so that there is less tendency for cake

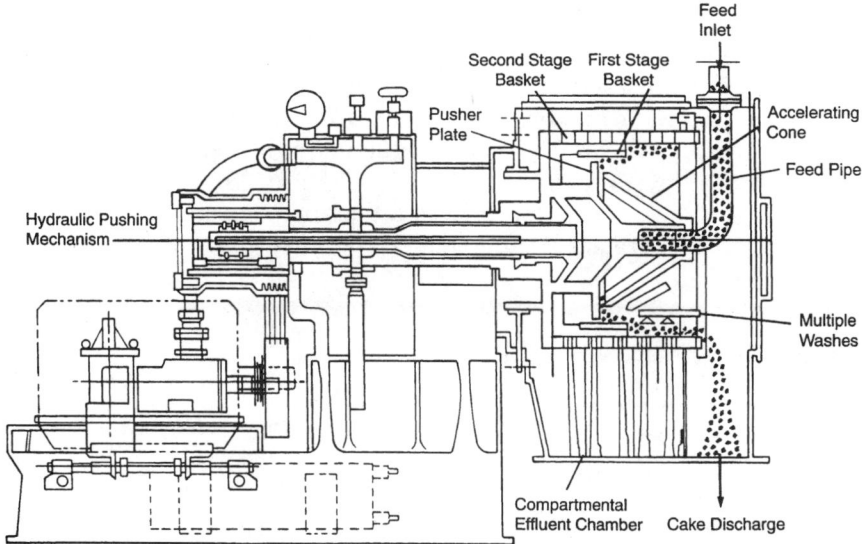

Figure 9.3 Two-stage pusher centrifuge. (*Courtesy of Bird Machine Company.*)

buckling despite a relatively long basket. Bulk filtration takes place at the feed zone in the small-diameter sections with minimum power consumption, and film drainage takes place in the large-diameter sections under maximum G force. In one design the second-stage basket is conically shaped to facilitate cake transport with increasing G force for better dewatering. Concurrently the cake also thins out, providing further enhancement to drainage. In addition the centrifugal force component along the conical basket facilitates cake transport, reducing the "push" requirement of the centrifuge. This is especially advantageous for high solids throughput.

Cake transport

Below a limiting feed rate, the cake thickness is constant for a given basket length. This can be seen by balancing the push pressure F_p on the cake with height h against the frictional shear stress τ_y the basket exerts to retard the cake movement,

$$F_p(2\pi R_b h)\left(1 - \frac{h}{2R_b}\right) = \tau_y 2\pi R_b L \tag{9.1}$$

where R_b is the basket radius, and L the length of the basket for a given stage. With h/R_b being small, this further reduces to

$$\frac{h}{L} = \frac{\tau_y}{F_p} = C_f \tag{9.2}$$

where the effective frictional coefficient C_f is typically about 0.1. Table 9.1 shows the values of C_f for different materials. For example, a basket 250 mm (10 in) long with $C_f = 0.1$ has a cake height of $h = 0.1 \times 250 = 25$ mm (1 in). At a higher solids feed rate exceeding the aforementioned critical feed rate the cake height increases once more. This is shown by the top curve in Fig. 9.4.

For each stroke ℓ_s the cake moves a distance ℓ_c (see Fig. 9.5). Thus the velocity of the cake V_c and that of the push plate V_p are, respec-

TABLE 9.1 C_f Values for Various Materials

Cake material	C_f
Hard crystals (e.g., quartz)	0.075
Medium-hard crystals (e.g., $ZnSO_4$, Fe_2SO_4)	0.088
Plastic	0.088
Sand, silica	0.10
Cellulose fiber	0.15

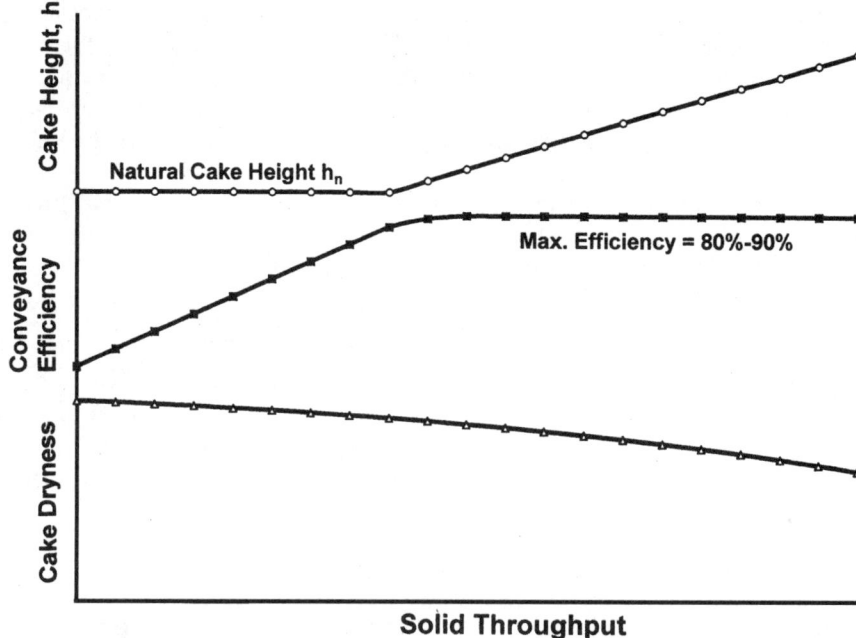

Figure 9.4 Pusher characteristics. Top—cake height versus throughput; center—conveyance efficiency versus throughput; bottom—cake dryness versus throughput.

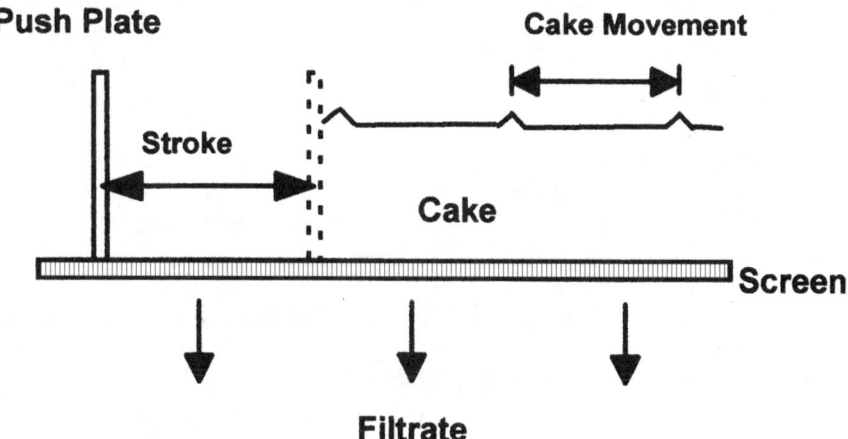

Figure 9.5 Cake conveyance on pusher screen.

tively, $V_c = \ell_c f$ and $V_p = \ell_s f$, where f is the push frequency. The ratio of the two velocities is the conveyance efficiency,

$$\eta_c = \frac{V_c}{V_p} = \frac{\ell_c}{\ell_s} \tag{9.3}$$

Depending on the moisture of the cake as well as on the frictional characteristics of the screen, a 90% efficiency can be attained for certain materials with proper operation. The typical characteristics of the cake conveyance efficiency are shown by the middle curve in Fig. 9.4. At low throughput the conveyance efficiency is poor and the cake supposedly to be conveyed as a solid plug is still semifluid. However, at high throughput the cake reaches a solids consistency where it will act as a semisolid, with the efficiency reaching 80 to 90%. The cake solids (dry basis) throughput m_{ss} is given by

$$m_{ss} = V_c A_c \rho_c W_s = 2\pi R_b \ell_s \eta_c f \left(1 - \frac{h}{2R_b}\right) h \rho_c W_s \qquad (9.4)$$

where ρ_c is the density of the cake, A_c the cross section area of the cake, and W_s the cake solids fraction by weight. Therefore, from (Eq.) 9.4, the cake height can be determined knowing m_{ss} and the other quantities. Thus

$$\frac{h}{R_b} = \frac{1}{2\left(1 - \sqrt{1 - 2c}\right)} \approx \frac{c}{2} \qquad (9.5)$$

where $c = m_{ss}/(2\pi \eta_c f \ell_s W_s \rho_c R_b^2)$. The approximate form, that is the foremost right side of (Eq.) 9.5, holds for small cake heights that are much less than the basket radius (planar approximation), in which case the cake height increases linearly with the solids throughput. At large cake heights the increase shows more like a square root behavior, with a less-than-proportional increase with the solids rate (see top curve in Fig. 9.4). Almost all pushers operate approximately at the increase portion of the height versus throughput curve.

As discussed, the cake-advance velocity increases with increasing feed rate. At higher feed rates, the velocity reaches a maximum and stays constant thereafter. Any further increase in rate only results in a thicker cake at the same maximum conveyance velocity, which has already reached its maximum. As the feed rate increases further, it reaches a point at which it is greater than the bulk filtration rate, resulting in a slurry pool forming above the cake surface, which in effect drives a higher filtration rate due to the extra liquid head. If this increase in filtration rate does not balance the increase in feed rate, a swallowing limit is reached, at which the liquid slurry runs over the product cake, which in proper operation acts as a dam holding back the slurry.

In another instance, when the cake at the feed zone has not reached a consistency to be conveyed, it slops over on the preceding formed cake in the basket. This is especially true for cake with a slow filtration rate in which case the stroke frequency has to be reduced in order to increase the retention time.

An important consideration for pusher centrifuges, especially at high solids throughput, is the discharge cake moisture. As shown in Fig. 9.4, at high solids throughput the cake height also increases, resulting in a smaller effective dewatering time t_d since $t_d \propto 1/h$. This produces a wetter cake, which often becomes the limiting factor at high throughput.

Retention time

The time available for cake dewatering is the retention time of the cake in various stages of the pusher. The retention time for a given stage is the length of the basket divided by the cake velocity. Thus given that the cake retention time is cumulative as the cake gets transported from one stage to the next, we have

$$t_r = \sum_i \frac{L_i}{\ell_s f \eta_{ci}} = \frac{1}{\ell_s f} \sum_i \frac{L_i}{\eta_{ci}} \qquad (9.6)$$

where L_i and η_{ci} are the axial length and the conveyance efficiency, respectively, of each basket stage i.

Example: Given the following values for a two-stage pusher: $L_1 = 12$ in (300 mm), $L_2 = 14$ in (356 mm), $f = 60$ cycles/min $= 1$ cycle/s, $\eta_{c1} = \eta_{c2} = 0.8$, and $R_s = 1.5$ in (38 mm). Then

$$t_r = \frac{12 \text{ in}}{(1.5 \text{ in})(1)(0.8)} + \frac{14 \text{ in}}{(1.5 \text{ in})(1)(0.8)} = 22 \text{ s}$$

Typically, the retention time of the cake in a pusher is on the order of minutes.

Single-stage versus multistage pushers

By having a short basket in each stage of a multistage pusher, the possibility of cake buckling is reduced. Cake buckling has been a problem, especially for long single-stage pushers. Because the cake thickness in each stage is smaller (the cake thickness is proportional to the basket length), a thinner cake results, which proves to be more effective in washing and dewatering. In addition, there is no cake heel as the cake tumbles in transition from one stage to the next. This also helps eliminate moisture retained by the capillary rise in an otherwise undisturbed cake. Effective cake washing usually takes place at the transition between stages (or at a step change in basket diameters). Despite these advantages for the multistage pusher, the single-stage pusher can better handle bulk filtration than its multistage counterpart. Since bulk filtration has to be accomplished in the first stage, a thicker cake is formed on a single long basket. Therefore solids recovery is higher for the single-stage pusher than for the multistage one. With either design, it is important that the cake thickness be uniform both circumferentially and axially, with no axial ridges or valleys on the cake surface, which would result in poor washing and dewatering.

Pusher centrifuges are used for dewatering and washing crystals and other particulate solids, including short fibers. The average size of the crystals in the feed should be 200 mesh (75 μm) or larger for good operation and to minimize the loss of fine solids to filtrate. Feed-slurry concentration is usually 35–60% (up to 95%) w/w for conventional pushers. Lower concentration can be tolerated on the cone-screen and multistage types, especially the ones with a more efficient feed accelerator, which can handle higher liquid filtration. The washing efficiency of pusher centrifuges can be excellent, frequently in excess of 95% displacement of mother-liquor impurities with a wash-to-crystal-weight ratio of only 1:10. Industrial pusher centrifuges range in size from 250 to 1250 mm (10 to 50 in) in single, two, and four stages. They are operated at 600–2000g. The upper range corresponds to smaller units. The solids throughput is between 0.2–80 t/h (dry solids).

Double-acting pusher

A double-acting pusher[2] has been developed recently with a push plate dividing the pusher basket into two effective pushers. The feed pipe is positioned so that it always feeds onto a fresh screen area without cake on either side of the push plate. This avoids laying feed on the preformed cake, which offers an additional filtration resistance, as is found when operating conventional pushers. Consequently the bulk filtration rate and thus the feed rate are significantly higher for the double-acting pusher. There are drawbacks: (1) The loss of solids through the screen is higher because of the absence of the cake precoat on the screen. (2) Washing cannot be implemented in this design. Thus it is only appropriate for applications that do not require cake wash. (3) The overhang length of the basket is limited by design.

Screen-Bowl Centrifuges

The screen-bowl centrifuge consists of a solid-bowl decanter to which at the smaller conical end a cylindrical screen has been added (see Fig. 9.6). The scroll spans the entire bowl, conforming to the profile of the bowl. It combines a sedimenting centrifuge together with a filtering centrifuge. Therefore the solids that are processed are typically larger than 20–45 μm (0.00080–0.002 in).

As in a decanter, an accelerated feed is introduced to the separation pool. A more effective accelerator design is preferred due to the re-

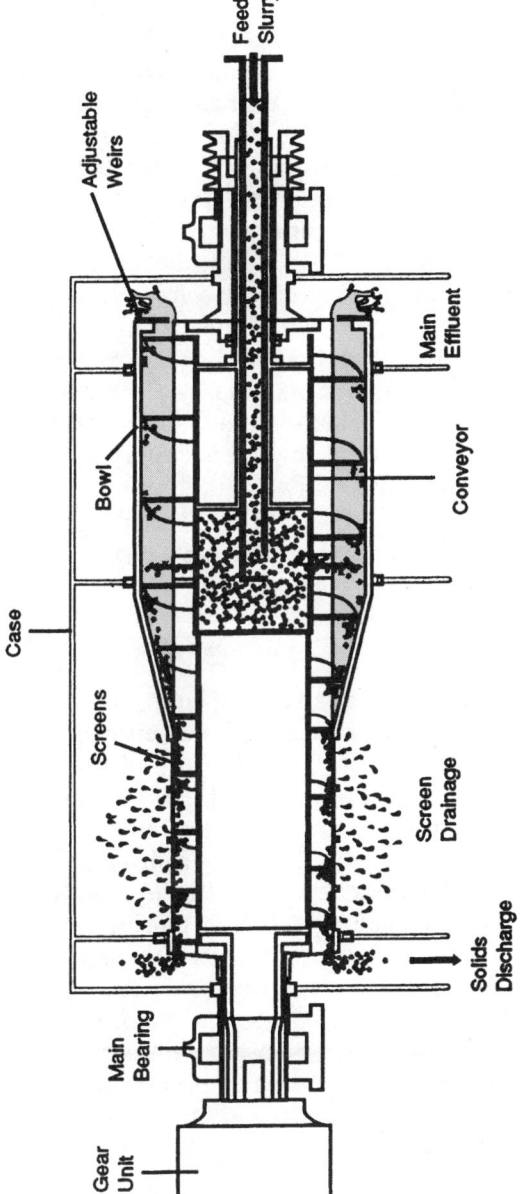

Figure 9.6 Screen bowl. (*Courtesy of Bird Machine Company.*)

duced clarifier length and lower operating G force with screen bowls. This applies especially at high throughput, where centrate solids are kept to a minimum. The denser solids settle toward the bowl wall, and the effluent liquid overflows the ports at the large end of the machine. The sediment is scrolled toward the beach, typically at a steeper angle (up to 15–20°) compared to the decanter centrifuge. (For hard-to-convey cake, the beach angle should be lowered to 10° to avoid high torque. This longer conical beach unfortunately further compromises the length of the screen and the cylindrical clarifier.) As the solids are conveyed to the cylindrical screen section, the liquid in the cake further drains through the screen, resulting in additional dryness. The first half of the screen section is very effective for cake washing to remove impurities, whereas the second half of the screen section is reserved for ultimate dewatering of any residual mother liquor and the wash liquid.

The screen is constructed of an assembly of wedge bars or cut plate supported by a cage, with apertures between adjacent bars or "land" of the cut plate, which open up toward a larger radius (for self-cleaning purposes). This prevents solids from blinding the screen, thus reducing the filtration area and adding conveyance torque. For abrasive materials such as coal, the screens are made of wear-resistant materials such as tungsten carbide.

Screen recycle

The screen openings come in various sizes, typically from 0.1 to 0.5 mm (0.004 to 0.020 in). In general the screen opening should be as wide as possible (2 to 3 times the mean size of the particles) so that liquid from the cake or wash liquid can drain out freely. The percent of open area ranges between 5 and 15% depending on the design and opening of the screen. Typically the screen drain flow rate can be as high as 10% of the volumetric feed rate. In the event that a significant amount of solids is carried along, the screen drain is recycled back and mixed with the fresh feed making a second pass to the screen bowl. The finer solids, which had passed through the screen the first time, will eventually get caught in the cake and discharge off the machine. Figure 9.7a shows the the recycle setup, whereas Fig. 9.7b shows the case without the recycle. The appropriate material balance should be used in each case to come up with the total solids recovery, based on what enters and exits the dotted boxes in these figures.

Dewatering Models

Pusher. In Chap. 7 the dimensionless dewatering time t_d was used to correlate the cake moisture to drainage under centrifugal force. It can

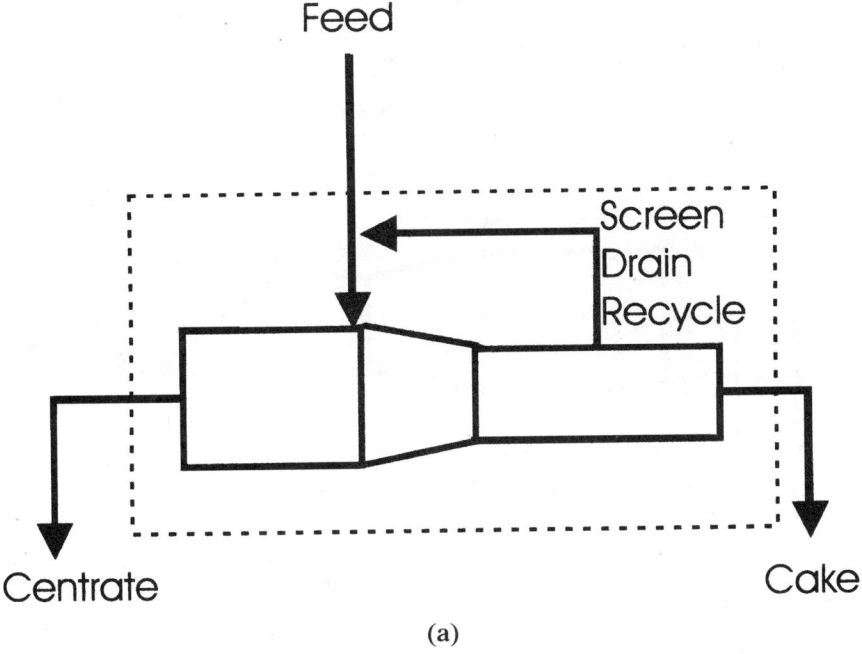

(a)

Figure 9.7 Schematic of screen bowl. (*a*) With screen drain recycle.

be shown that the dimensionless dewatering time t_d for the pusher is given by

$$t_d = \sum_i \frac{d_h^2 \Omega^2 R_i L_i}{v h_i f \eta_{cils}} \quad (9.7)$$

where h_i is the cake height for stage i. Cake height can be determined from either measurements or prediction,

$$h_i = c_f L_i \quad (9.8a)$$

$$h_i = \frac{m_{ss}}{2\pi \eta_{ci} f \ell_s W_s \rho_c R_b} \quad (9.8b)$$

In Eqs. (9.7) and (9.8) Ω is the rotational speed of the basket, R_i the radius of stage i, m_{ss} the solids throughput (dry basis), c_f the cake frictional coefficient, η_{ci} the cake conveyance efficiency, f the push frequency, ℓ_s the push stroke, L_i the basket stage length, W_s the cake solids fraction by weight, and ρ_c the cake bulk density.

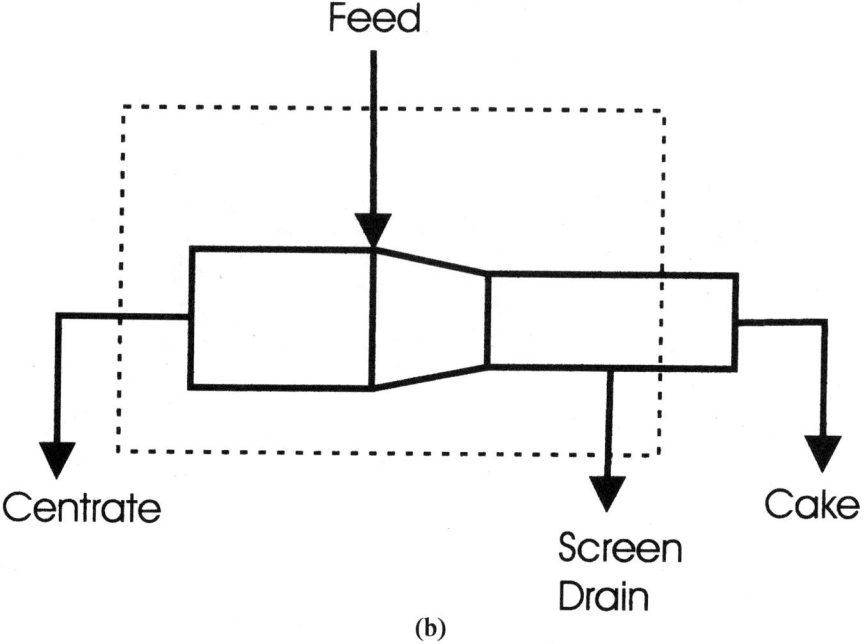

Figure 9.7 Schematic of screen bowl. (b) Without screen drain recycle.

Screen bowl and screen scroll. The dimensionless dewatering time for the screen bowl and the screen scroll can be determined from

$$t_d = \frac{\pi d_h^2}{\sqrt{2}\nu} \sqrt{\frac{n\rho_s \varepsilon_s \, \text{gr}}{\tan \phi \, m_{ss}}} (\Omega \overline{R})^{3/2} \left(\frac{L_{\text{screen}}}{L_{\text{lead}}}\right) \quad (9.9a)$$

The geometric mean radius is defined by

$$\overline{R} = \sqrt{R_1 R_2} \quad (9.9b)$$

where for screen scrolls R_1 and R_2 are the entrance and exit radii, respectively. For screen bowls the effective radius is the geometric radius of the screen section. d_h is the hydraulic diameter of the particles, ν the kinematic viscosity of the liquid, ρ_s the solids density, ε_s the cake solids volume fraction, ϕ the angle of repose of the moist cake, L_{screen} the longitudinal distance along the screen, L_{lead} the conveyor lead, n the number of leads, and gr the gear ratio. Low cake moisture demands lower solids throughput m_{ss}, higher gear ratio gr, multiple leads n, high rim speed ΩR, longer screen L_{screen}, shorter conveyor lead L_{lead}, and dewatering carried out at elevated process temperature where liquid viscosity is significantly reduced. On the other hand, high chatter (material depen-

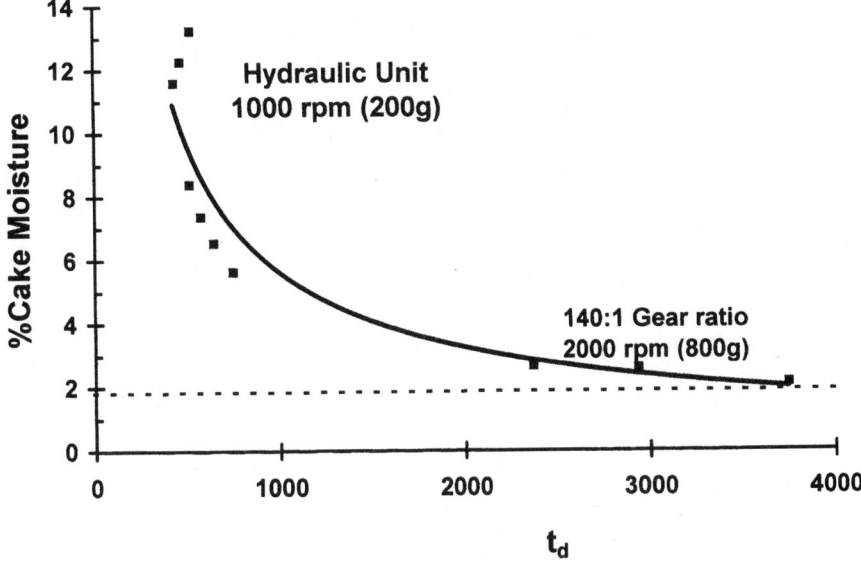

Figure 9.8 Dewatering test results of glass beads using a screen-bowl centrifuge.

dent) and conveyance torque may limit operating the machine at a high gear ratio or at very low differential speed. Also, when the screen bowl is operated at too high a G force, this may result in excessive loss of fines through the screen. Figure 9.8 shows the test results for cake moisture from the screen-bowl tests of glass beads with a mean size 70 μm, which portray the familiar dewatering curve. The tests were carried out at 200–800g using a 457-mm (18-in)-diameter screen bowl.

References

1. W. W. F. Leung et al., "Feed Accelerator Improves Coal Dewatering Centrifuge Capacity and Basket Life with Improved Cakes and Recovery," presented at the 10th Pittsburgh Coal Conference, Sept. 20–24, 1993.
2. W. Stahl et al., "The Application of the New Double-Action Pusher Centrifuge," presented at the 7th World Filtration Congress, Budapest, Hungary, May 20–23, 1996.

Chapter

10

Applications of Filtering Centrifuges

Filtering centrifuges have been applied to the dewatering of granular materials. In this chapter a selection of applications of both batch and continuous filtering centrifuges will be presented.

Before discussing specific examples it is useful to classify various applications according to the properties of the slurries and their separation requirements. The types of centrifuges were listed in Table 7.1 with their ranges of operating G levels in accordance with the applications specified by minimum feed solids concentrations of 10 to 50% w/w and mean particle sizes of 5 to 300 μm (0.0002 to 0.012 in).

Types of Filtering Centrifuges

Vibrating conical screens. For easiest separation with concentrated feed and coarse particles larger than 300–500 μm (48–32 mesh), vibrating, oscillatory, and tumbler centrifuges at low G levels of about 50–300g can be used for dewatering prethickened feed slurries in order to dry cake to its minimum surface moisture.

Screen scrolls and pushers. For particles of finer sizes or with a rough surface, both of which require longer dewatering times, screen scrolls are used with controlled retention time through the use of an appropriate gear ratio for the gear box. For materials that command better washing procedures, pushers are employed using displacement wash to remove impurities from cake forming on the basket with uniform thickness. Typically pushers and screen scrolls are used to dewater particles above 75–150 μm (200–100 mesh). Wedge wire screens with longitudinal slots along the axis are used in pushers with openings of 100–300 μm (0.004–0.012 in). Some laser-cut slots can provide even finer openings down to 20–40 μm (0.0008–0.0016 in), which are employed for dewatering finer crystals. In general screen-scroll centrifuges are used with wedge wires having openings of 300 to 750 μm (0.012 to 0.03 in). The wedge wires are woven to conform to a conical geometry. For finer openings, less expensive perforated plates supported by a basket cage design are often used. Perforations of various sizes and percent open areas are available. (Unfortunately the sieves with smaller openings also give a smaller percent open area, which further reduces the filtration rate.) In addition to the sieve openings and better controlled retention times, the screen scrolls have a range of G forces of 500–1500+g. The higher end of the range corresponds to small pilot screen-scroll units.

Screen bowls. With dilute materials of only 10–30%, a large portion of the slurry is in the form of a liquid, and the screen bowl, with sedimentation and thickening capabilities, becomes more applicable. Separation of solid from liquid takes place in the cylindrical section of the screen bowl. The sediment is further thickened in the dry beach of the screen bowl to form a cake with a consistency suitable for dewatering on the screen deck with minimum loss of fines through the screen. The G levels are between 500 and 2000g. For mineral applications, where erosion of mechanical parts becomes a serious maintenance concern which can increase the operating expense, all screen bowls operate below 1000g and typically at 400–700g. For chemical applications where abrasion is not of concern the G levels can go up to 2000g.

Baskets. Pendulum, vertical and horizontal peelers, and inverting filtering centrifuges are either semiautomatic or fully automatic baskets, which are used to dewater higher-value cake products measured on a dollar per gram basis instead of dollars per ton, as with the centrifuges mentioned. Filter cloth with various fine openings is used as a filter medium. The G levels are generally higher, between 700 and 2000g. The dewatering time is a few seconds for the vibratory, oscillatory, and tumbler centrifuges processing coarse materials; about 10 s or less with screen scrolls; and about 20–40 s with pushers. On the other hand, the dewatering time for batch baskets is more on the order of minutes, anywhere between 5 and 30 min, depending on the difficulty of cake dewatering. For difficult-to-dewater cake it could stretch to several hours. Therefore high-value solids of 5–20 μm (0.0002–0.0008 in) are processed by these automatic basket centrifuges.

Flue-Gas Desulfurization

Basket centrifuges have been used to dewater gypsum (calcium sulfate) to below 10% cake moisture. Typically, the cake moisture can decrease to as little as 6–8%. The dewatering time is on the order of 5–10 min, depending again on the difficulty of cake dewatering. The total cycle time, including feeding, washing, and dewatering, is between 10 and 12 min for 8% cake moisture target, and increases to 14–16 min for 6% cake moisture target. Washing is required to remove incoming feed chloride from 20,000–40,000 ppm to less than 100 ppm. For wallboard-grade gypsum, the chloride should be less than 50 ppm. More often 20–50 ppm chloride content is achieved with a good displacement wash.

In operation, the basket is often overfed with slurry at 25–40% w/w so that part of the feed that has been partially clarified overflows the lip ring or overflow weir. This overflow usually contains 4–5% w/w solids and is returned back to the feed tank. The feed takes several minutes until the cake in the basket has reached a depth of 150–230 mm (6–9 in), depending on the diameter of the basket. A typical basket is 1100–1300 mm (44–52 in) in diameter, and the cake fills to 80–90% of the maximum depth as set by the weir.

Figure 10.1 shows a basket designed especially for flue-gas desulfurization with overflow capability. The overflow is captured in a stationary gutter at the weir end of the rotating basket and subsequently redirected back to the feed tank. The basket is held by three heavy-duty linkages anchored to a firm base support. The flow sheet is simple. The slurry containing the suspended gypsum particles is stored in a feed tank from which it is pumped to a bank of 6–10 basket centrifuges in parallel. A 1300-mm (44-in)-diameter basket takes 10–15 ton/h dry solids and operates at 900g. For an 80-ton/h plant there are typically 8–10 baskets with 6–8 baskets operating at one point and 1 or 2 baskets used as spares. Overflow from a basket, typically with 4–5% w/w solids, is returned back to the feed tank. When the feed condition changes, such as when the feed-slurry solids concentration drops to 15–20%, the feeding-filtration cycle time needs to increase to filter the additional liquid prior to dewatering. In consequence, the solids throughput may decrease slightly as compared to the nominal operating condition. Also, when the amount of fly ash increases, it takes longer to dewater the cake as the smaller fly ash particles render the gypsum cake less permeable.

Fully Automatic Basket Applications

For fine chemicals and pharmaceuticals where the product is measured by the gram the horizontal peeler, the vertical peeler, and the

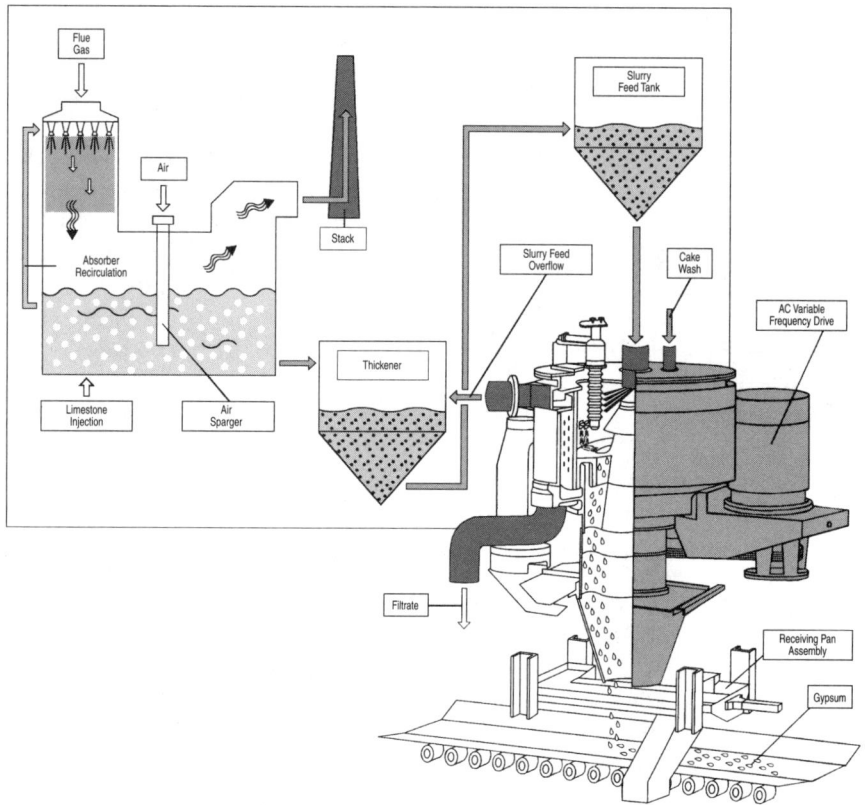

Figure 10.1 Flue-gas desulfurization circuit with cutaway of gypsum basket. (*Courtesy of Bird/Ketema.*)

inverting basket (sometimes referred to as inverting filtering centrifuge) are good candidates for dewatering to the desired dryness. For materials that are shear-sensitive, both the vertical and the horizontal peeler discharge the cake under reduced speed, typically under 100 rpm, regardless of the fact that the horizontal peeler is capable of discharging under full operating speed (which is applicable for cake materials that can tolerate high shear). The inverting basket, as the name implies, discharges cake by inverting the filter medium in the form of a bag pulled axially to empty out the cake deposited inside the bag. This removes the cake in its entirety and is also suitable for shear-sensitive material. The horizontal peeler uses a back flush at regular intervals to remove the residual cake heel, which might glaze over a period of several cycles, forming an impermeable layer and thus reducing the filtration rate. The vertical peeler uses an air gun at the knife tip as well as back pulsation of air from the basket's outer diameter to remove the cake heel. For a 1270–1524-mm (50–60-in)-di-

Applications of Filtering Centrifuges

TABLE 10.1 Applications for Fully Automatic Baskets

Specialty chemicals and pharmaceuticals	Bulk chemicals	Petrochemicals	Food processing
Antibiotics	Aluminum fluoride	ABS	Modified starch
Fumaric acid	Ammonium chloride	Bisphenol	Molasses
Nicotinic acid	Ammonium sulfate	Polyacrylonitrile	Starch
Pesticides	Iron sulfate	Polyethylene	
Salicylic acid	Naphthalene	Polypropylene	
Sodium hydrogen carbonate	Paraffin	Polyvinyl chloride	
Stearate	Sodium chloride		
Sulfonic acid	Sodium perborate		

ameter basket, the centrifugal gravity is nominally between 1000 and 1200g, whereas for the smaller units with a 625-mm (25-in) diameter all three types of baskets often achieve as much as 2000g.

The vertical peeler, the horizontal peeler, and the inverting filter centrifuge all have their unique features targeting similar applications, which are listed in Table 10.1. Note that clean-in-place and sanitary-in-place features are important for pharmaceuticals and food processing, respectively.

Conical-Screen Centrifuges in Coal Prep Applications

Conical-screen centrifuges are being used in coal prep plants. The axis of rotation can be either horizontal or vertical. These centrifuges are used in all applications discussed in this section.

Vibrating conical screens

Vibrating screen centrifuges are used to dewater coal in coal prep plants. One possible arrangement in the separation step is to have the incoming plant feed first classified by the jig, with the good coal sent downstream to a screen. The screen-retained portion is sent to a vibrating screen centrifuge, and the cake coming off the centrifuge becomes a clean product coal. The screen filtrate contains finer clean coal, which is processed further downstream (see Fig. 10.2a). Another possible arrangement is to have the midsize coal feed first classified by a hydrocyclone with the underflow fed through a screen, where the screen-retained portion is processed downstream by a vibrating centrifuge. The filtrate stream containing the fine clean coal is further processed, whereas the cake is sent to refuse (see Fig. 10.2b). Another possible application is to have the midsize coal feed passed through a shanking table, the filtrate of which goes to a finer screen. The screen filtrate has finer coal fractions to be further thickened into a valuable

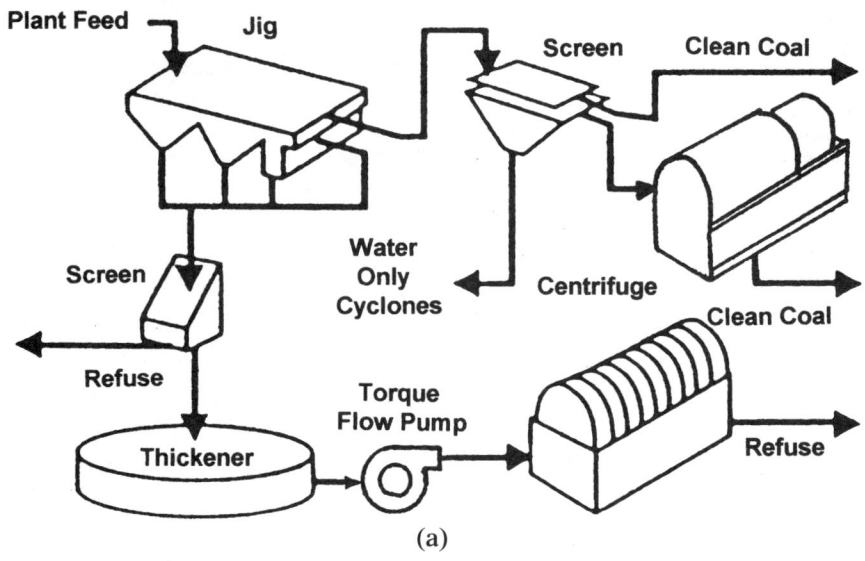

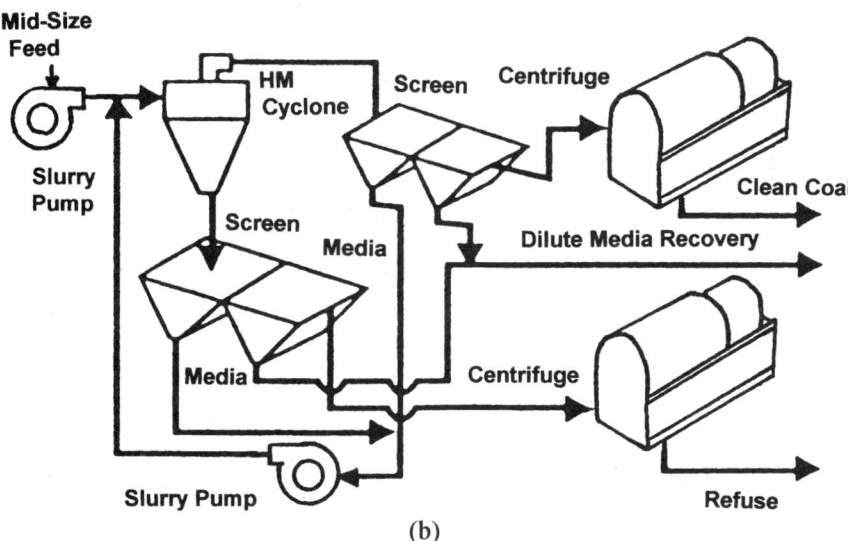

Figure 10.2 Coal prep circuits.

product and the screen retain is dewatered in a vibrating centrifuge with the cake going into refuse (see Fig. 10.2c).

In the applications described, the coal is further divided into two categories—coarse and fine coal. The coarse coal has a size range of 1.24 in × 8 mesh (30 mm × 2 mm) and a median size of ¾ in (18 mm) with 6–10% feed moisture producing cake at 3% surface moisture. On

Applications of Filtering Centrifuges 235

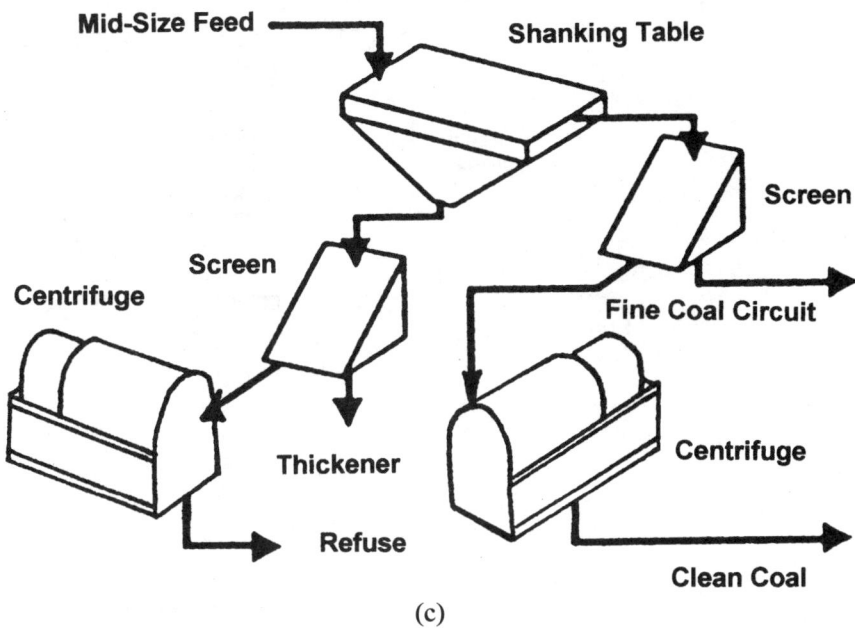

(c)

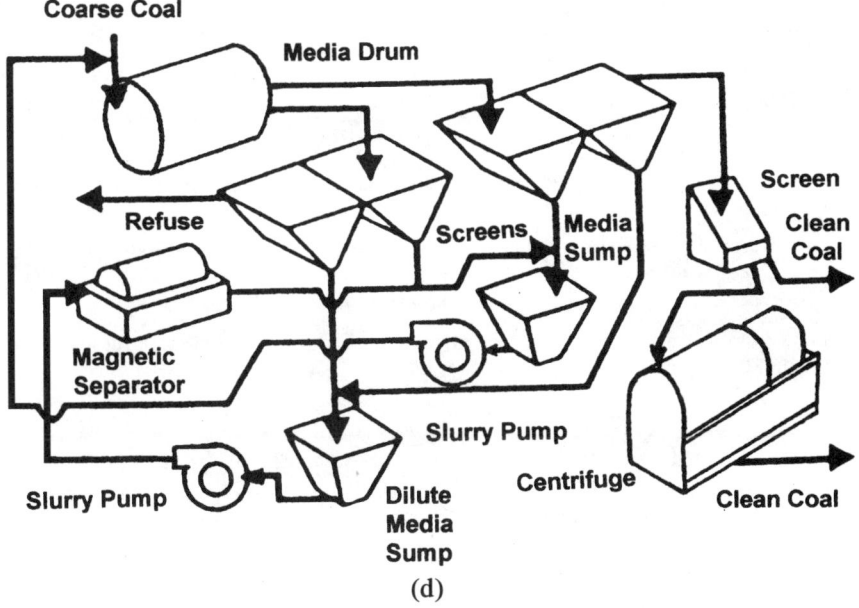

(d)

Figure 10.2 (*Continued*)

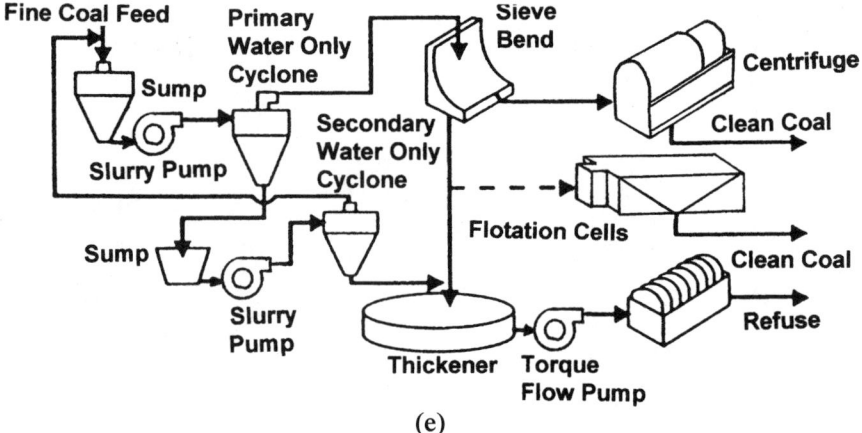

Figure 10.2 (*Continued*)

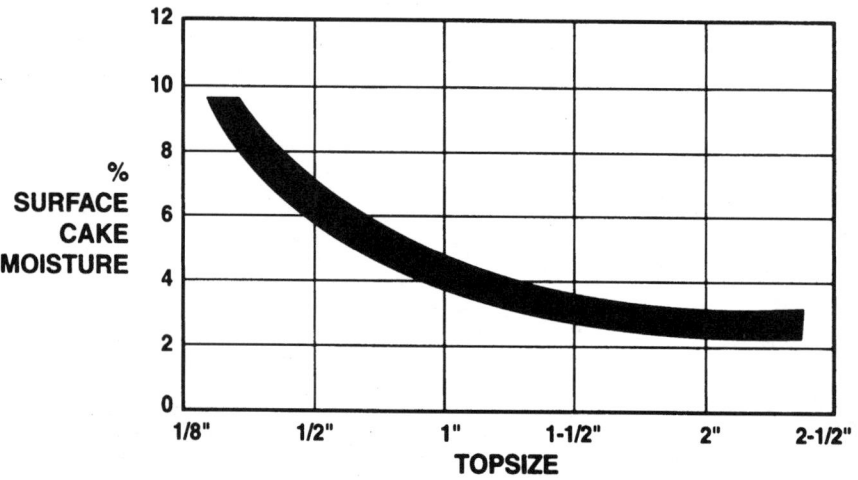

Figure 10.3 Vibrating conical-screen centrifuge used for coal feed containing less than 25% moisture.

the other hand, the fine coal has a size range of ¼ in × 28 mesh (6 mm × 0.5 mm) and a median size of 1.5 mm with 20% feed moisture producing cake at 7–8% surface moisture. This is summarized in Fig. 10.3, where the surface cake moisture is plotted against the top size with feed moisture less than 25%.

The capacity depends on the size of the machine or the large diameter of the conical basket. The dry solids throughput for 1100-, 1200-, and 1300-mm (44-, 48-, and 52-in) baskets is 80–200, 165–220, and 160–350 ton/h (73–182, 150–200, and 145–318 t/h), respectively. The exact capacity depends on the specific coal application. The highest efficiency is obtained by minimizing the fraction less than 28 mesh (600 μm) fines in the feed. For best results the undersize fine material (28 mesh × 0 mesh, 600 × 0 μm) should be dewatered by screen-bowl centrifuges.

Screen scrolls

Screen-scroll centrifuges are used to process coarse and fine coal in coal prep plants. For example, in Fig. 10.2b the overflow from the hydrocyclone downstream of the midsize feed is taken to a screen with the screen-retained coarse coal fed to a 900-mm (36-in)-diameter screen scroll for drying. In Fig. 10.2c the retain from the shanking table is fed to a screen with the screen retain, a coarse coal, sent to a screen scroll for drying to obtain a clean coal product at 6% moisture. In Fig. 10.2d the coarse coal is fed through a media drum and a series of screens downstream with the output stream of the circuit fed ultimately to a screen scroll for dewatering. In Fig. 10.2e the fine-coal feed is taken to a two-stage hydrocyclone where the underflow of the first cyclone is fed to the second cyclone, the overflow of which is sent to a sieve bend. The retain on the sieve bend is sent to screen scrolls to obtain a fine-coal product with 7% surface moisture.

The coarse coal has a size range of $3/8$ in × 28 mesh (6 mm × 0.6 mm) and a median size of $1/8$ in (3 mm) with 35% feed moisture producing a cake at 6% surface moisture. Another slightly finer grade of coarse coal has a size range of 5 mesh × 100 mesh (4 mm × 0.15 mm) and a median size at 14 mesh (1.2 mm). This has feed moisture of 28%, yielding a cake at 7% moisture. With the fine coal the size range is between 16 mesh and 325 mesh (1 mm and 45 μm). The cake depends strongly on the feed particle size. With 28-mesh (0.6-mm) median-size feeds, the surface cake moisture can be reduced to 8+%, whereas with 48-mesh (0.3-mm) median-size feed the moisture jumps to 11+%. The feed moisture for the fine coal is generally between 35 and 52%. This is summarized in Fig. 10.4, where the percent surface cake moisture is graphed against the top size of coal feed.

The capacity for the coarse coal for a 900-mm screen scroll is about 100 ton/h when operating at 250g, whereas for finer coal, the same size machine at twice as high a G force (that is, 500g) can only handle up to half as much capacity (namely, 50 ton/h).

Wedge wires are often used for this application. Slot openings between wires can range between 0.75 and 0.3 mm (20 and 48 mesh), depending on the application. Thin perforated plates are also used. The life of the basket is about several hundred hours, depending on the abrasiveness of the coal processed, and whether the feed is accelerated to proper speed when it is introduced into the basket. The

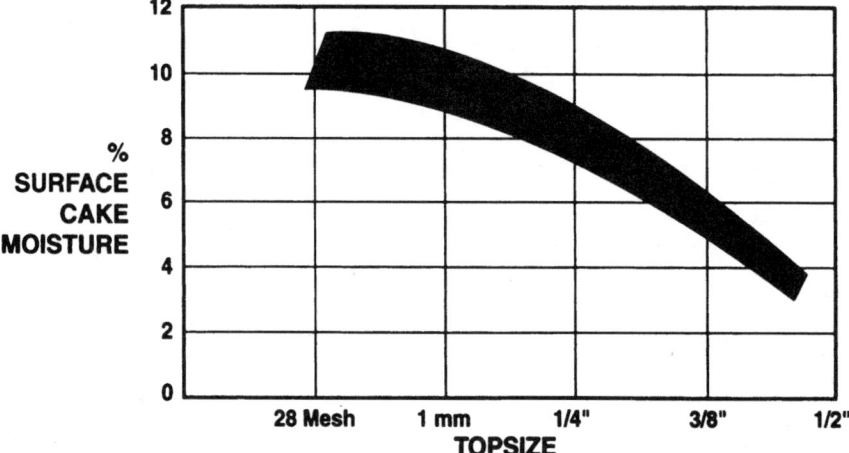

Figure 10.4 Screen-scroll centrifuge used for coal dewatering and drying.

scroll carries a discrete number of helical scraper blades, which are spaced out around the circumference to control the retention of solids on the screen. Another scroll design uses complete 360° blades, similar to those of the scroll decanter. This provides a better control of solids movement in the basket and thus better utilization of the dewatering screen area. However, this benefit is at the expense of higher blade tip wear, which may be undesirable for processing abrasive coal, especially at high tonnage.

Screen Scrolls for Chemical, Mineral, and Industrial Applications

Unlike the vibrating centrifuge, where processed materials spend at most a few seconds in the conical basket, the processed materials can stay in a screen scroll for a longer time. This is achieved by selecting an appropriate gear ratio so as to provide a low differential speed between the scroll and the basket. Both cyclo gear and a planetary gear box are used. Table 10.2 gives a list of materials in chemical, mineral, and industrial applications which have been processed by screen-scroll centrifuges.

One of the many applications is in processing potash, which is primarily potassium and some sodium chloride. The potash is dewatered to 5% moisture using a 1000-mm (40-in)-diameter screen-scroll centrifuge at a solids capacity of 25 ton/h. The conveyor scroll has multiple wrap-around leads to provide effective dewatering of potash crystals. Some screen-scroll centrifuges have also included the advanced feed acceleration designs. With such a design, the feed slurry acceler-

TABLE 10.2 Screen-Scroll Applications

Chemical applications	
Adipic acid	Pearl Polymerizate
Amino acetic acid	Polyethylene
Ammonium persulfate	Polystyrene
Ammonium sulfate	Polyvinyl alcohol
Ammonium thiosulfate	Potassium bichromate
Barium chloride	Potassium chloride
Bisodium sulfate	Potassium phosphate
Bisphenol	Potassium sulfate
Borax	Prussic acid salts
Calcium chloride	Silver nitrate
Calcium formate	Sodium acetate
Cobalt acetate	Sodium bicarbonate
Cobalt sulfate	Sodium carbonate
Crystal soda	Sodium chloride
Dimethyl terephthalate	Sodium formate
Dinitro methylanilin	Sodium phosphate
Disodium phosphate	Sodium sulfate
Ferrocyanide salts	Thiocyanate salts
Iron sulfate heptahydrate	Thiourea
	Tin sulfate
Manganese sulfate	Trisodium phosphate
Monophosphate	Trisodium sulfate
Nickel sulfate	Zinc sulfate
Oxalic acid	Others

Food and pharmaceutical applications	
Aspirin	Glacial acetic acid
Carboxymethylcellulose	Glauber salt
Chocolate (broken)	Lactose
Citric acid	Methylcellulose
Coffee grounds	Nuts (broken)
Finely chopped onions	Vegetables
Fruit juices	Vegetable extracts
Fungal mycelia	Others

Mineral applications	
Anthracene	Potash
Coal	Rock salt
Copper sulfate	Others

Industrial applications	
Celluloid and cellulose wool	Nylon chips
	Plastic granules
Cotton linters	Plexiglas beads
Gunpowder	Rubber regenerate
Gypsum	Others
Ion exchange resin	

ates promptly to speed, thus acquiring the needed centrifugal gravity to effect bulk filtration and subsequent cake dewatering as the feed is introduced to the feed zone at the small-diameter section of the conical basket. This contrasts with the conventional design with ineffective accelerators, where the feed undergoes slipping and sliding on the screen surface at the feed zone in an effort to get accelerated to the same tangential speed as the basket. This is inefficient, causing undesirable erosion on the basket in the feed section. For a 1000-mm (40-in) screen scroll, a much higher solids throughput of 44 ton/h (versus 25 ton/h) is achieved with the cake moisture further reduced to 4.5%.

The screen-scroll centrifuge has also been used in food processing. For finer solids perforated screens are often used. Standard perforations of say 0.13–0.81 mm (0.005–0.032 in) are readily available. Finer-precision perforations can be obtained such as by laser cutting. The centrifuge is used to recover lactose crystals from concentrated whey at a feed of 20% w/w and at particle sizes of 10–150 mesh (1.7 mm–100 μm). The cake containing 4% moisture is sent downstream for further refining, whereas the filtrate liquid containing a trace amount of some fine crystals is separated out by a hydrocyclone, with the overflow containing very fine crystals sent for downstream processing and the underflow containing coarser crystals recycled back to the screen scroll. Other similar crystalline materials processed include citric acid with a cake moisture of less than 0.5%, sugar with 1.5% moisture, and polyethylene crumb of 5% cake moisture.

Screen scrolls have also been used in the corn wet-milling process. After wet milling, the screen scroll is used to dewater the fibers after passing through a series of staggered screens using countercurrent wash to remove the starch by the wash liquid. The recovery of fiber is typically 95% for 1–10% w/w solid feed. The cake moisture can be reduced down to 75% for fine fibers and 60% for coarse fibers with sizes of between 100 mesh and ½ in (0.150 and 12 mm). Depending on the sizes of the machines, up to 15 ton/h solids throughput is not uncommon. Screen scroll is excellent for dewatering pulp materials. For example, potato pulp can be dewatered to 80% cake moisture and paper pulp to 65% moisture. Other fibrous materials are soybean, spent brewery grains, whole and milled wheat, apple, orange, and other fruit pulp.

Pushers for Salts, Soda Ash, Potash, and Polymers

The pusher centrifuge is most appropriate for washing and dewatering crystalline materials above 100 μm (150 mesh) given that the cake deposited on the screen surface is uniform in thickness in both the circumferential and the axial directions. Pushers can also process finer crystals above 325 mesh (45 μm) provided finer sieves, such as 40-μm

(400 mesh) openings, are used. A higher loss of fines is expected when there is a significant fraction less than 325 mesh (45 μm), or when the feed solids concentration drops below 30% by bulk volume. At the other extreme, for practical economic reasons the maximum particle size is limited to 5 mm (0.2 in). The dry solids rate varies from 0.5 to 50 ton/h. Washing capability is very good with good filtrate quality or complete solids capture.

A 400-mm (16-in)-diameter pusher can process 3–5 ton/h of sodium chlorate at a feed consistency of 25%, resulting in cake moisture of 1–2%. On the other hand, a 500-mm (20-in)-diameter pusher can process 15–20 ton/h of salt to 2% cake moisture. If required, washing can be implemented to reduce the sulfate content. The bulk cake density of salt is about 80 lb/ft^3 with feed solids typically about 40–50% w/w. On the other hand, a 700-mm (28-in) pusher can process 20–35 ton/h of sodium bicarbonate monohydrate (soda ash) with feed solids of 35–50% to yield a dry cake with 2–3% moisture. In addition, a 900-mm (36-in)-diameter pusher can process 40–50 ton/h of potash crystals to yield cake with 3–5% moisture. Finally, polymers are processed with 300–900-mm (12–36 in) pushers at rates of 0.5–10 ton/h to yield cake moisture in the range of 15–35%, depending on the polymer size and shape. These include polyethylene, polypropylene, ABS, and polystyrene among others. Table 10.3 shows a more comprehensive list of materials that have been processed by pusher centrifuges.

The pusher centrifuge is an ideal equipment for solid-liquid separation when the process requires continuous operation, high solids purity, cake dryness, minimal crystal attrition, separation of the wash liquid from the mother liquor, and a piece of equipment to be used in combination with filters.

Screen-Bowl Applications

Fine-coal dewatering

Screen bowls have commonly been used to dewater coal at the downstream end in a coal prep plant. The cake moisture obtained from a screen bowl depends on (1) the size of feed, (2) the operating condition of the centrifuge such as feed rate, G, differential speed, and pool depth, and (3) the design geometry, such as screen length, lead, number of leads, screen opening. Under optimal operation the total cake moisture (inherent and surface moisture with the inherent moisture limited to less than 1–1.5%) directly correlates with the fines fraction (that is the fine particle fraction less than 325 mesh or 45 μm) in the feed. Specifically, the percent total cake moisture increases with increasing -325 fines fractions, as shown in Fig. 10.5. With 10% of

TABLE 10.3 Pusher Applications

Adipic acid	Nickel sulfate
Amino acid	Nitrocellulose
Ammonium bicarbonate	Oxalic acid
Ammonium chloride	Potash
Ammonium nitrate	Potassium bisulfate
Ammonium sulfate	Potassium chlorate
Aspirin	Potassium chloride
Bark	Potassium nitrate
Borax	Potassium persulfate
Boric acid	Potassium phosphate
Caprolactum	Quartz
Cellulose	Sea salt
Cellulose acetate	Sodium acetate
Coffee powder	Sodium bicarbonate
Copper sulfate	Sodium bichromate
Disodium phosphate	Sodium chlorate
Ferrous sulfate	Sodium chloride
Foundry sand	Sodium formate
Glycerine salt	Sodium nitrate
Guanidine nitrate	Sodium nitrite
Hexachlorocyclohexane	Sodium phosphate
Hexamethylene tetramine	Sodium sulfate
	Sodium sulfide
Lactose	Sugar
Lead nitrate	Trichloroacetic acid
Linters	Urea
Naphthalene	Zinc sulfate
	Others

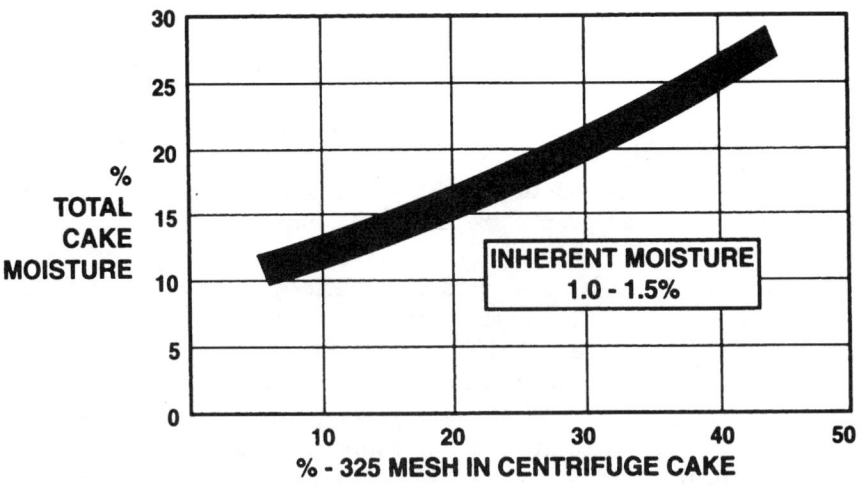

Figure 10.5 Screen bowl for fine-coal dewatering with 28 mesh×0 (0.5 mm×0 mm) feed.

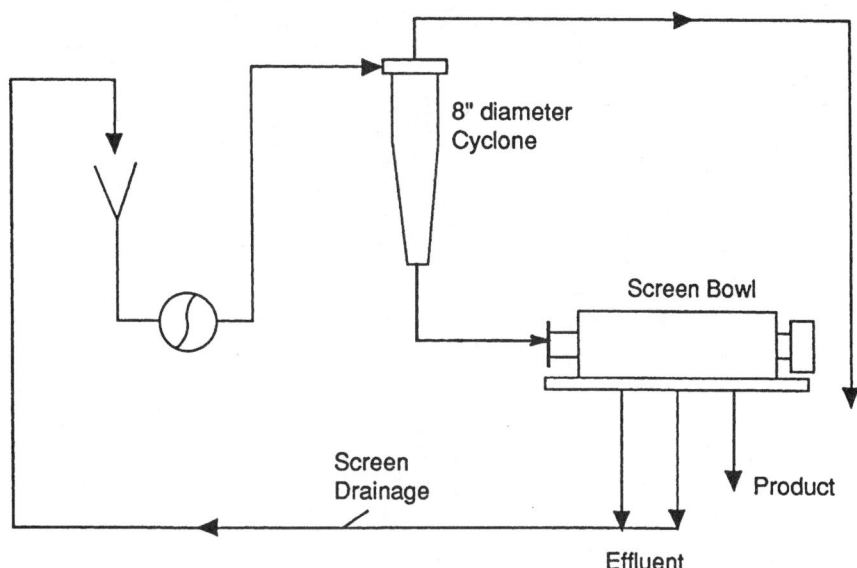

Figure 10.6 Screen bowl fed from hydrocyclone underflow.

−325 mesh fines, the total cake moisture is about 10–12%. Furthermore with 20, 30, and 40% of −325 mesh, the cake moisture jumps to 15–17, 20–22, and 25–28%, respectively. Obviously, the screen bowl performs best when the −325 mesh fines are minimized. These figures need to be adjusted for nonoptimal design and for cake with inherent moisture higher than 1–2%.

Figure 10.6 shows a typical arrangement where a hydrocyclone is installed upstream of a screen bowl, 910 × 1830 mm (36 × 72 in). The cyclone makes a cut such that the underflow of the cyclone with 28 mesh × 200 mesh (0.6 mm × 150 μm) is fed to a screen bowl. The feed rate is 180–260 gal/min, 20–37 ton/h (dry solids), 13–15% ash and 39–53% w/w suspended solids in the feed. The cake moisture obtained is between 11 and 13%. The product cake averages out to 25 ton/h (dry solids), with a further reduction in ash amounting to 10–12%. This is 3% less than that of the feed, revealing that finer particles less than 325 mesh carry a heavy amount of ash and are classified out in the centrate. The centrate is at a flow rate of 175–200 gal/min with 0.3–1.5% w/w suspended solids. The screen drain is about 15–20 gal/min, which is 8–10% of the volumetric feed rate. It has 10–15% w/w suspended solids and 11–14% ash, which is the same as the feed. Given the screen drain is recycled back to the feeder and subsequently back to the screen bowl, the 10–15% solids in the 15–20-gal/min screen drain would eventually be captured and should not be considered as a loss item.

TABLE 10.4 Analysis of Screen Bowl for 28 Mesh×200 Mesh Coal Feed

Screen analysis		% feed	% cake product	% centrate	% screen drain
Mesh	μm				
+28	(600)	0.71	0.36	—	—
28×48	(600×300)	65.65	62.60	—	1.55
48×100	(300×150)	18.13	17.88	—	10.18
100×200	(150×75)	10.75	12.22	—	18.91
200×325	(75×45)	2.15	2.80	53.26*	12.94
−325	(−45)	2.39	5.04	46.74*	56.42

*The fine solids fraction lost in the screen drain contains essentially useless nonburnable clay or ash.

Therefore the solids recovery based on loss of fines in the centrate alone is determined to be 98–99%. However, the fines in the centrate with 53% between 200 mesh and 325 mesh (75 μm and 45 μm) and 47% under 325 mesh (less than 45 μm) is essentially clay which has no burning value (Btu or kWh) or use. As a matter of interest, the particle-size fractions corresponding to the feed, product cake, effluent, and screen drainage are given in Table 10.4 for further reference.

Chemical applications

Screen bowls have been used in many chemical applications. Dimethyl terephthalate (DMT) is obtained by crystallization from methanol. The solid crystals are separated in a screen bowl, where the crystals are further washed by methanol to remove any impurities.[1]

In manufacturing of bisphenol-A (BPA) from phenol and acetone, one of the steps involves the crude product stream containing BPA, phenol, and impurities sent to a crystallizer. After the crystallizer, the BPA needle-shaped crystals are washed by pure phenol in a screen-bowl centrifuge to a high degree of purity before downstream processing.[1]

Screen bowls have also been used for dewatering of polyvinyl chloride (PVC) slurry to dry cake suitable for downstream drier.

Application guidelines

Screen bowls are generally applicable for dilute (2% w/w ss) to thickened (60% w/w ss) feed slurries. The range of particles varies between 2 μm and 5 mm (practical limit). The solids rate (dry) can range between 1 and 150 ton/h from a small 150-mm (6-in)-diameter bowl to a 1370-mm (54-in)-diameter bowl. Washing is good with clear liquid discharge. The screen drain can be recycled if the solids content becomes significant. It is best suited for continuous operation, handling solids greater than 325 mesh (45 μm), which provides fast draining and fast settling and with less escape of fines through the screen, with openings typically 150 μm (0.006 in) and larger. It has several advantages over other competitive equipment for higher capacity with wash capability, ability to handle dilute feeds, and when conventional filter cake becomes too wet. Heat trace can be applied on the housing and peripherals and with proper insulation, the machine can operate with seal at elevated temperatures, which is useful for chemical operations.

References

1. *Hydrocarbon Processing,* Gulf Publishing, Houston, TX, 1997, pp. 103–173.

Chapter 11

Feed Acceleration

It is well appreciated that centrifuges rotate at high speed by which a centrifugal force is generated to produce solid-liquid separation, either by sedimentation or by filtration. By contrast, the feed delivery system is often thought of only as a means of somehow getting the feed slurry into the machine. That it may have a significant effect on performance is often overlooked. In this chapter the discussion will be on feed acceleration, whereby the feed stream from an initial condition without rotation is accelerated to a significant rotational speed, thus generating the required centrifugal gravity to effect separation.[1-6] This is most important for continuous feeding centrifuges.

Ideal Feed Accelerator

Upon entry to the sedimenting centrifuge pool surface or to the filtering basket, the desirable features of feed behavior include the following:

1. The tangential speed of the feed should match the solid-body tangential speed of the pool or basket, which has a value of ΩR. Here Ω is the angular speed of the centrifuge and R is the radius of the pool or basket. This assures the feed to attain the proper centrifugal force, and avoids shearing forces from underacceleration, which might lead to instability, turbulence, and resuspension. The performance of an actual feed accelerator may be characterized by an accelerator efficiency defined as

$$\eta_a = \frac{u}{\Omega R} \qquad (11.1)$$

where u is the average tangential speed of the feed at a given radius R, and this is compared to the tangential speed of solid-body rotation. The centrifugal force is given by

$$G = \frac{u^2}{R} \qquad (11.2)$$

It is obvious from Eqs. (11.1) and (11.2) that $G = \Omega^2 R \eta_a^2$. When $\eta_a = 100\%$, $G = \Omega^2 R$. Therefore another measure of the established G force is the G efficiency, where the gravity obtained from the tangential velocity u in Eq. (11.2) in a circular trajectory is compared to that under solid-body rotation,

$$\eta_G = \frac{u^2/R}{\Omega^2 R} = \eta_a^2 \qquad (11.3)$$

When referring to the accelerator alone, the efficiencies are evaluated with the values of u and R at the exit of the accelerator. However, the performance of the machine is in practice determined by the value of the G efficiency reckoned with the values of u and R at the entry of the feed to the pool or basket.

Figure 11.1 shows a feed stream leaving the accelerator at radius R_1 with tangential speed $u_1 = \Omega R_1$ (assuming it is fully accelerated) and radial speed v_1. It enters the pool or basket at a larger radius R_2. In the absence of external force and torque, the absolute velocity remains constant. However, the tangential velocity component reduces only to a smaller value of $u_1(R_1/R_2)$ by the law of conservation of angular momentum. This is also demonstrated through the simple vector construction shown in Fig. 11.1. However, the tangential speed of the pool or basket under solid-body rotation is at ΩR_2. Therefore the accelerator efficiency at R_2 becomes

$$\eta_{a2} = \frac{u_1(R_1/R_2)}{\Omega R_2} = \frac{u_1(R_1/R_2)^2}{\Omega R_1} = \eta_{a1}\left(\frac{R_1}{R_2}\right)^2 \qquad (11.4)$$

and by Eq. (11.3),

$$\eta_{G2} = \eta_{G1}\left(\frac{R_1}{R_2}\right)^4 \qquad (11.5)$$

In particular the G efficiency is significantly affected by a change in radius between the accelerator discharge and the pool or basket. Unfortunately the loss in tangential speed contributes to the gain in the undesirable radial velocity, as depicted in Fig. 11.1.

2. The radial speed and the associated momentum should be small. This avoids a plunging of the feed into the pool or basket, with accom-

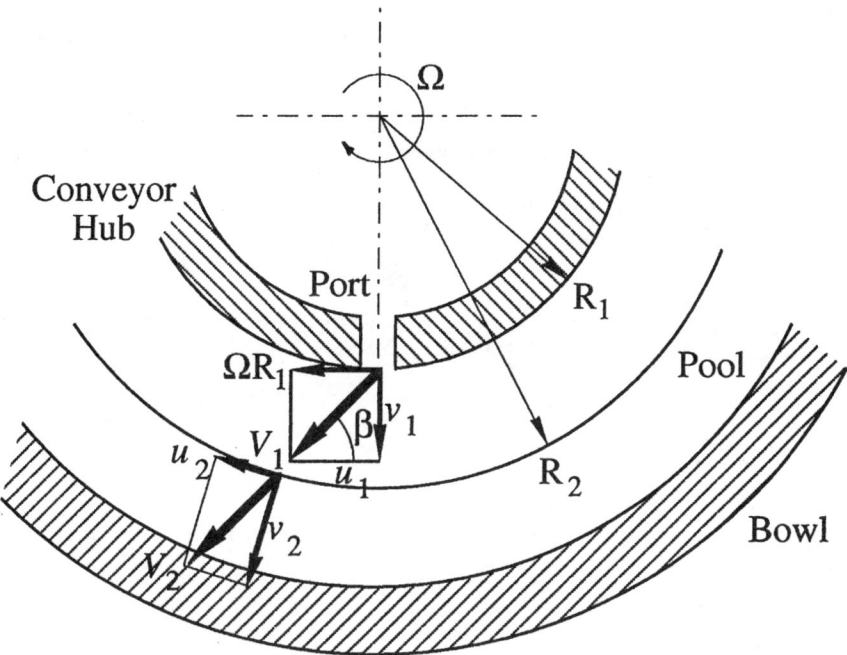

Figure 11.1 Kinematics of feed stream as it passes from accelerator exit to pool entry.

panying disturbances in the former case, and with loss of fines and wear in the latter.

3. The feed flow should be uniform in the circumferential direction to avoid disturbance from concentrated jets.

Conventional Accelerators

Conventional hub accelerator

In solid and screen bowls the feed compartment is formed by blocking the interior of the conveyor hub by two baffles, which are perpendicular to the machine axis and spaced some distance apart. One of the two baffles has a solid wall whereas the other has an opening at the center through which the feed is delivered. This is depicted in Fig. 11.2a. The feed slurry is delivered to the feed compartment by a stationary pipe. Through contact with the rotating baffles and the inner wall of the feed compartment, it gets accelerated before discharging to the annular pool through feed ports in the conveyor hub. Unfortunately the acceleration to the speed of rotation by means of liquid viscosity via diffusion of momentum is rather ineffective. Also, the feed ports in most designs follow the helical blade, and given that they are not at

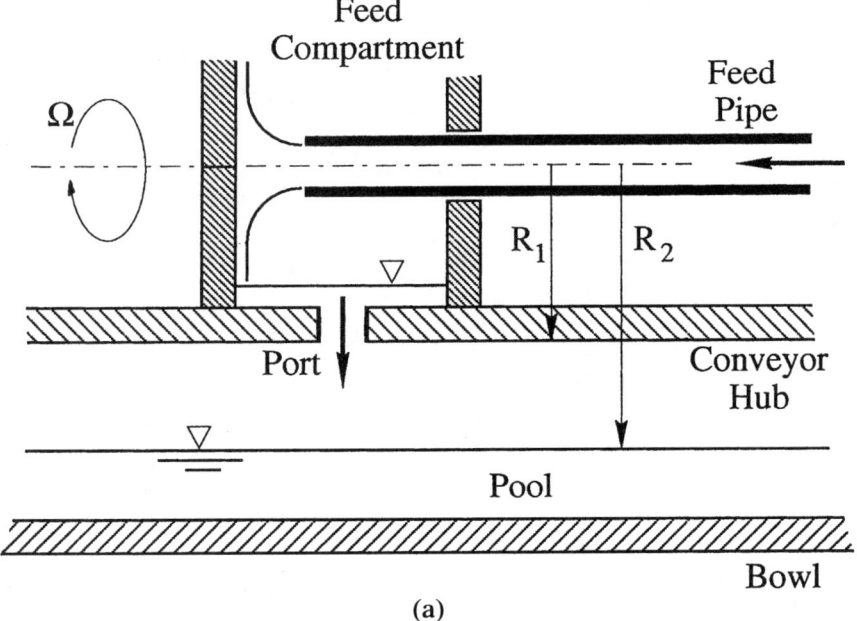

Figure 11.2 (a) Hub accelerator.

the same axial location, they receive an unequal amount of flow. Furthermore, the feed after exiting the feed compartment usually has to plunge over a radial distance to the annular pool. This reduces the already deficient tangential speed and increases the undesirable radial momentum as the feed enters the pool in discrete jets, resulting often in turbulence, mixing, and wear.

Conventional cone accelerator

Figure 11.2b shows a schematic of a conventional cone-shaped accelerator.[1] Appropriately adapted, it may be applied to either a decanter or a pusher centrifuge. The feed is directed to a distributor which lays down the liquid in a thin conical layer at the small end of the cone. As shown in Fig. 11.3a, viscous forces exerted by the cone wall in the tangential direction cause the liquid layer to accelerate in the tangential direction. By the no-slip condition, the liquid layer immediately adjacent to the wall of the cone acquires a tangential speed ΩR, whereas the layer further out has a much lower tangential speed. This is all accredited to the slow transfer of momentum by diffusion via liquid viscosity. The larger the liquid viscosity, the more effective is the transfer mechanism; however, it is still limited.

As shown in Fig. 11.3b, the radially outward G field, as a result of the tangential velocity distribution in the liquid layer, has both a nor-

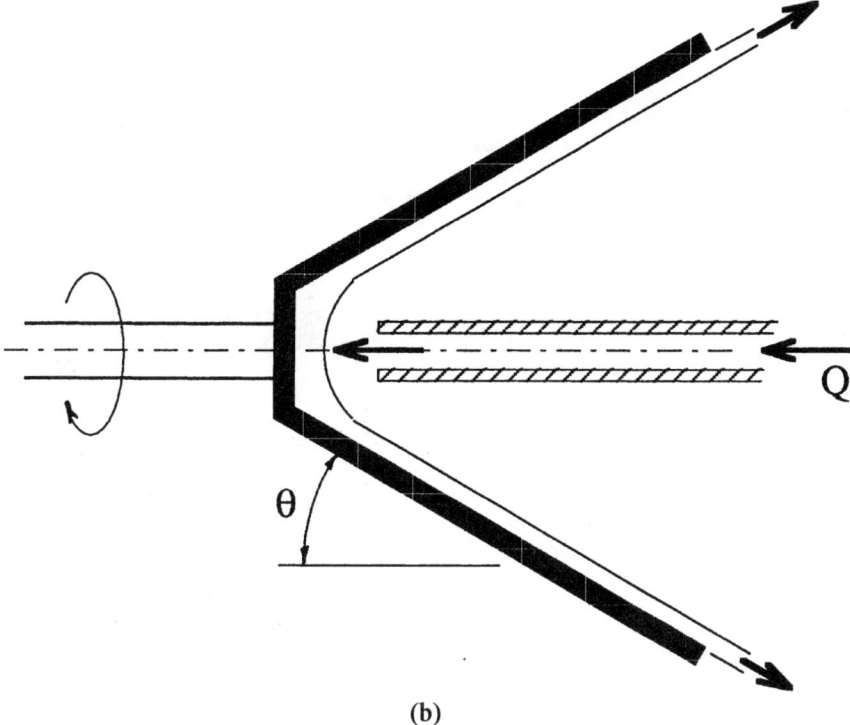

Figure 11.2 (b) Cone accelerator.

mal component perpendicular to the cone surface and a longitudinal component directed along the cone wall. The latter drives the liquid down the cone with a longitudinal velocity, which starts from zero at the wall due to the no-slip condition and reaches a smaller finite velocity at the liquid surface after acquiring a local maximum velocity in the layer in response to the tangential velocity profile shown in Fig. 11.3a. As the liquid layer reaches the cone at a larger diameter, the liquid layer thickness further decreases because of (1) increasing longitudinal velocity and (2) increasing circumference of the cone. The foregoing trio of events actually occur interactively and simultaneously, and it is evident that a steady equilibrium condition can never be attained because the radius $R(x)$ is constantly increasing with distance x along the cone. Accordingly, the local tangential speed always lags the local value of ΩR, and η_a, and thus η_G, cannot possibly reach 100%.

Conventional double-disk accelerator

Double-disk geometry[2] has been used in feed acceleration for pushers. As shown in Fig. 11.2c, it consists of two parallel disks, a target disk and a cover disk with an annular opening to accept the feed delivered

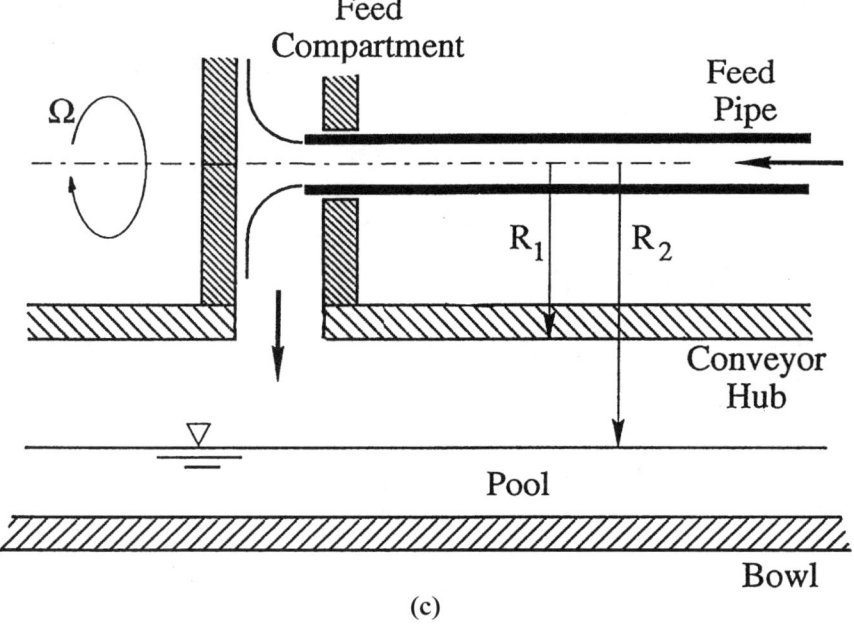

Figure 11.2 (c) Double-disk accelerator.

from a stationary feed pipe. The double disk provides a good distribution of feed to the basket. However, as with the cone accelerator, it has poor efficiency because the feed slurry is accelerated by contact with the smooth surface of the rotating disks, the momentum transfer of which depends on the viscosity of the slurry, which is rather ineffective. In fact, a double disk is a cone with a 90° semicone angle. Therefore the mechanism of feed acceleration follows that of the cone with consequential poor performance.

Elaborate analysis further reveals that the efficiency of the cone and double-disk accelerator is increased by higher angular speed Ω, larger exit radius R_{exit}, higher liquid viscosity μ, smaller flow rate Q, and smaller semiangle for the conical accelerator.

Conventional radial vanes in disk and tubular centrifuges

Radial vanes have been used in decanters and disk and tubular centrifuges. Feed slurry is forced to move at the same tangential speed ΩR as a solid body at a given radius R. The major disadvantage is that the feed discharges off the driving face of the accelerating vanes in a finite number of discrete jets carrying also a very high longitudinal velocity (which is comparable in magnitude to the tangential speed). When the jets plunge into the liquid pool, this causes localized distur-

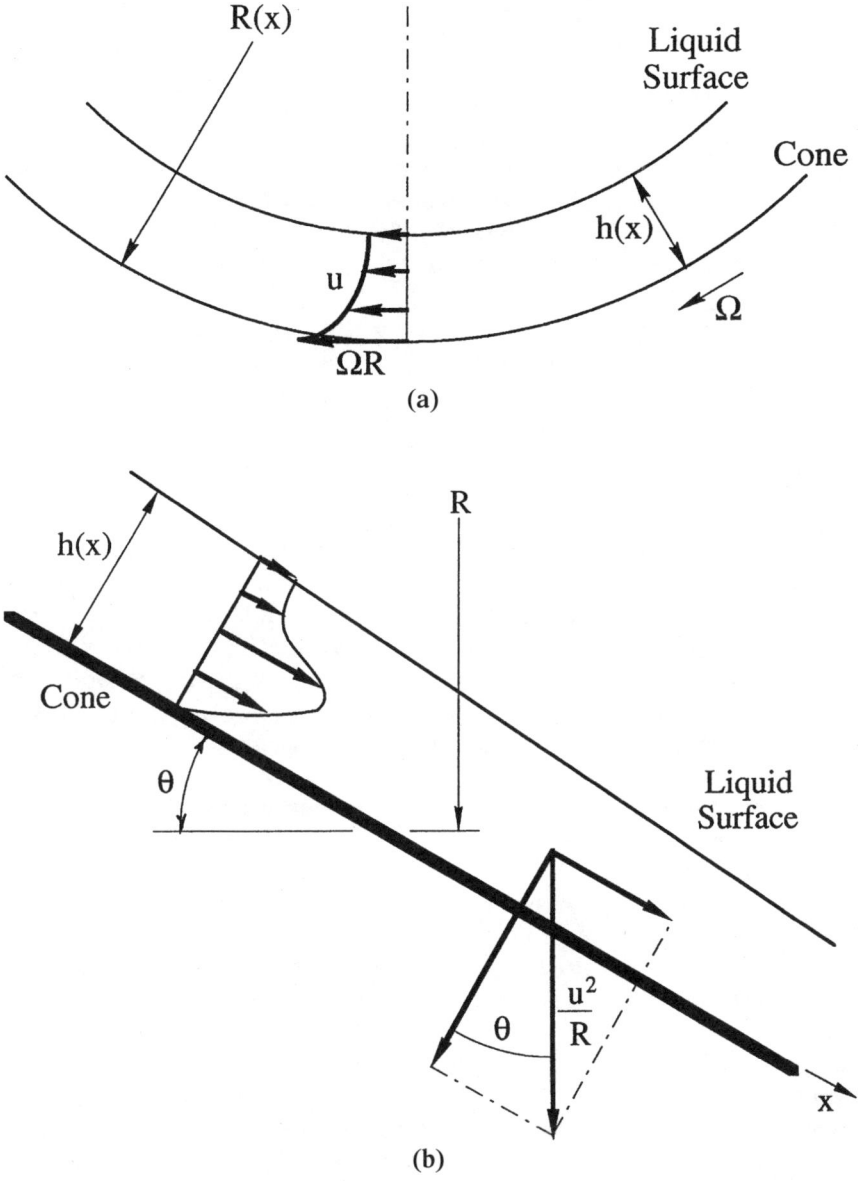

Figure 11.3 Mechanics of acceleration for cone accelerator. (a) Liquid film in cross-sectional view taken normal to axis. (b) Liquid film running down inside of cone as seen in diametric cross section.

bance to the solids already settled in the pool. More importantly, for shear-sensitive solids such as flocculated particles (floc in short), the solids get significant shear as the feed is abruptly accelerated to speed at the starting radius of the accelerating vanes.

Improved Feed Accelerators

Improved hub accelerator

1. Viscosity is a poor means of momentum transfer. Therefore acceleration of the feed in the conveyor hub is poor. The feed is largely accelerated as it is discharged through the ports by contact with the driving face of the ports. Consequently the thicker hub, which provides more driving surface area than the thinner hub, has better feed acceleration. This can be further improved with rectangular or elliptical ports with the longer axis of the opening aligned with the axis of rotation, thus maximizing the driving surface area on the feed slurry.

2. The feed ports should be located at the same axial position (that is aligned feed ports) such that each feed port receives the same amount of feed.

3. As the feed flows through the feed ports, it is subject to the Coriolis force, which directs the feed to move opposite to the direction of rotation. Instead of flowing through the port, the feed tends to slip backward with respect to rotation. This increases the resistance to flow through the feed ports, resulting in the liquid "backing off," forming a standing pool inside the feed compartment. On the other hand, the pressure head from the standing liquid pool provides additional force to drive the liquid flow through the ports. For a given flow rate, an equilibrium pool level in the feed compartment may be reached whereby the driving force from the buildup pool balances the resistance from the Coriolis force to flow thorugh the ports. Such an equilibrium pool level increases with increasing feed rate due to the additional flow resistance. When the level of the liquid pool reaches the baffle inner diameter in the feed compartment, the liquid overflows and leaks out of the machine through the feed-pipe trunnion. This restricts the maximum flow capacity through the machine and poses a serious limitation to high-throughput applications.

The problem can be resolved by installing an anti-Coriolis baffle at the pressure or driving face of the feed port, as shown in Fig. 11.4. The baffle extends inside the conveyor hub into the slurry pool of the feed compartment to counteract the Coriolis force, facilitating the liquid to flow through the port with reduced flow resistance. Figure 11.5 compares the accelerator efficiency for a range of feed rates up to 400 gal/min (1540 L/min) among three different cases: (1) no baffle; (b) anti-Coriolis baffle 0.75 in (19 mm) tall, extending midpoint into the slurry pool formed inside the hub; and (c) anti-Coriolis baffle 1.5 in (38 mm) tall, extending across the entire slurry pool. In the absence of the baffle, the accelerator efficiency drops off to 50% at 400 gal/min (1540 L/min). When a "full" baffle is installed, which effectively counteracts the undesirable Coriolis force at high flow rate and thus high

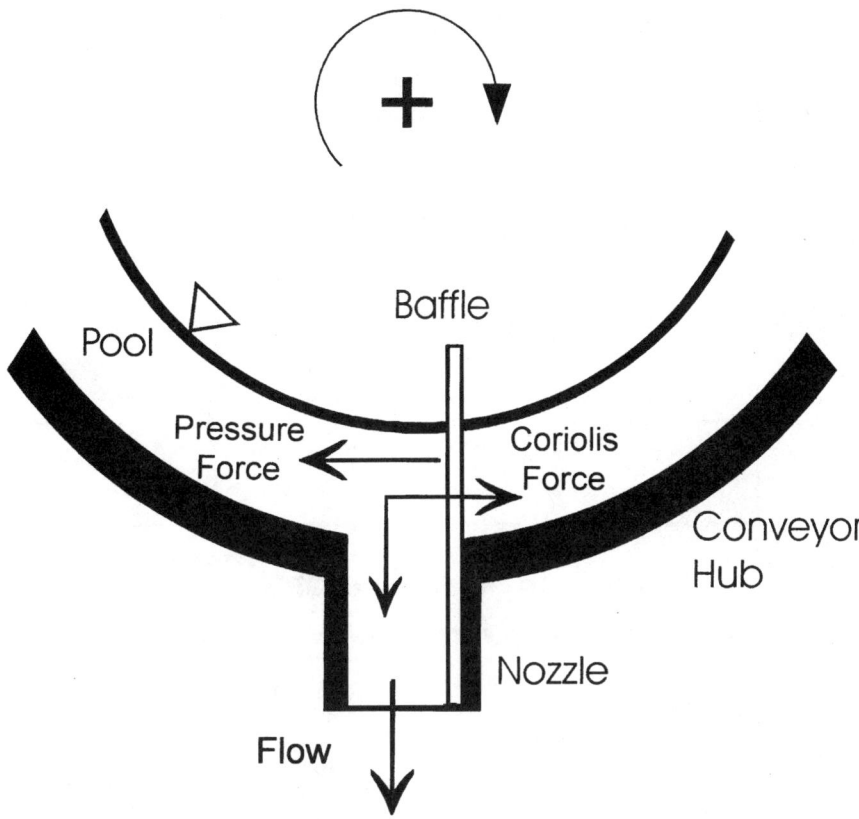

Figure 11.4 Anti-Coriolis baffle inserted in feed port to counteract Coriolis force.

through-flow velocity, the efficiency is restored back to 100%. With a partial baffle protruding into the pool, the benefit is compromised as there is still rotational slip in the pool above the baffle due to the Coriolis force at high flow rate.

4. Although the feed may attain the tangential speed corresponding to the discharge radius (that is, 100% efficiency) as it leaves the ports of the feed compartment, the acceleration and G efficiencies decrease further as the feed drops onto the pool at a larger radius by virtue of Eqs. (11.4) and (11.5). An overspeeding vane, which extends radially and circumferentially outward from the feed port, is attached to the pressure face of the port to form a continuous passageway. The overspeeding vane provides a tangential speed to the feed greater than the local ΩR at the discharge radius R of the feed port to compensate for (1) the loss of tangential speed of the feed due to the radial drop from the accelerator to the pool, and (2) higher tangential speed at the pool which is at a larger radius.

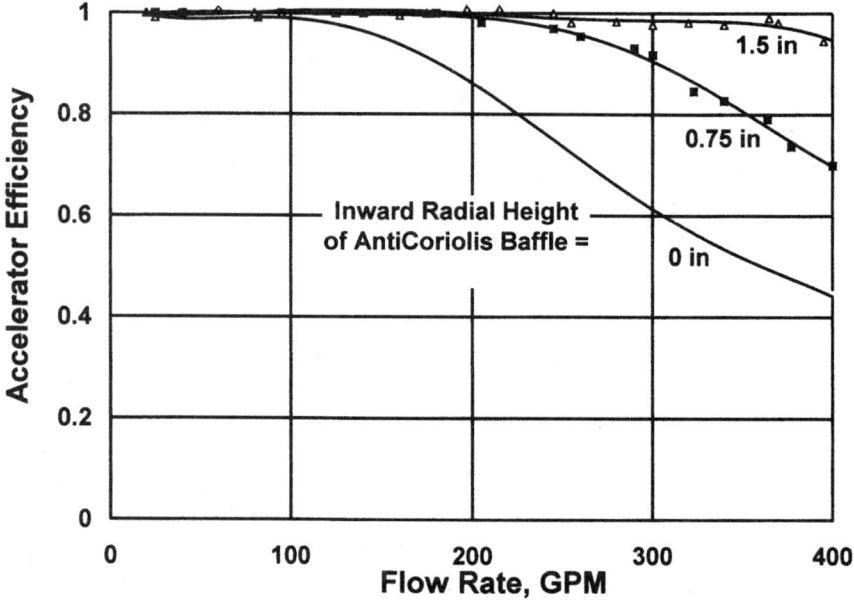

Figure 11.5 Experiment with anti-Coriolis baffle for three different protrusion lengths.

Improved cone accelerator

Given viscosity is a poor transfer of momentum, the cone accelerator can be equipped with a set of vanes assisting feed acceleration. Unfortunately this results in highly concentrated individual streams leaving the vanes. To eliminate this undesirable effect, the exit section of the cone does not have vanes (that is, it is unvaned) so that it is used as a smoothener[1] (Fig. 11.6a). The movement of the flow from the radius of the vane exit to the larger radius at the smoothener exit, in combination with tangential friction and the velocity profile in the liquid streams, acts to smear out the individual streams into a conical sheet that is approximately uniform in the circumferential direction. This distributing action is shown in Fig. 11.6b by the streamlines sketched as seen in the laboratory frame.

The smoothening action in Fig. 11.6b is at the expense of η_a. This is because viscous friction is not effective in accelerating the flow from the vane exit to the cone exit, which is at a larger diameter and higher rim speed ΩR. This deficiency in ΩR may however be compensated by curving the accelerating vanes forward in the direction of rotation, as shown in Fig. 11.7. This "overspeeding" strategy gives rise to a velocity triangle in which at the vane exit u is larger than ΩR, which implies that $\eta_a > 100\%$ at the vane discharge. This overspeeding strategy may therefore be used to compensate not only for the loss in η_a that occurs at

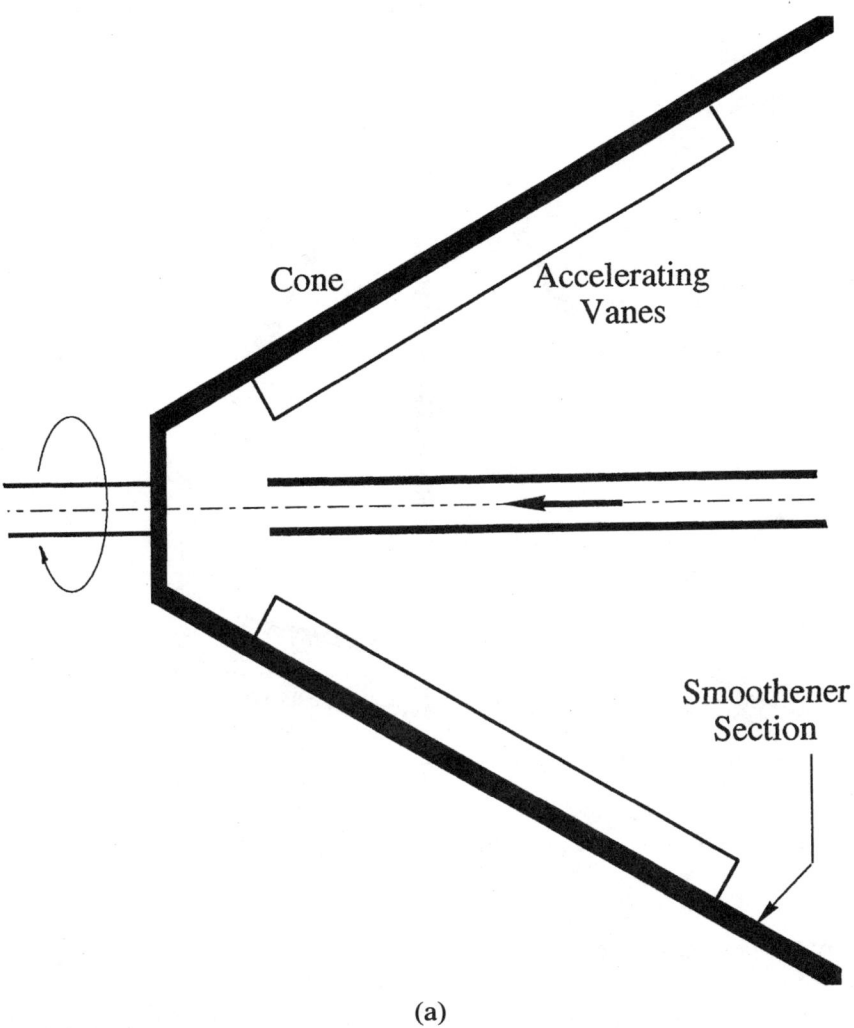

Figure 11.6 Cone accelerator with straight vane and smoothener. (*a*) Diametric cross section.

the smoothener, but also for the loss in η_a as the feed travels from the radius of the smoothener exit to the larger radius of the pool or basket.

Improved double-disk accelerator

Figures 11.7 and 11.8 meet the ideal-accelerator criteria discussed and further incorporate the features of overspeeding and smoothening.

The improved double-disk accelerator[2] consists of a compact design with two parallel disks, a solid target disk and a cover disk that has a

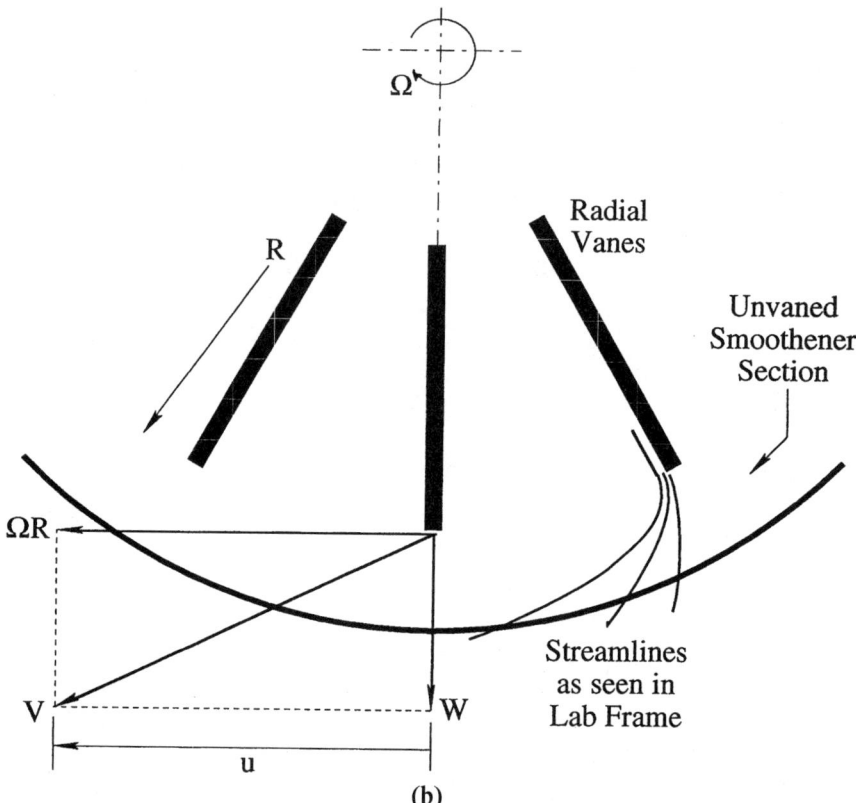

Figure 11.6 Cone accelerator with straight vane and smoothener. (*b*) Cross section normal to axis showing smoothening in unvaned cone surface.

center opening to accept a feed pipe. Located between the target and cover disks are multiple accelerating vanes, which are curved forward in the direction of rotation. The feed is directed against the target disk and is distributed to the vanes, where it accumulates on the driving face of the vanes. Through centrifugal force the vanes accelerate the feed so that it leaves the vanes with a relative speed W_1, which is somewhat less than ΩR_1, as seen in Fig. 11.7. Vectorial addition of these two velocities yields a greater absolute velocity V_1 of the feed stream as it exits the vaned accelerator in the reference frame of the laboratory. The tangential component of V_1 is seen to be greater than ΩR_1 at the vane exit. This overspeeding strategy again yields $\eta_a > 100\%$ at the vane exit, which is similar to the improved conical accelerator.

The accelerator efficiency at the vane exits is determined mainly by the vane exit angle θ and by the frictional losses of the feed as it travels along the driving faces of the vanes. Figure 11.9 shows the acceleration efficiency (expressed in fractions above 1 or 100% in the figure) as a

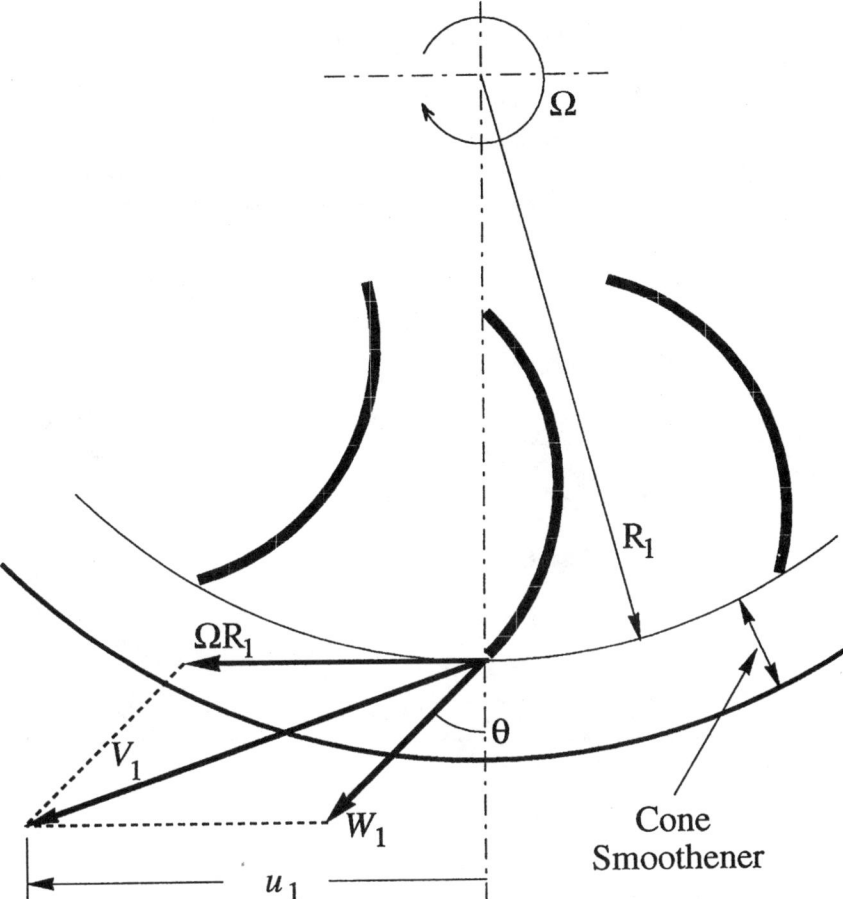

Figure 11.7 Overspeeding vanes and smoothener.

function of the exit angle and the loss coefficient C_{vane}, where the latter is the ratio of the vane head loss to the kinetic energy $W_1^2/2$. (W_1 is the longitudinal velocity component along the vane at discharge.) It is evident that the accelerator efficiency can go well above 100%.

The flow leaving the vanes is highly concentrated into individual streams. In order to eliminate this undesirable formation, a smoothener is installed, for example, in the form of a cone or a disk. The functional aspect is identical to that of the improved cone accelerator.

Improved accelerator vane

The multiple accelerating vanes in the double-disk as well as in the disk and tubular centrifuges start at a small yet finite radius. They are preferably curved forward in the direction of rotation, as in Fig.

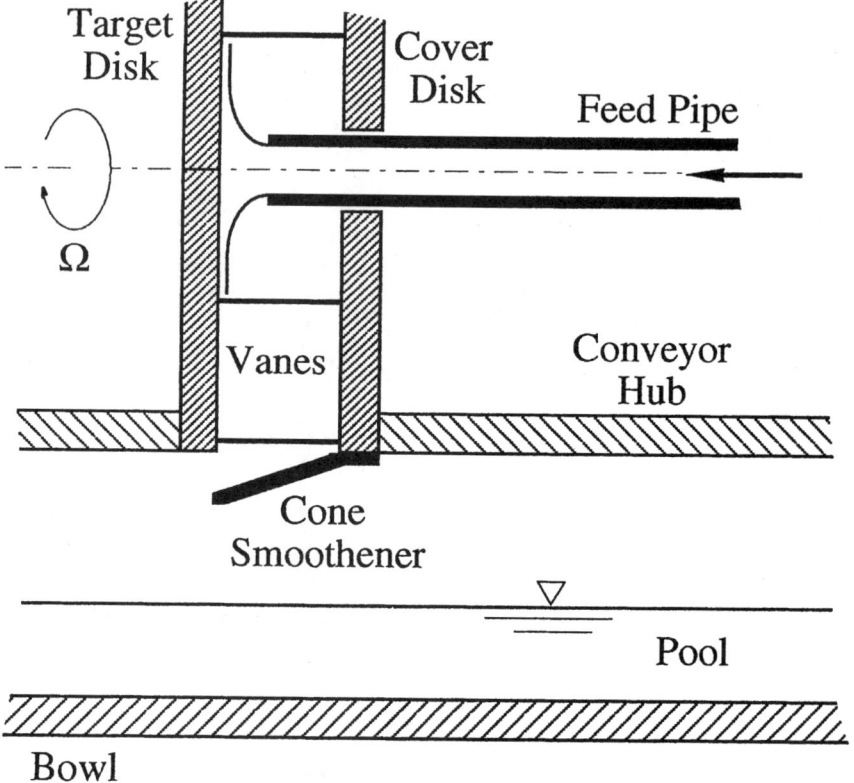

Figure 11.8 Double-disk accelerator with vanes and smoothener.

11.7, such that in the frame of rotation, the feed stream in the rotating frame of the accelerator is directed at a small angle of attack into the channels formed by the curved vanes. This reduces the shearing and backsplash associated with impact at a large angle of attack, especially when the vanes are oriented radially. This is beneficial for processing shear-sensitive slurries using high-G centrifuges. It also provides an alternative to the sealed hermetic design discussed in Chap. 4, where the feed area is flooded to the axis, permitting the feed to be accelerated by a vane assembly starting from the axis.

While the overspeeding vanes discussed here appear superficially to show a centrifugal pump or fan, it is in fact a very different mechanism. The vanes of pumps and fans are curved backward for good reasons, whereas the overspeeding vanes are curved forward in the direction of rotation.

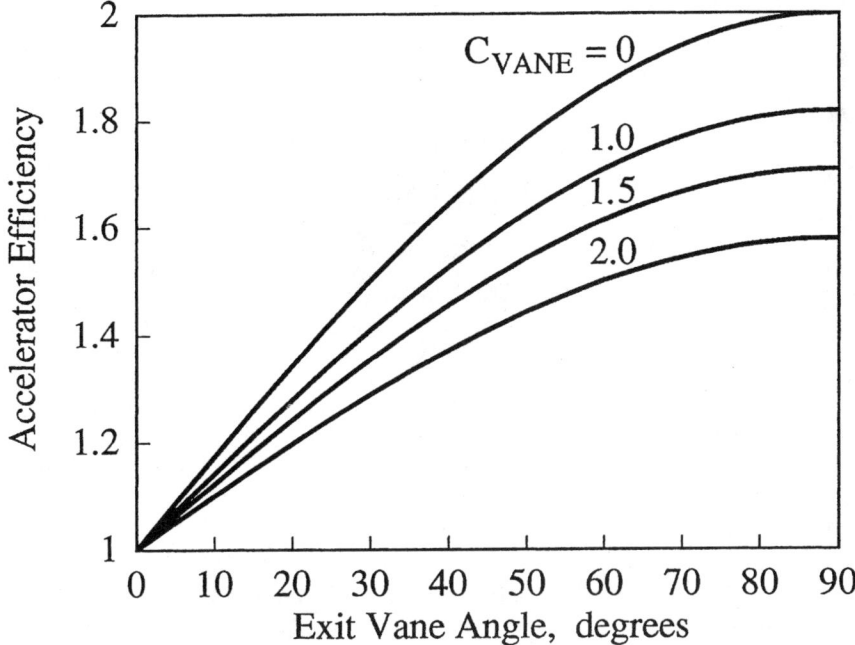

Figure 11.9 Effect of vane loss coefficient and exit angle on accelerator efficiency.

Practical Considerations

In applying the concepts of overspeeding and smoothening, it is important that the designer consider many factors. Among these are the number of vanes, the vane height normal to the cone-disk surface, the location and geometry at the start and end points of the vanes, the speed and gravitational droop of the feed jet, the geometry of the feed distributor, splash, erosion, and plugging.

Gravitational Droop

The feed leaving the stationary feed pipe should have adequate velocity, which depends on the flow rate and the exit diameter of the feed pipe so as to aim toward the center of the distributor surface without incurring much droop due to the earth's gravity. Otherwise the feed is poorly distributed on the distributor or target surface, resulting in splashing and poor subsequent acceleration. Figure 11.10 shows a schematic of the droop due to a low-velocity feed. (The direction of gravity is pointing downward in the figure.) Gravitational droop δ_{droop} should be less than

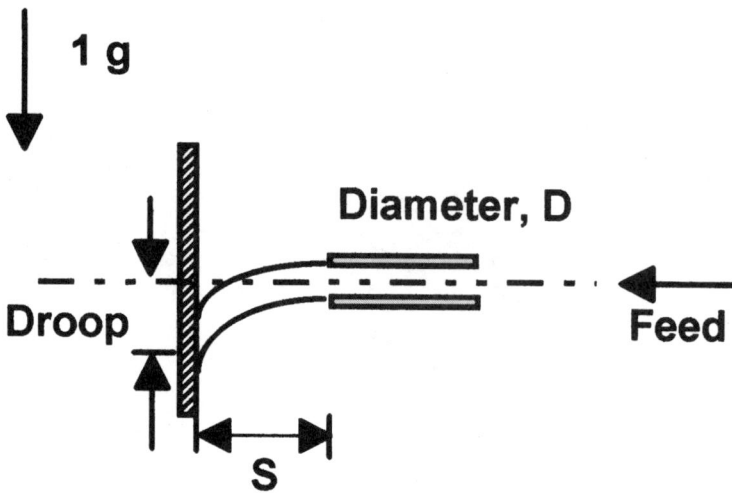

Figure 11.10 Gravitational droop during feeding.

3 mm (1/8 in) from the centerline. It can be determined based on the feed pipe standoff distance from the distributor target surface s, the feed pipe exit diameter D_{fp}, and the volumetric flow rate Q, using

$$\delta_{droop} = \frac{\pi^2}{32} g \left(\frac{s D_{fp}^2}{Q} \right)^2 \qquad (11.6)$$

For example, when $Q = 385$ L/min (100 gal/min), $D_{fp} = 76.2$ mm (3 in), and $s = 76.2$ mm (3 in), then $\delta_{droop} = 14.88$ mm (0.586 in). When D_{fp} is reduced to 63 mm (2.5 in) and subsequently to 50 mm (2 in), δ_{droop} decreases to 7.16 and 2.95 mm (0.282 and 0.116 in), respectively. The standoff can also be decreased accordingly. A reduction in D_{fp} and s may incur a higher pumping pressure loss, which needs to be factored into the design, taking into consideration how the feed is being delivered—by pump or by gravity head.

Side-by-Side Testing

As shown in Fig. 11.11, two rotating units were set up side by side. They were identical except for their accelerators. One has a conventional hub accelerator and the other a double-disk accelerator with forward curved vanes and a smoothener. The bearing and water feed were arranged so that the views from the open ends were unobstructed. The two bowls had identical dimensions of 254 mm diameter by 305 mm length (10 by 12 in) and were operated at the same speed of 1000 rpm, volumetric feed rates, and pool depths. Free-wheeling rotation meters, mounted with paddles slightly immersed in the pools, rotated at the

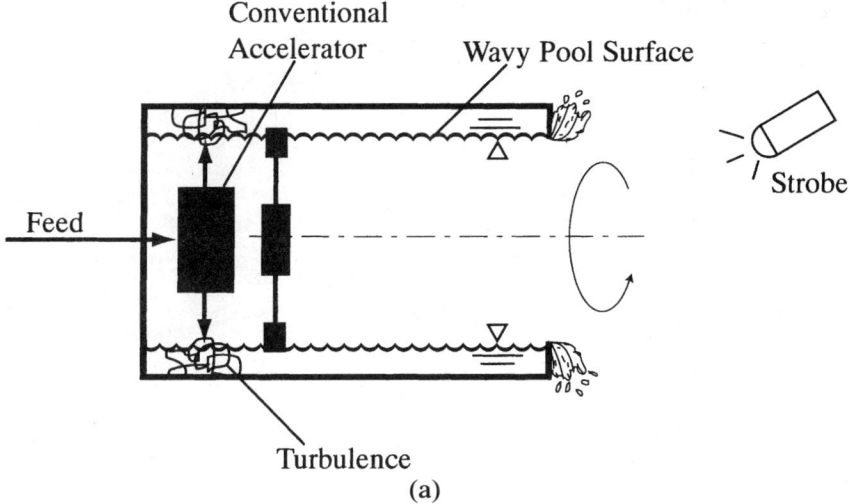

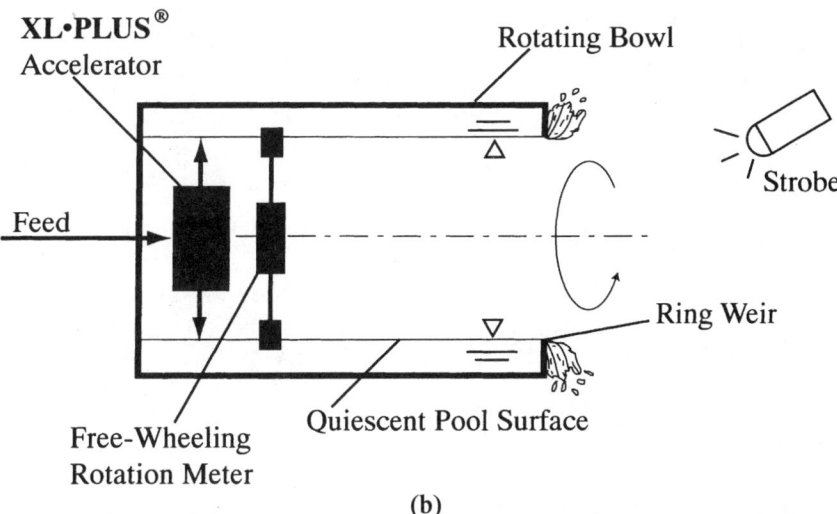

Figure 11.11 Setup for side-by-side tests. (*a*) Conventional accelerator. (*b*) Improved feed accelerator.

same angular speed as the pool. A strobe on each unit was used to measure the angular speeds of the pool and the bowl, and also to facilitate observation of the pool surface.

The measured η_a and η_G are seen from Fig. 11.12 to be 100%, that is, the pool rotates at the angular speed of the bowl. Moreover, the pool surface (Fig. 11.11*b*) is smooth and quiescent, indicating that there is little disturbance from the significantly reduced radial momentum of the entering feed. In contrast, Fig. 11.11*a* shows that the conventional

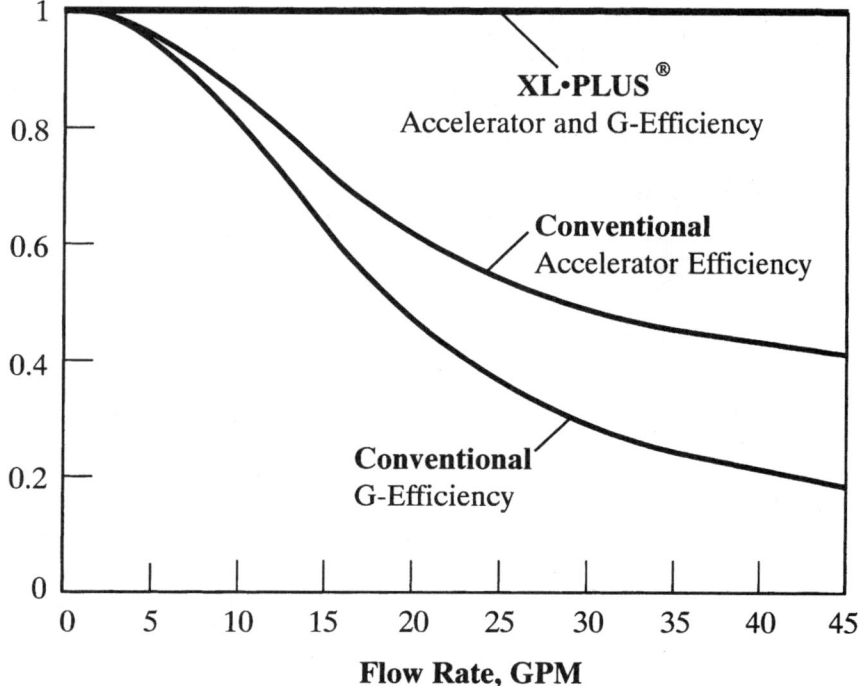

Figure 11.12 Results of side-by-side tests.

accelerator has a relatively poor accelerator efficiency, especially as the flow rate increases. The G efficiency is poorer still. Also, waves on the free surface and a turbulent appearance where the feed enters the pool (Fig. 11.11a) indicate that the entering feed has a large radial momentum, which causes undesirable disturbances.

Experimental Tests

The average velocity of the feed leaving the accelerator can be determined from the torque data T_{net}, given that average velocity, from momentum balance, is given by $u = T_{net}/\rho Q R_{exit}$ and the accelerator efficiency is thus $\eta_a = T_{net}/\rho Q \Omega R_{exit}^2$, where ρ is the liquid density, Q the flow rate, Ω the rotational speed, and R_{exit} the exit radius of the accelerator. All the measurements are made with this arrangement.

Hub accelerator. Typical results for a hub accelerator are shown in Fig. 11.13 for a configuration in which there are rectangular slots in the hub, each with circumferential versus axial dimensions of 50.8 by 76.2 mm (2 by 3 in). The outer diameter of the hub is 249 mm (9.8 in) with an axial length of the feed compartment of 146 mm (5.75 in). Results are shown for both a thin-wall hub and a thick-wall hub. With a thicker hub

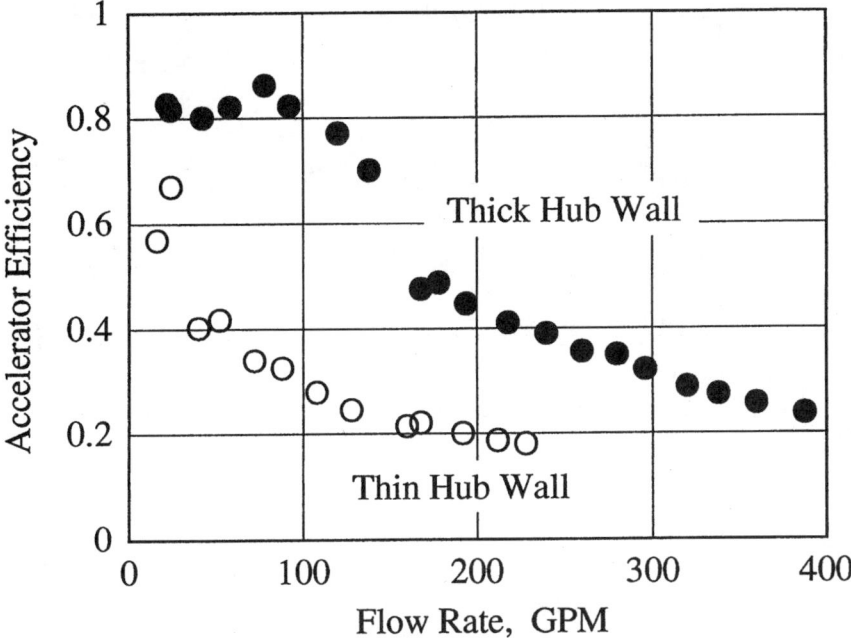

Figure 11.13 Effect of hub wall thickness on accelerator efficiency.

η_a is about 85% for flows up to about 379 L/min (100 gal/min) and drops off rapidly to about 30% at 1323 L/min (350 gal/min). In accordance with Eq. (11.4), η_a will deteriorate further as the feed passes from the port exits to the radius of the decanter pool.

Double disk. Results are given for a pair of double disks with a 152-mm (6-in) radius and an axial spacing between disks of 31.8 mm (1.25 in). When there were no vanes between the disks, η_a was too low. When 16 forward curved vanes were installed with a starting radius of 83.8 mm (3.3 in) and extending to the disk diameter, where $\theta = 40°$, the upper set of data in Fig. 11.14 was obtained at 1900 rpm. The efficiency exceeds 100% even at small flow rates and keeps climbing as the flow increases because the relative speed W_1 increases with the flow rate. At 1140 L/min (300 gal/min) η_a reaches 130%.

Installation Experience

Solid-bowl decanter on sewage sludge

Dewatering of mixed sludges. In a municipal sewage installation the solid-bowl decanter was tested in its original configuration with a conventional accelerator. Subsequently it was modified by the installation

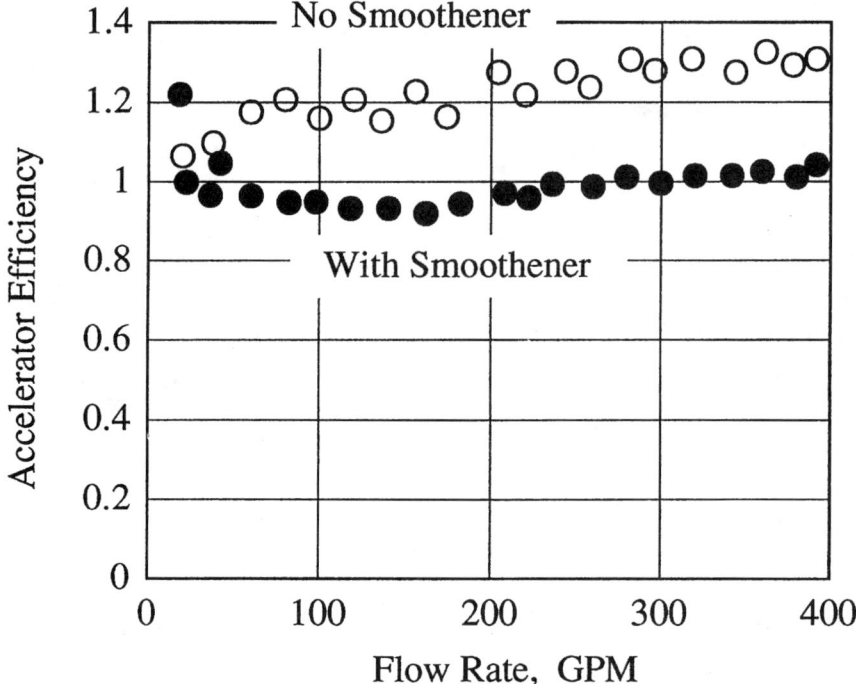

Figure 11.14 Effect of smoothener on efficiency of improved feed accelerator with forward curved vanes.

TABLE 11.1 Performance Data at 3200 rpm for Dewatering Municipal Sludge in Solid-Bowl Decanter

	Before modification	After modification
Capacity, gal/min	45	60
Polymer pump setting, %	75	85
% cake solids	15–16	15–16
% recovery	86	96

of a vaned cone assembly with a semicone half-angle of 30°, an exit radius of 175 mm (6.9 in), eight longitudinal vanes of 12.7 mm (0.5 in) height each and without a smoothener. Table 11.1 shows the performance of the unit at 3200 rpm. While there was a 13% increase in polymer use, the capacity was increased by 30–40%, which resulted in a net reduction of 15% in polymer consumption per pound of sludge. The results indicate that the modification improved performance substantially. With a smaller semicone angle and a cone smoothener, even better performance would be expected.

Dewatering of heat-treated sludges. A solid bowl was used to process heat-treated sewage sludge taking 10% w/w solids of feed and dewatering to above 42% cake solids at a rate of 570–646 L/min (150–170 gal/min). The solid recovery was 86–90%. When the conventional hub accelerator was converted to an XL•PLUS®* feed accelerator system, the same machine could process 25–40% more feed rate, while the polymer consumption was reduced from 3.1 to 1.7 kg/t (6.25 to 3.4 lb/ton). This suggests that with shear-sensitive solids such as flocculated solids, it is important to provide gentle acceleration by bringing the feed to the same tangential speed as the rotating pool with circumferential uniformity and minimal radial disturbance. Otherwise this can be at the expense of higher polymer consumption due to shearing of the flocculated solids. This is further confirmed by an increase in solids recovery from 86–90% to 95% despite the polymer dosage being cut by half.

Solid-bowl kaolin classification

The objective was to classify kaolin slurry from a feed with 78% less than 1 μm to obtain a product with 88% solids less than 1 μm using a solid-bowl centrifuge. The machine originally processed 15 ton/h with a 71% yield (or size recovery) of the 1-μm product. After retrofitting with an improved hub accelerator, incorporating overspeeding and anti-Coriolis baffles, the modified machine processed 80% more feed solids throughput capacity with a yield increased to 79%. This was attributed to prompt acceleration of the fine particles to the slurry pool speed and making the proper separation.

Screen-bowl clarification

For the same overall machine length, a screen-bowl centrifuge has a shorter clarifier than a solid bowl due to the presence of a screen section. Therefore clarification may be limiting, especially for slurry with fine solids, short clarifier, and for a poor feed acceleration resulting in a much lower G force. This is partly compensated with an improved feed accelerator. A screen bowl had to dewater coal slurry, which was subsequently fed to a boiler for power generation. When the conventional hub accelerator was replaced by an XL•PLUS® feed accelerator system, the recovery of coal was increased from 94 to 96% while the cake moisture was maintained at 21–22%. The median particle size lost in the effluent decreased from 5 to 2 μm. The additional 2 percentage points of coal recovered provides substantial savings to the 1500-MW coal-fired power plant on an annual basis.

---

*XL•PLUS® is a registered trademark of Bird Machine Company.

Screen scroll

Coal plant. A screen-scroll centrifuge was used to dewater coarse coal of ¼ in × 28 mesh (6 mm × 0.5 mm) in a coal preparation plant. The cake was found to be distributed nonuniformly onto the basket with almost half the basket area without coal.[3] This poor distribution was rectified with an improved XL•PLUS® feed accelerator system. As a result, the coal is distributed uniformly onto the basket surface and at the same time accelerated to match the tangential speed of the basket at the feed zone. As shown in Fig. 11.15, the solids throughput increased almost threefold while keeping the cake moisture relatively constant. An adjunct benefit is that by accelerating the feed to the basket circumferential speed, there is less wear on the basket and the scraper at the feed zone. The life of several baskets tested using the improved feed accelerator was doubled. This increases production significantly while it reduces maintenance cost in a coal preparation plant.

Salt plant. The screen scroll was also tested in a potash plant. By modifying an existing screen-scroll centrifuge with an XL•PLUS® feed ac-

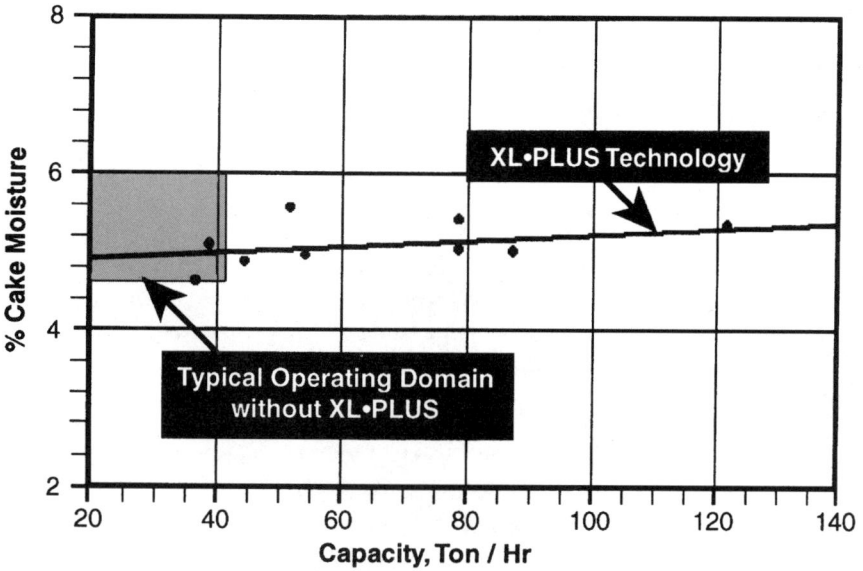

Figure 11.15 Cake moisture versus capacity for coal dewatering; improved XL•PLUS® feed accelerator used for screen-scroll centrifuge.

celerator design, the solids capacity was increased by 76% while at the same time reducing the cake moisture from 5 to 4.5%.

Pusher centrifuge in dewatering and washing of sodium chloride

The objectives were to reduce the moisture and the impurities (that is, sodium sulfate) of the sodium chloride crystals in the cake by washing and subsequently dewatering using a two-stage pusher. Initially the machine had an unvaned conical accelerator with a semicone angle of 18° and an exit radius of 130 mm (5.1 in). It was modified by installing 16 longitudinal vanes, each 31.8 mm (1.25 in) tall and 82.6 mm (3.25 in) long. This left a smoothener section about 25 mm (1 in) long. Table 11.2 shows average results obtained from 12 tests prior to, and 6 tests following, the modification. The capacity was increased by a factor of 2.5, with reduced levels of both moisture and sulfate in the product crystals. Visual observation with a strobe further showed that notwithstanding the accumulation of the feed at the driving faces of the vanes, the smoothener section effectively restored circumferential uniformity at the basket entry. The uniformity was confirmed by the absence of longitudinally running ridges and valleys in the cake. The low sulfates in the salt crystals were further evidence of uniform washing associated with circumferential uniformity of the feed.

Pusher centrifuges on sodium bicarbonate

Two pushers were tested side by side for direct comparison at about 475 L/min (125 gal/min) of a slurry with a particle-size range of 50–150 μm, yielding a cake product at a rate of 4 dry ton/h. Centrifuge A had a conventional accelerator made of a pair of parallel disks without accelerating vanes. Centrifuge B was fitted with three different modified accelerators. Useful information was obtained from videotape recordings under stroboscopic lighting; namely, the shape of the slurry

TABLE 11.2 Average Results from 12 Tests Prior to and 6 Tests Following Modification

	Before modification	After modification
Capacity, ton/h	2.07	5.2
% cake (product) moisture	1.43	1.39
% cake sulfate	2.3	1.6

lines exiting the accelerator provided estimates of η_a, and the degree of circumferential uniformity was inferred from inspection of the solid cake on the baskets.

1. In centrifuge A, which had a conventional accelerator, the efficiency was perceived to be poor, but the cake was of good circumferential uniformity.
2. Centrifuge B, with 16 curved vanes between parallel disks (semicone angle = 40°) and a cone smoothener, had an efficiency greater than 100%, and the cake was of good circumferential uniformity.
3. In centrifuge B, with 32 straight radial vanes between parallel disks and no smoothener, the efficiency approached 100%, but the distribution of solids on the first-stage basket was distinctly nonuniform circumferentially. There were 32 sharp and distinct axially extending ridges, alternating with valleys, which extended around the circumference. This irregularity is adverse to the quality of the cake and interferes with uniform washing of the product to remove unwanted impurities.
4. Same as item 3, but with a smoothener installed. Here η_a was slightly less than 100%, but the cake was of good circumferential uniformity.

These tests made it clear that when accelerating vanes are used to obtain high η_a, a smoothener is essential to the prevention of a high degree of circumferential nonuniformity. With a smoothener and forward curved vanes, η_a in excess of 100% may be obtained without loss of circumferential uniformity.

Pusher centrifuge on sodium sulfate

Figure 11.16 shows the capacities measured in the tests. With a conventional accelerator, the machine delivered a maximum product rate of 8.6 kg/min (19 lb/min) of sodium sulfate at a feed rate of 57 L/min (15 gal/min). Washout of the cake occurred at a slightly higher feed rate, with consequent degradation of dryness. When the accelerator was replaced by an XL•PLUS® design, there was no washout, even at 110 L/min (29 gal/min). At this feed rate 37 kg/min (81 lb/min) of product was obtained. The data further indicate that the recovery percentage, which is proportional to the ratio of product to feed rate, roughly doubled.

Pusher centrifuge on polystyrene beads

After a retrofit with a double-disk accelerator based on forward curved vanes and smoothener, the improvements noted in Table 11.3 were determined. The performance improved in every aspect.

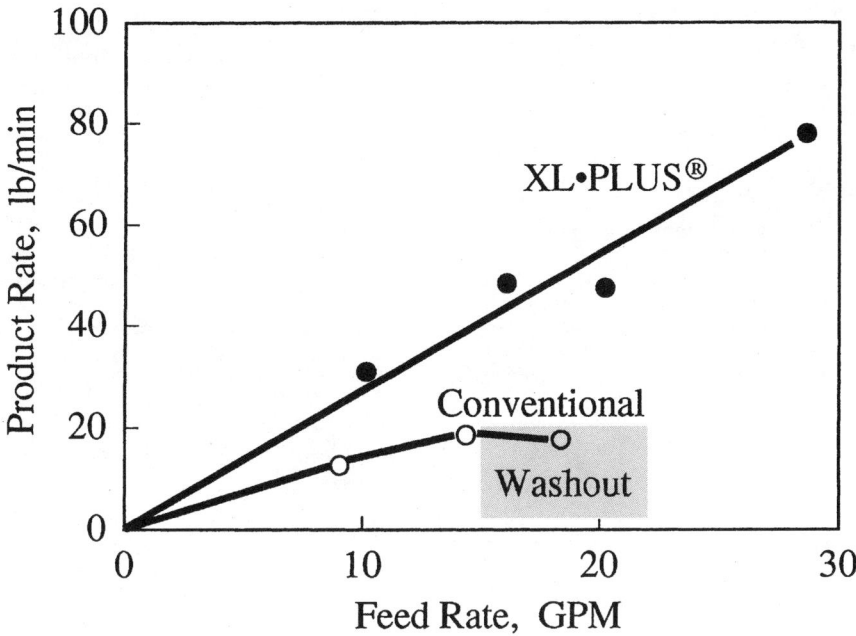

Figure 11.16 Comparing performance of pusher centrifuges for sodium sulfate, conventional and XL•PLUS® feed accelerators.

TABLE 11.3 Small Pusher Centrifuge for Dewatering Fine Polystyrene Beads

	Before modification	After modification
Capacity, gal/min	4.2	20
% cake solids	87	95
% solids recovery	82	99

Benefits

The improved feed acceleration system provides, especially for continuous-feed centrifuges, a feed stream with (1) a tangential speed matching that of the pool or basket and causing reduced disturbance or wear, (2) uniform circumferential feed distribution, and (3) minimized radial penetration into the pool or basket. This is achieved by the use of anti-Coriolis baffles, forward curved vanes, and a cone smoothener in conjunction with an improved feed-distribution system.

With sedimenting centrifuges the improved feed accelerator increases feed capacity, separation, and centrate clarity, especially on fine slurries and hard-to-separate and shear-sensitive materials; increases yield and finer cut for classification; increases the swallowing capacity of the feed accelerator; and reduces wear in the feed zone.

With filtering centrifuges, the improved feed accelerator increases feed capacity and cake dryness; reduces washout of the cake for pushers; increases the swallowing capacity of the feed accelerator; and reduces wear in the basket feed zone, especially with abrasive materials.

References

1. W. W. Leung and A. Shapiro, "Improved Design of Conical Accelerators for Decanter and Pusher Centrifuges," *Filtration and Separation,* Sept. 1996.
2. W. W. Leung and A. Shapiro, "Efficient Double-Disk Accelerator for Continuous-Feed Centrifuges," *Filtration and Separation,* Oct. 1996.
3. W. W. Leung and A. Shapiro, "Feed Accelerator System Including Accelerator Cone," U.S. patents 5,380,266, Jan. 10, 1995; 5,527,258, June 18, 1996.
4. W. W. Leung and A. Shapiro, "Feed Accelerator System Including Accelerator Disk," U.S. patent 5,401,423, Mar. 28, 1995.
5. W. W. Leung, "Feed Accelerator System Including Feed Slurry Accelerating Nozzle Apparatus," U.S. patent 5,423,734, June 13, 1995, U.S. patent 5,651,756, July 29, 1997, U.S. patent 5,658,232, August 19, 1997, and U.S. patent 5,683,343, November 4, 1997.
6. W. W. Leung and A. Shapiro, "Feed Accelerator System Including Accelerating Vane Apparatus," U.S. patent 5,520,605, May 28, 1996, U.S. patent 5,551,943, Sept. 3, 1996, and U.S. patent 5,632,714, May 27, 1997.

Chapter 12

Lab, Pilot, and Production Tests

Preliminary Screening

Screening tests are carried out to determine whether a given application is suitable for centrifugation. Furthermore, if centrifugation is indeed applicable, which is the most suitable centrifuge to use and under what condition should the machine be operated at—low, moderate, or high G? Screening tests are frequently performed to provide answers to these questions.

In conjunction with the screening tests, it is important to determine the following:

1. Nature of liquid phase
   a. Temperature: (i) nominal operating temperature, (ii) temperature variation, (iii) possible range of temperature adjustment
   b. Viscosity at conditions set by a
   c. Density at conditions set by a
   d. Vapor pressure at conditions set by a
   e. Corrosive characteristics
   f. Are fumes noxious, toxic, inflammable, or none of these?
   g. Is contact with air important, desirable, or must it be avoided?
   h. Are there dissolved solids and impurities?
2. Nature of solid phase
   a. Particle-size distribution (measurement method and instrument)
   b. Are particles amorphous, flocculant, soft, friable, crystalline, fibrous, or abrasive?
   c. Particle shape

    d. Is particle-size degradation unimportant, undesirable, or highly critical?
    e. Feed-solids concentration
    f. Density of solids particles
    g. Cake-retained mother liquor content: amount tolerable (maximum), amount desirable?
    h. Is cake washing required inside machine or by repulping outside machine to reduce soluble mother liquor impurities?
3. Does slurry settle under gravity? If so, how long (seconds, minutes, or hours) or at what rate?
4. Does slurry settle in a spin tube? If so, how long and under what G force? (Repeat trials at several Gs and times.)
5. Is supernatant liquid from tests 3 and 4 satisfactorily clear?
6. What is the nature of the sediments? Are they soft and fluid or plastic, or are they firm and granular? Can they be handled by the centrifuge solids conveyance or discharge mechanisms?
7. With filtration tests using a Buchner funnel, does slurry filter rapidly, slowly, or not at all due to the fine-solids fraction [such as sizes less than 45 μm (325 mesh)]?
8. What is the amount of feed which requires separation, batch quantity and batch time, or continuous volumetric rate of slurry and dry solids rate?

Any process data pertaining to existing situations versus desirable situations and results would be valuable. Also, it is reasonable to address these issues for a given slurry, given that centrifugal sedimentation or filtration may be possible to separate the solids from the liquid based on satisfactory answers to questions 4 and 5, or 7.

Spin-Tube Tests

Clarification

Spin-tube tests can be carried out to answer some of the questions regarding the settleability of a given suspension. After proper preparation, slurry samples are poured into two 15-mL transparent plastic tubes up to the 10-mL mark for convenience. The tubes are secured in metal holders placed diametrically opposite in a rotating-head assembly. They are spun promptly to speed so that the G force developed acts on the density difference between the particles and the liquid to make separation between these two phases. Both G and t are important variables in the tests. When polymer is used to agglomerate the fine solids, the dosage D of the candidate polymer, measured in kilograms per dry metric ton (or pounds per ton), is an additional variable, and so is the feed-solids concentration W_f. If the tube is insulated in a thermal jacket and heat loss is significant during the test, the temperature effect on separation can be studied. A systematic set of tests can be conducted with various G, t, D_p, W_f, and T. After each test, the supernatant is decanted off the tube and the clarity (qualitative observation) as well as both the suspended and the dissolved solids are measured. The suspended solids can be determined by filtering a known weight of the slurry sample and measuring the oven-dried solids retained on the filter surface. On the other hand, the fraction of total solids (suspended and dissolved) can be determined from the known weight of the original slurry sample and by weighing the total oven-dried solids of the same sample. The difference between total and suspended solids based on these two measurements is the dissolved solids in solution. The solids concentration from the feed and from the sediment cake, provided the quantity is sufficient for the latter, is also measured.

The solids (suspended or total) recovery by centrifugation is thus

$$\text{Rec}_s = \frac{1 - W_e/W_f}{1 - W_e/W_s} \qquad (12.1)$$

where W_e and W_s are the solids concentration of the supernatant and of the cake, respectively. Note W_f, W_e, and W_s can correspond to either the suspended or the total solids concentration, depending on which recovery—suspended solids or total solids recovery—is being used. Centrifugation only separates suspended solids. In any case, the solids recovery is a function of G, t, D_p, and W_f. Figure 12.1a shows the solids recovery as a function of t for several different Gs. For a given G level, the recovery is low at small t and approaches 100% at large t. Likewise, at higher G, the recovery increases for the same t. On the other hand, Fig. 12.1b shows recovery plotted against G for various ts. Most slurries tested behave in the manner depicted in Fig. 12.1a and b. Of interest is that the recovery depends only on the dimensionless time parameter $\xi = (V_{sg}t/R_b)(G/g)$ defined in Eq. (3.3). Hence the recovery should vary as the product Gt, as delineated in Fig. 12.1c.

For a system of well-defined particles with median (or mean) size d and density ρ_s in a suspension phase with density ρ_L and viscosity μ to be separated in a centrifuge rotating at speed Ω, the appropriate dimensionless time becomes $\zeta = [(\Omega d)^2(\rho_s - \rho_L)/\mu]t$. If $\Omega = 100$ rad/s (955 rpm), $d = 10^{-3}$ cm (10 μm), $(\rho_s - \rho_L) = 0.1$ g/cm^3, $\mu = 0.01$ g/cm · s, and $t = 10$ s, then ξ is calculated to be 1.0.

When the solids do not settle readily or the cake is soft and can easily get resuspended, a dilute flocculant solution is often used with concentration typically less than 0.1% w/w to yield an agglomerate of the feed solids for better separation. The flocculated solids (floc in short) formed

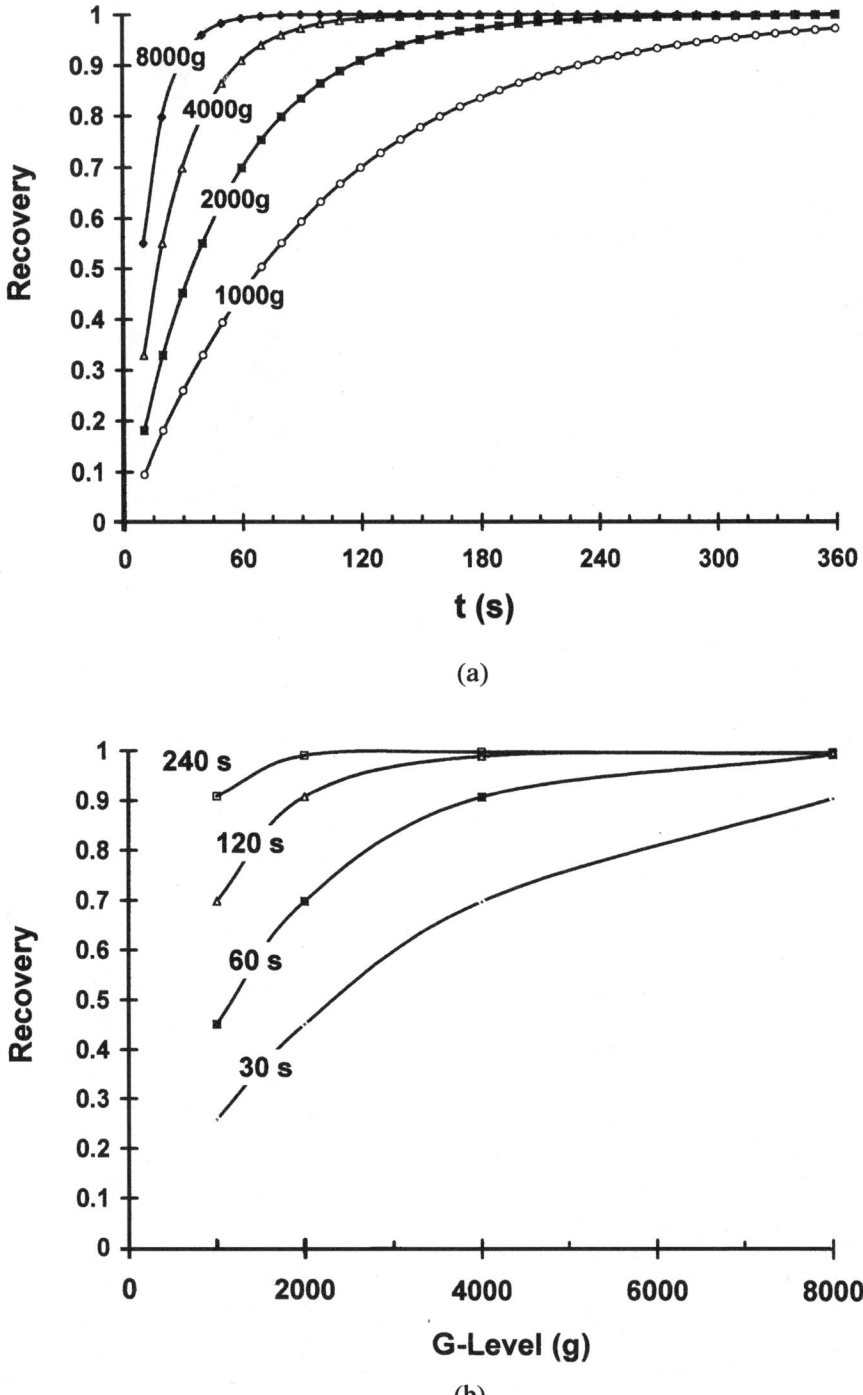

Figure 12.1 (a) Solids recovery versus t for different G. (b) Solids recovery versus G for different t.

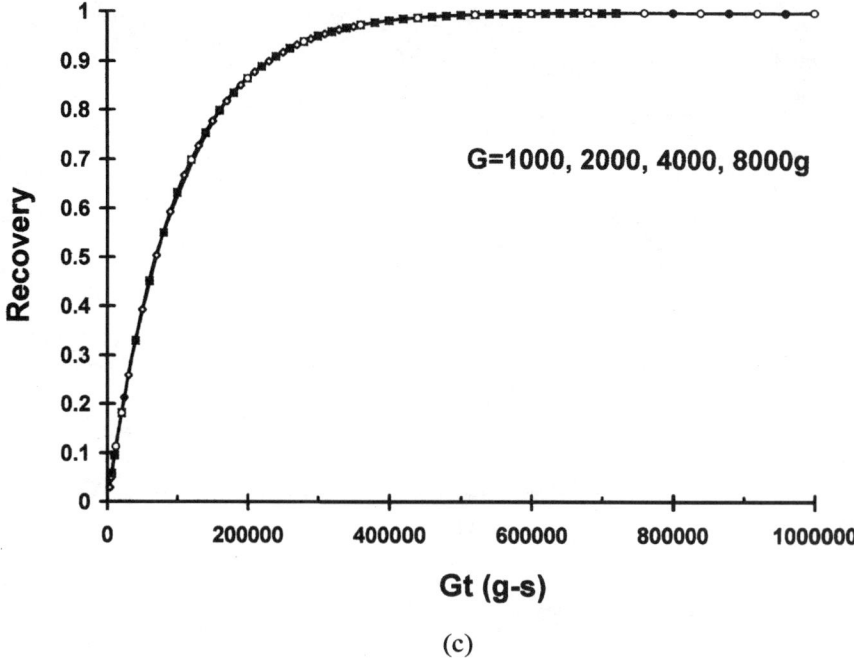

Figure 12.1 (c) Solids recovery versus Gt (for various centrifugal gravities).

are subsequently centrifuged to determine the supernatant clarity. The amount of flocculant is increased gradually in the test until both a clean supernatant and a firm cake are obtained within a reasonable test period, say 30–60 s. Figure 12.1d shows the effect of polymer dosage on solids recovery. By diluting the feed in some applications the undesirable hindered settling between particles can be further reduced.

Material balance

When the test sample quantity is small, measurement is limited to the supernatant and the original feed. The cake that settles out in the test tube is in such a minute quantity that it renders measurement inaccurate and impractical. A simple material balance before and after separation can be made to deduce the cake solids.

1. Solids balance

$$v_f \rho_f W_f = (v_f - v_s)\rho_e W_e + v_s \rho_s W_s \qquad (12.2)$$

2. Solids and liquid balance

$$v_f \rho_f = (v_f - v_s)\rho_e + v_s \rho_s \qquad (12.3)$$

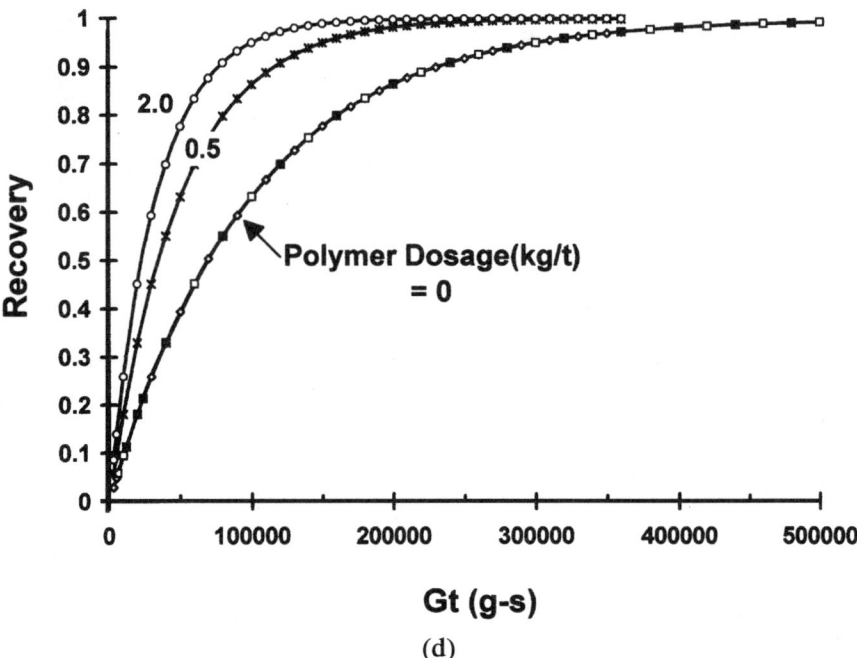

Figure 12.1 (d) Solids recovery versus Gt for various polymer dosages.

Combining these two equations, the cake solids can be determined from

$$W_s = \frac{v_f \rho_f W_f - (v_f - v_s)\rho_e W_e}{v_f \rho_f - (v_f - v_s)\rho_e} \qquad (12.4)$$

where v_f is the original bulk volume of the suspension in the test tube, and v_e and v_f are the bulk volumes of the centrate and the cake after separation. For example, if $W_f = 0.176$, $\rho_f = 1.145$ g/mL, $v_f = 10$ mL, $W_e = 0$, $\rho_L = 1.0$ g/mL, and $v_s = 1.2$ mL, then from Eq. (12.4), $W_s = 0.76$. The bulk cake volume and the % cake solids as a function of Gt are shown in Fig. 12.2a and b. The cake bulk volume increases during formation and subsequently decreases slightly due to compaction under G force, (see Fig. 12.2a). The % cake solids increases rapidly upon increasing Gt during cake formation, but subsequently increases only gradually during cake compaction (see Fig. 12.2b). The handleability of the cake can be determined from measuring the yield stress. In a semiquantitative test, a rod is placed in the cake after the supernatant liquid has been decanted off. The weight (under 1 g) of the rod is supported by the yield stress of the cake acting on the sur-

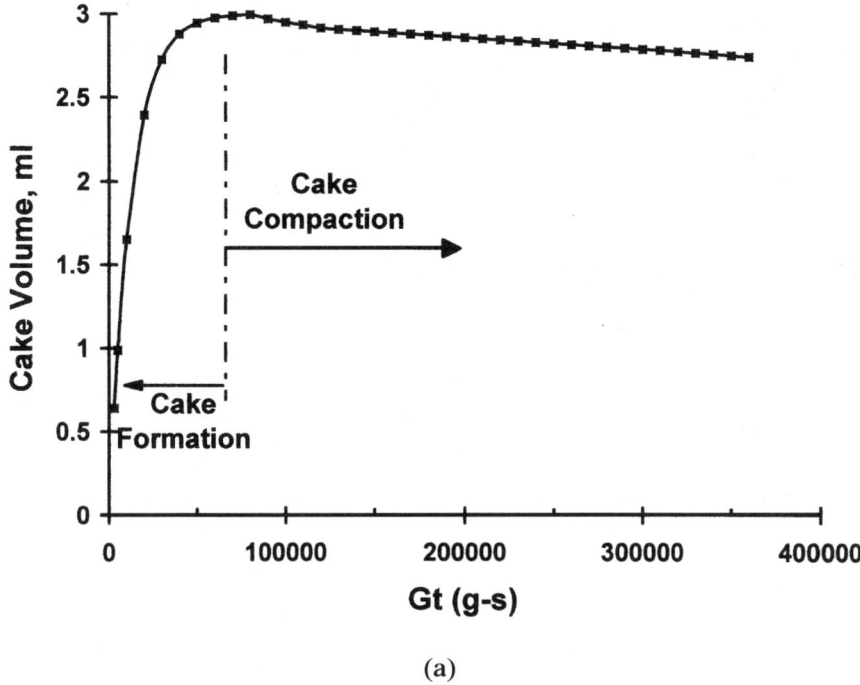

(a)

Figure 12.2 (a) Cake volume change with time in spin tube.

face area of the rod segment that has penetrated into the cake. A force balance can be used to infer the yield stress. Thus

$$\tau_y = \frac{mg}{2\pi r h_{pe}} \quad (12.5)$$

where mg and r are the weight and the radius of the rod, respectively, and h_{pe} is the length of penetration of the rod into the cake. The smaller the h_{pe} value, the higher is the cake yield stress and the better the cake can withstand handling such as conveyance. Figure 12.2c shows the measured yield stress τ_y, as a function of Gt. In summary, the cake properties change dramatically from small Gt when the cake is in the formation stage to larger Gt when the cake undergoes consolidation. This is evident in Fig. 12.2.

Spin-tube classification

A sample of the slurry for the classification test is prepared. The feed particle-size distribution $F_f(d)$, especially the fraction less than 45 μm (325 mesh), is measured using a particle-size analyzer. About 40 mL

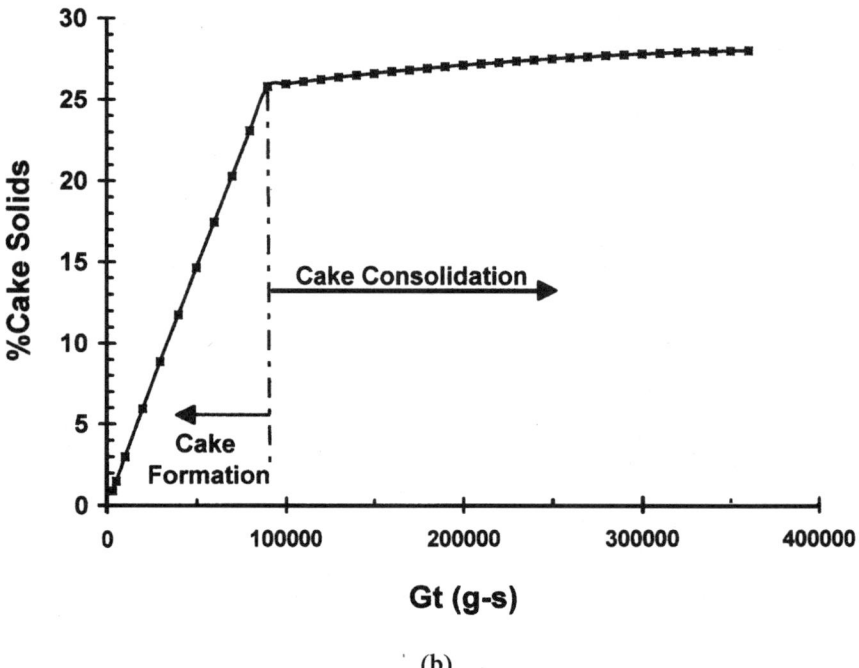

Figure 12.2 (b) Cake solids percent change with time.

of the sample v_f is poured into a 50-mL test tube. The tube and its contents are spun to 1000–3000g for 10–20 s. As the supernatant having volume v_e and density ρ_e is decanted off, the particle-size measurement $F_e(d)$ is taken. The remaining cake has volume v_s and density ρ_s. A material balance before and after the separation gives

$$\rho_e v_e + \rho_s (v - v_e) = \rho_f v \qquad (12.6)$$

The recovery of the total solids in the centrate, or in this case the supernatant, is

$$\%\text{Rec}_e = \frac{\rho_e v_e}{\rho_f v} \approx \frac{v_e}{v} \qquad (12.7)$$

and the size recovery, or the yield on size d, from Eq. (2.17b) is

$$Y(d) = \frac{\rho_e v_e F_e(d)}{\rho_f v F_f(d)} \approx \%\text{Rec}_e \frac{F_e(d)}{F_f(d)} \qquad (12.8)$$

Tests can be conducted at various G, such as 1000g, 2000g, and 3000g, and at different ts. The results of $\%\text{Rec}_e$, $F_e(d)$, and $Y(d)$ can be graphed as a function of Gt or $\mathcal{L}e$ number, as discussed in Chap. 6.

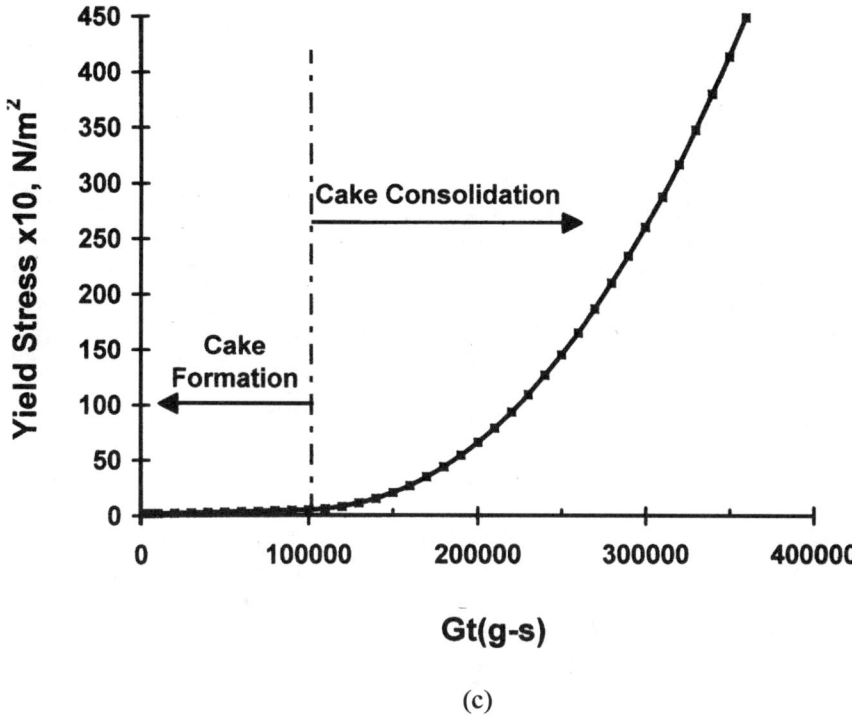

(c)

Figure 12.2 (c) Yield stress of cake as a function of centrifugation time.

Degritting test

Degritting tests can be run in a similar way as the classification tests. In addition to determining %Rec_e and %yield in the centrate, the grit level is also measured. The grit in the supernatant product cannot exceed a maximum allowable concentration while still meeting an acceptable recovery of the fine-particle product in the centrate. Typically, a moderate to high G force, usually between 1000 and 2500g, is required for the separation. Otherwise when the G force is too high, finer product particles also settle with the grit, yielding poor recovery.

Batch Solid-Bowl Tests

The batch solid-bowl centrifuge shown in Fig. 1.2a can be used to run semicontinuous clarification, classification, and degritting tests. A small feed sample is introduced batchwise or in a semicontinuous manner into a rotating solid bowl, say 100 mm (4 in) in diameter. The effluent is collected continuously, and measurements are made of the particle-size distribution and the solids concentration. The test terminates as the cake height reaches about 75–80% of the weir height, when the centrate quality starts to deteriorate due to entrainment of the sediment by the fast-flowing effluent stream. From the centrate solids weight fraction W_e the total solids recovery by the centrifuge for clarification can be estimated as $\%\text{Rec}_s \approx 1 - W_e/W_f$. Tests can be conducted at various Q, G, and for a limited pool range. Results are presented as $\%\text{Rec}_s$ versus Q for various G and pool levels. This is similar to that shown in Fig. 2.12, but with the abscissa expressed in terms of volumetric feed rate. The flow dynamics, which affect separation in a continuous-feed centrifuge, can be realized in a batch solid bowl, but unfortunately not in spin tubes. Thus the batch solid bowl comes somewhat closer to testing a continuous sedimenting centrifuge than a batch spin tube. Classification and degritting tests can be carried out similarly and the particle-size distribution needs to be monitored for all of the inflow and outflow streams of test centrifuge.

Basket Tests

There are two useful tests—a filtration test and a desaturation test. The former is concerned with the measurement of the cake resistance (or permeability) and the filter medium resistance to determine the bulk filtration rate, which might limit feeding of a filtering centrifuge. The latter evaluates the cake desaturation characteristics, which dictate the final cake moisture and consequently affect also the throughput. From Eq. (7.6) we get

$$\frac{\frac{1}{2}\rho\Omega^2(R_b^2 - R_p^2)}{q} = \frac{\mu}{2\pi K} \ln\left(\frac{R_b}{R_c}\right) + \frac{\mu r_m}{2\pi R_b} \qquad (12.9)$$

The numerator on the left-hand side of Eq. (12.9) represents the driving pressure, and the denominator the flow rate per unit basket length. The ratio thus represents the overall resistance to filtration. As shown in Fig. 12.3, a graph of the overall resistance $\frac{1}{2}\rho\Omega^2(R_b^2 - R_p^2)/q$ versus cake thickness, as measured indirectly by $\ln(R_b/R_c)$ ($= \ln[1/(1 - h/R_b)]$), gives a straight line intersecting the y axis at $\mu r_m/2\pi R_b$ with slope $\mu/2\pi K$. There are two approaches by which these quantities can be inferred.

Constant-head tests

A constant driving head for filtration can be obtained by an overflow arrangement. The basket is fed such that a constant yet small overflow from the basket is maintained. The overflow from the rotating basket is caught by a stationary gutter located adjacent to the basket. Baffles are installed at the basket and casing to ensure that the overflow liquid does not leak into the filtrate collection bin. This arrange-

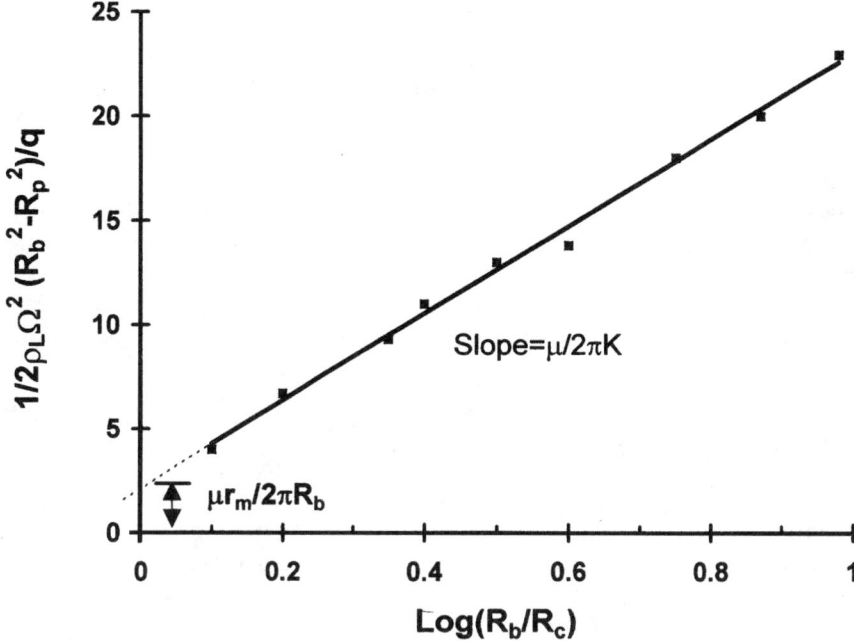

Figure 12.3 Total cake resistance versus effective cake thickness in filtration.

ment provides a reliable means of achieving a constant head with radius R_p for filtration tests.

Another arrangement can be made without the overflow arrangement, where the feeding rate just balances the filtration rate with the liquid-free surface holding stationary. When the feeding rate is greater than the filtration rate, the liquid level increases. Vice versa when the feeding rate is less than the filtration rate, the free surface of the liquid pool subsides.

In both arrangements the filtration rate is steady and can be measured at leisure. With the second arrangement, the filtration rate is equal to the feeding rate.

The test can be conducted for several cake heights. By increasing the cake height h (thereby also reducing the cake radius $R_c = R_b - h$) the resistance also increases nonlinearly, but appears linear in the form plotted in Fig. 12.3. The slope of the trend line through the data gives the cake permeability

$$K = \frac{\mu}{2\pi s} \qquad (12.10)$$

and the y intercept of the line provides the combined medium resistance and cake heel resistance,

$$r_m = \frac{2\pi R_b I}{\mu I} \qquad (12.11)$$

where s and I are the slope and the y intercept, respectively, of the linear trend for the test data in Fig. 12.3. As the medium resistance increases over time because of clogging and, more so, because of glazing of the cake heel due to high-speed peeling of the cake with a horizontal peeler, r_m increases, and so does the y intercept in Fig. 12.3. Also, the cake permeability is inversely related to the slope. If the slope of the linear trend is steep, it is indicative of high cake resistance. On the other hand, gradual slope is indicative of low cake resistance. Also, if the trend of the data set is linear regardless of the cake height, it can be inferred that the cake is incompactible. On the other hand, if the trend increases nonlinearly with the effective cake height $\ln(R_b/R_c)$, the cake is considered compactible with lower cake permeability under increased cake height and cake compaction pressure $p_s = (\rho_s - \rho)\varepsilon_s Gh$.

Variable-head tests

After a fixed amount of solids is introduced into the basket, the cake starts to form with a growing thickness h from the inner wall of the basket. Before the liquid level falls below the cake surface, a fixed known volume of liquid, preferably the mother liquor or a liquid compatible in property with the mother liquor, can be introduced to the cake to determine the filtration velocity. By noting the time it takes for the liquid pool surface above the cake to travel from an arbitrary initial radius R_{p1} to a larger final radius R_{p2}, the cake permeability can be determined. This is done by balancing the filtration rate with the decrease of liquid level. Recall the filtration rate per unit distance along the axis of a basket centrifuge,

$$q = 2\pi R_p \frac{dR_p}{dt} = \frac{\frac{1}{2}\rho\Omega^2(R_b^2 - R_p^2)}{\frac{\mu}{2\pi K}[\ln(R_b/R_c) - r_m K/R_b]} \qquad (12.12)$$

Integrating Eq. (12.12) between limits $t = 0$ and Δt, corresponding to the liquid-free surface at initial radius R_{p1} and final radius R_{p2}, we get

$$\Theta = \frac{\frac{1}{2}\rho_L \Omega^2 \Delta t}{\ln(R_b^2 - R_{p1}^2/R_b^2 - R_{p2}^2)} = \frac{\mu}{2\pi K}\ln\left(\frac{R_b}{R_c}\right) + \left[\frac{r_m \mu}{2\pi R_b}\right] \qquad (12.13)$$

where Θ denotes total cake resistance to filtration. It can be determined from the measurement of Δt for a given pair of initial and final radii R_{p1} and R_{p2}, a given rotational speed Ω, and cake height h or cake radius R_c ($= R_b - h$).

The measurement of the receding pool can be facilitated by laying down visible marked intervals along the radius of the basket "endwalls," and, with the help of stroboscoping, timing the duration Δt for the pool surface to pass between two arbitrary radii R_{p1} and R_{p2}. A plot of Θ versus $\ln(R_b/R_c)$, or $2.3023 \log_{10}(R_b/R_c)$, for various cake heights, as suggested by Eq. (12.13), gives a linear trend with the slope yielding the cake permeability or resistance, and the y intercept the combined filter medium and cake heel resistance.

An alternative approach is to introduce a known volume of the filtrate liquid above the cake prior to the cake surface drying out, that is, before the liquid level recedes below the cake surface. For filtrate volume v_{f0} and cake surface radius R_{p2}, the radius R_{p1} corresponding to the initial liquid surface by material balance is $[R_{p2}^2 - (v_{f0}/\pi b)]^{1/2}$, where b is the basket axial width. This can be used in conjunction with the variable-head tests discussed.

For most filterable materials the cake matrix is incompactible. For compactible cake, where compaction takes place due to increasing stress supported by the particles at contact with each other, the resistance versus height chart shows a nonlinear increase in cake resistance with the effective cake height $\ln(R_b/R_c)$.

Desaturation

For most crystalline solids, cake compaction is rather fast, followed by slow drainage or desaturation where the liquid saturation S drops below unity. Desaturation starts as the liquid pool above the cake disappears and the liquid level recedes below the cake surface. The process can be thought of as liquid draining through capillary tubes of various diameters under centrifugal gravity. The liquid drains out through the larger capillaries followed by the smaller ones. Although the bulk liquid has flowed through a given capillary, the inner wall of the tube is still coated with a liquid film, which takes a longer time to drain with progressively reduced film thickness over time. The drainage rate of the film depends on the roughness of the tube surface (that is, the particle surface roughness) and the viscosity of the liquid, which in turn depends on the temperature of the separation process (see Fig. 2.3).

Bucket Tests

The swinging-bucket (or beaker) centrifuge is a very versatile bench-scale equipment for carrying out cake desaturation measurements. Stainless-steel or titanium cylindrical tubes are used in making high-strength buckets to withstand the centrifugal stress. The collar of the tube rests on a sleeve, which is hinged from the swinging head of the centrifuge. Under centrifugal force, the cylinder assumes a horizontal position about the vertical axis of rotation. The large-diameter end of the bucket is equipped with a filter supported by a drainage screen. The filter opening is selected based on the particle size of the sample to be dewatered. Generally with a prethickened slurry, where fines are agglomerated in the cake, a coarser sieve opening of 150–210 µm (100–70 mesh) is usually adopted. If necessary, a finer cloth can also be used for finer-particle test samples. The liquid filters through the cake and is collected in the housing. In one design the screen is held in place by a cap nut from the large end (see Fig. 12.4). A free-filtering thickened slurry, typically 30–60% w/w (15–30% v/v) of solids, is used in the test. By introducing the thickened slurry in the bucket, a cake of a given thickness h forms promptly, followed by drainage, where the moisture from the cake is further removed or desaturated over a period of time t under a given centrifugal gravity G.

In the bucket test the volume of the cake can be measured accurately given the inner diameter of the bucket d_b is known and the cake height h can be measured by a depth caliper. The volume of the cake is then $v_c = 1/4 \pi d_b^2 h$. After centrifugation, the cap nut is removed and using a plunger the entire cake is taken out for analysis on both the wet and the dry mass, denoted by M_{wet} and M_{dry}. Subse-

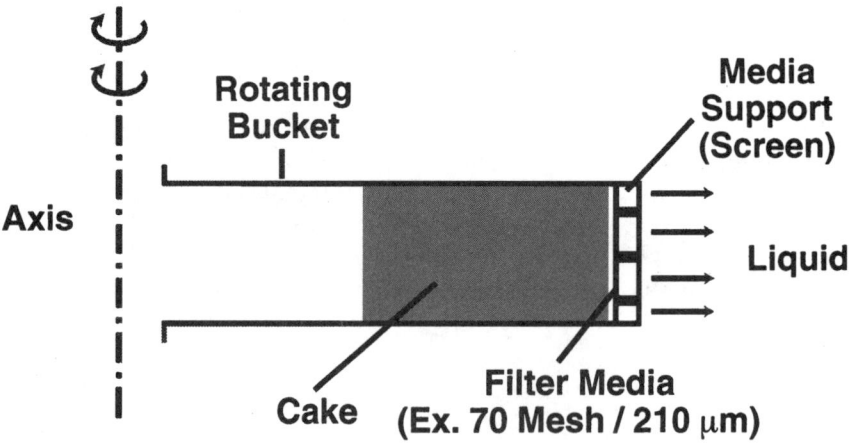

Figure 12.4 Bucket centrifuge.

quently the solids volume fraction ε_s (or the cake void fraction ε), the saturation S, and the cake solids weight fraction W_s (moisture by weight W_m) can all be determined using the equations

$$\varepsilon_s = 1 - \varepsilon = \frac{M_{dry}}{v_c \rho_s} \tag{12.14a}$$

$$S = \frac{M_{wet} - M_{dry}}{v_c \varepsilon \rho} \tag{12.14b}$$

$$W_s = 1 - W_m = \frac{M_{dry}}{M_{wet}} \tag{12.14c}$$

Example. Given: $v_c = 25$ cm^3, $M_{wet} = 30$ g, $M_{dry} = 16$ g, $\rho_s = 2.5$, and $\rho = 1.2$ g/cm^3, the following parameters are calculated based on Eqs. (12.14a) to (12.14c): solid volume fraction $\varepsilon_s = 0.26$; cake porosity $\varepsilon = 0.74$; liquid saturation $S = 0.63$; % cake solid $W_s = 53.3\%$; and % cake moisture $W_m = 46.7\%$. In Table 12.1 the pertinent data can be recorded for the bucket tests.

The liquid saturation S can be determined by another method using known values of W_s and ε_s. For a unit volume of cake with, respectively, solids volume fraction ε_s, cake void fraction ε, and liquid saturation S in the void, the solids fraction is $\rho_s \varepsilon_s$ and the liquid fraction $\rho(1 - \varepsilon_s)S$. W_s is thus related to both S and ε_s by

$$W_s = 1 - W_m = \frac{\rho_s \varepsilon_s}{\rho(1 - \varepsilon_s)S + \rho_s \varepsilon_s} \tag{12.15}$$

from which

$$S = \frac{\rho_s}{\rho_L} \frac{\frac{1}{W_s} - 1}{\frac{1}{\varepsilon_s} - 1} \tag{12.16}$$

It follows that if W_s and ε_s are both determined, S can be deduced from Eq. (12.16). Various combinations of G, h, and t can be used in the tests. For example, test conditions can be such as:

$G/g = 500, 1000, 2000$

$h = 12$ mm (0.5 in), 38 mm (1.5 in), 76 mm (3 in)

$t = 30, 60, 120, 240, 480$ s

The total number of tests becomes $3 \times 3 \times 5 = 45$ tests. Note that the shortest time should be much greater than the combined time for the

TABLE 12.1 Data Sheet for Bucket Test

Test	G/g	h, cm	t, seconds	tare, g	Wet sample weight + tare, g	Dry sample weight + tare, g	Wet sample weight M_{wet}, g	Dry sample weight M_{dry}, g	v_c^*	ε_s	ε	S	W_s	W_m
1														
2														
3														
4														
etc.														

*If cake bulk volume is v_c unknown, can use Eqs. (12.14d,e) knowing the bulk cake density bench-scale centrifuge to accelerate and decelerate for the test results to be meaningful. Typical bench units have 5–10-s acceleration times. Therefore any time less than 15 s would be meaningless. Also, maximum dewatering time should be selected so as to ensure that the final equilibrium cake moisture is reached. This is especially important when sample results are not known a priori and it takes at least several hours or overnight to dry off the cake moisture using a convection oven. The range of cake heights used for the test should encompass at least the expected minimum, the maximum, and an intermediate height. When possible, three different G forces should be selected, corresponding to the low, an intermediate, and the maximum values. It is not necessary that the maximum value selected correspond to the targeted maximum G force of the production machine because (1) this restricts exploring the higher G benefit on cake moisture reduction and precludes incorporating higher G in future designs, and (2) a higher G may possibly be used to simulate similar results, which at times can be achieved at a lower G with longer dewatering time or lower cake height.

Figure 12.5 shows some typical trends of cake dewatering with an incompactible material. In Fig. 12.5a the cake moisture decreases with increasing dewatering time for a given cake height and G level until an equilibrium value is reached. As the cake height h increases, the moisture versus time curve increases accordingly due to increasing resistance in liquid drainage. In most cases the data reach the same asymptote, yet it takes longer to reach the same equilibrium provided the capillary effect is insignificant. Otherwise when the capillary effect is significant, that is, the capillary rise height is comparable to the cake height, thinner cake actually yields higher cake moisture due to capillary saturation. In Fig. 12.5b the cake moisture is graphed against time with the G level as parameter for each curve. The moisture versus time curve decreases as the G force increases. While the solid curve shows a much lower cake moisture as G increases (capillary number $N_c > 5$), some filter cake behaves similar to the broken curve ($N_c < 5$), where the final equilibrium cake moisture is identical to that of the case at lower G, yet the benefit is such that it takes a much shorter time at higher G to reach the equilibrium cake moisture. This points to the benefit that a shorter dewatering time is realized for a batch centrifugal filter or, equivalently, a higher throughput capacity for a continuous-feed centrifuge.

Another form of the same result is shown in Fig. 12.5c and d, where cake moisture is plotted against cake height. In both charts the cake moisture increases with the cake height. As the dewatering time or the G force increases, the moisture versus height curve decreases as shown in Fig. 12.5c and d, respectively. Finally, the cake moisture versus G curve is shown in Fig. 12.5e and f for varying cake heights

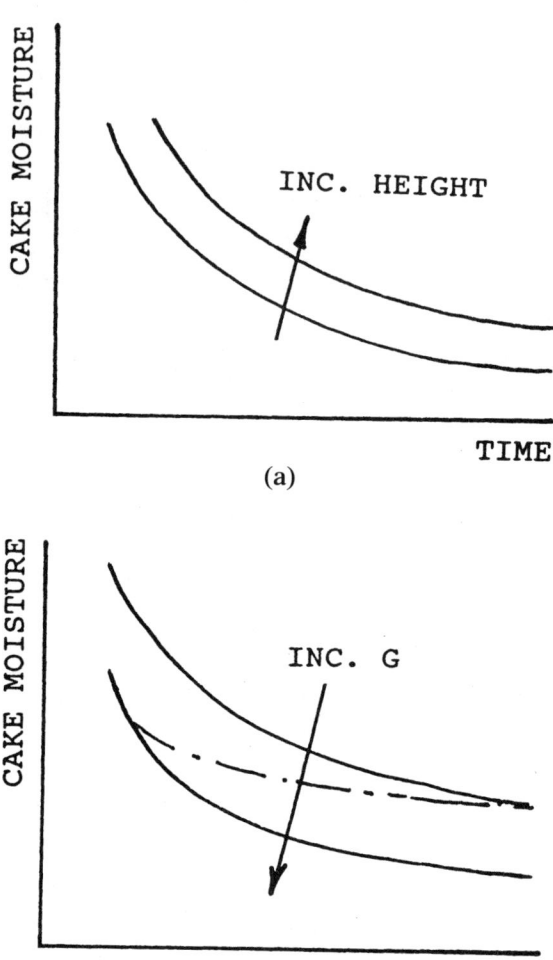

Figure 12.5 Typical characteristics of cake dewatering during desaturation.

and dewatering times. As the cake height increases, the moisture curve shifts upward. As the dewatering time increases, the moisture versus G curve shifts downward. Note that as the G force increases, the cake moisture decreases sharply initially, and then gradually. Only after the G force has reached a critical value does a further drop in cake moisture result. This corresponds to desaturating the pendular moisture where the capillary number N_c exceeds 5 to 10.

For n_G different Gs used in the tests, n_h different hs, and n_t different ts, the total number of charts and curves that can be generated is

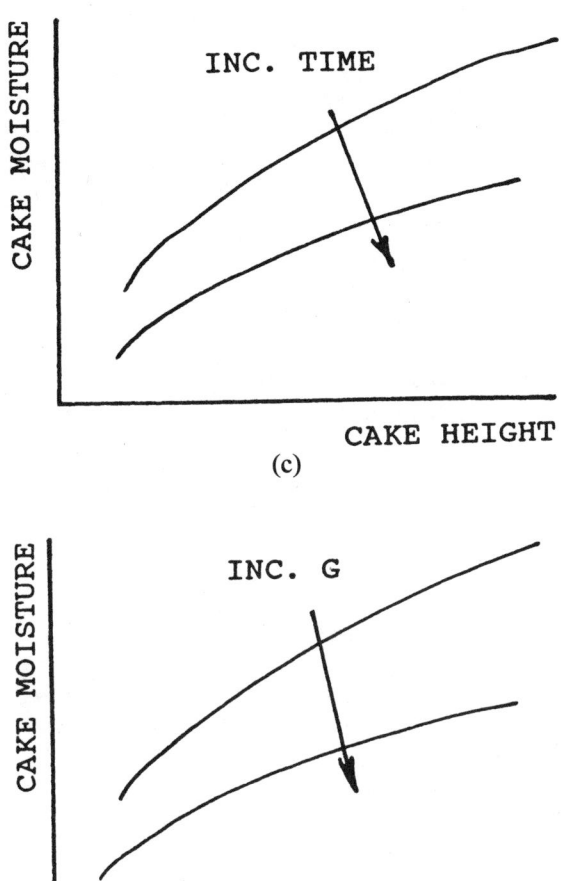

Figure 12.5 (*Continued*)

$2(n_G + n_h + n_t)$ and $2(n_G n_h + n_t n_G + n_G n_h)$. It is obvious that these can be excessive.

From Chap. 7, the best way of correlating the results is to plot % cake moisture W_m (or liquid saturation S) versus the dimensionless dewatering time t_d, which incorporates all the governing variables and is defined by Eq. (7.15b). Using this approach, a single curve is generated from repeated testing in the laboratory, as shown in Fig. 12.6 for % cake moisture by weight and liquid saturation versus t_d for a granular test material. In certain cases a family of curves is obtained, based on the departure of the residual cake moisture under different G forces and different cake heights.

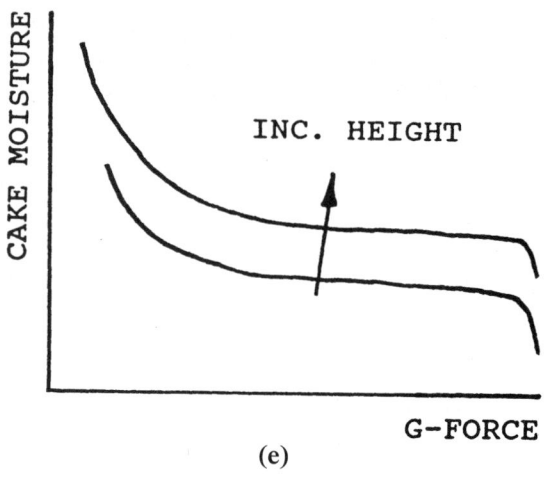

(e)

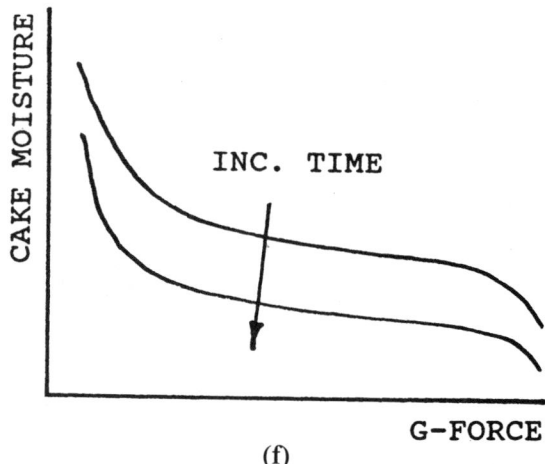

(f)

Figure 12.5 (*Continued*)

Cake Void Fraction

For incompactible cake the cake solids fraction or volume fraction is practically constant, independent of the compacting centrifugal force. This is not the case with compactible cake. Figure 12.7 shows the void fraction ε determined from the bucket data for the same incompactible material, with moisture data given by Fig. 12.6. As can be seen, despite the increasing centrifugal stress and long compaction time by a factor of over 1000 (3 log cycles), the cake void fraction stays very much constant for incompactible cake.

While Figs. 12.6 and 12.7 give results for larger granular materials of between 50 and 200 μm, Fig. 12.8 shows test results for dewatering

Lab, Pilot, and Production Tests 293

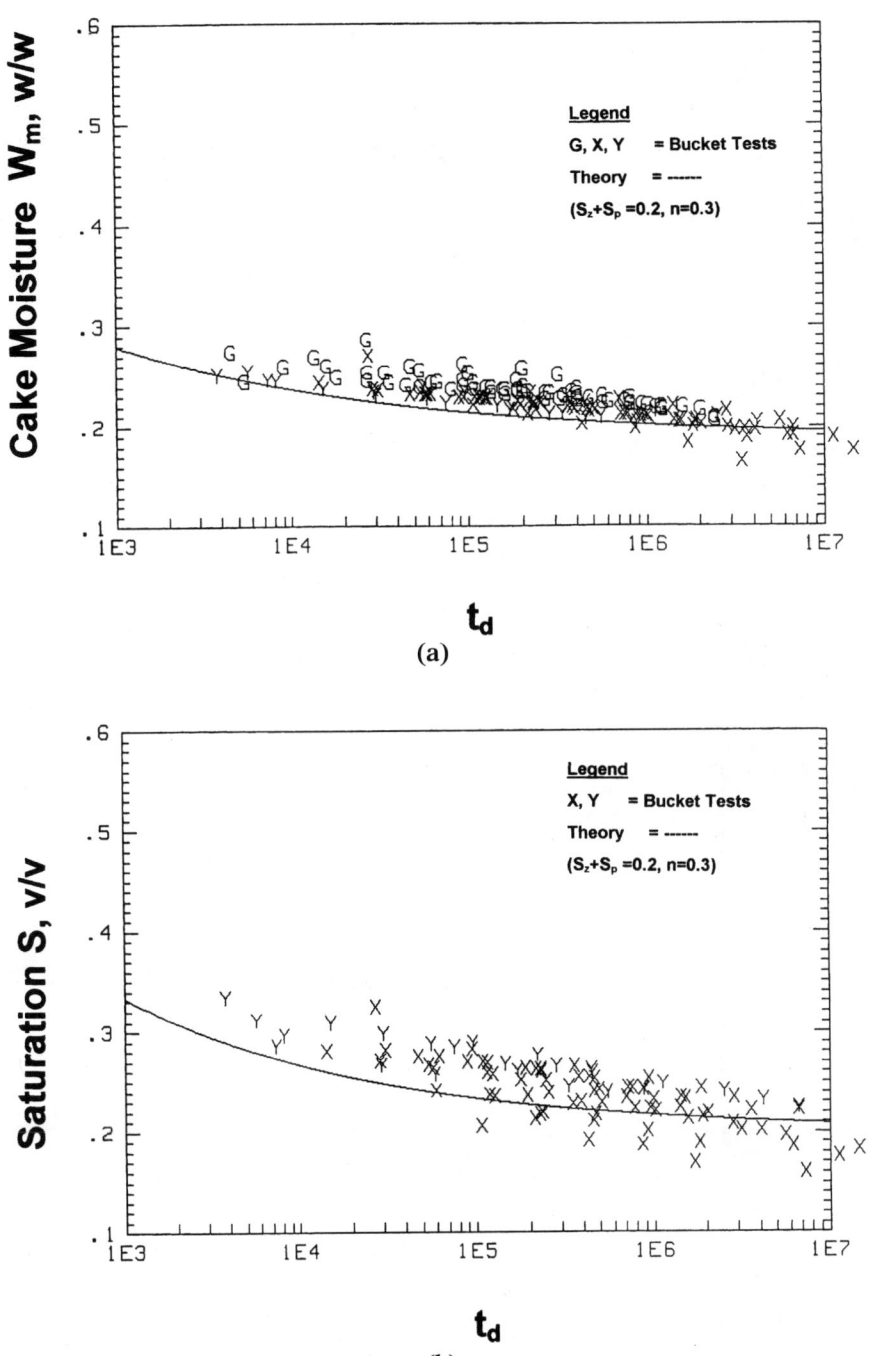

Figure 12.6 Dewatering characteristics of a given resin. (a) Percent cake moisture versus t_d. (b) Saturation versus t_d.

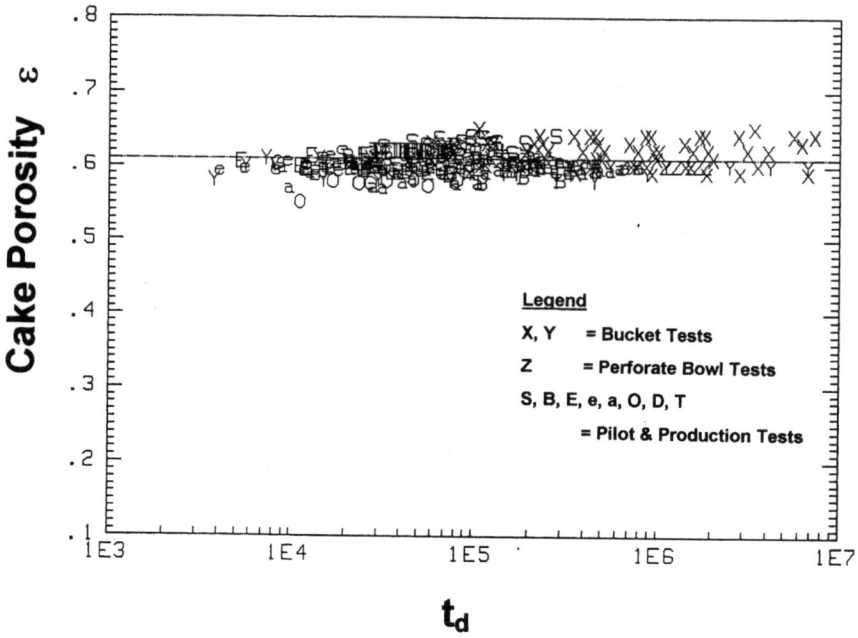

Figure 12.7 Deduced cake void fraction of resin in Fig. 12.6.

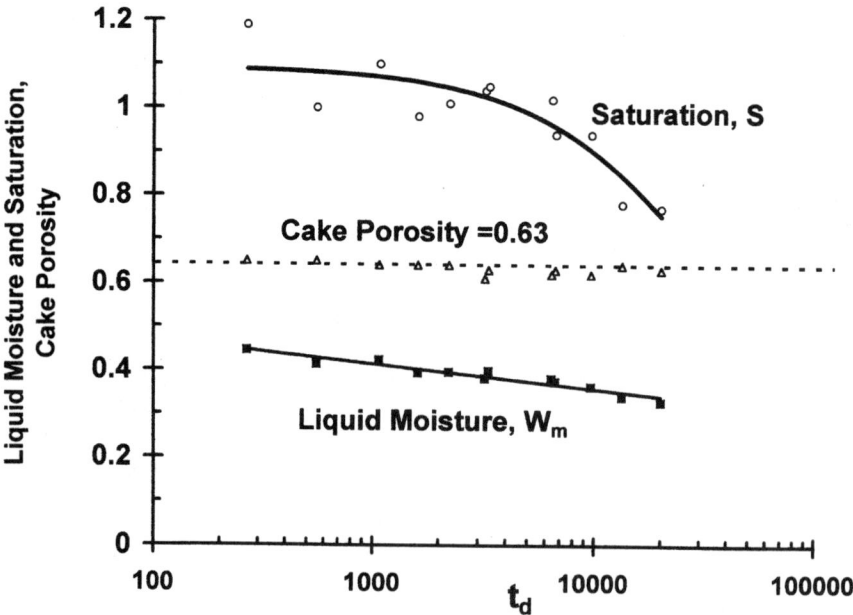

Figure 12.8 Cake dewatering for finer-particle slurry.

a fine mineral slurry with particle sizes of about 10 μm. Here the cake does desaturate, but at a very slow rate, as S never dropped below 80% during the tests. The porosity is maintained at 63%. The porous structure is quite open. Despite this, for cake pores presumably in the micrometer range, there are significant capillary effects and perhaps electrical forces in holding the moisture to the particle surface. High centrifugal gravity and long dewatering time are required for desaturation, where t_d needs to get beyond 100,000. For the tests the magnitude of t_d is reduced because of the low cake permeability and small particle sizes. Therefore the cake moisture stays at a high level of 30–40% w/w.

Pilot Tests

Pilot tests should be treated like smaller versions of production runs, yet there is much more flexibility as compared to production testing. Some pilot tests very often consist of some temporary or even permanent setups of the equipment, including accessories such as tanks, mixers, pumps, flowmeters, feed lines, centrate or filtrate lines, cake discharge lines, and hoppers. In some instances the plumbing can be changed easily by using quick-disconnect piping and valve arrangements with regard to the location of feed, the introduction of polymer into the feed lines before feeding the centrifuge, and wash location.

It is important to ensure in the pilot tests that the feed is representative of the properties of what will be running on a production scale. It is also prudent to test the feed under upset conditions due to upstream processes, such as reactors, crystallizers, grinding mills, flotation columns, thickeners, hydrocyclone, and so on, so as to determine how the upset condition affects the separation process in the centrifuge. The variables in the tests, where applicable, are:

- Volumetric and solid rate
- Speed or G
- Pool and interface levels
- Conveyance and solids discharge rate (differential speed between bowl and conveyor for scroll typed centrifuge)
- Process temperature
- Feed solids concentration
- Size of screen or filter medium openings
- Wash rate
- Polymer concentration and dosage

The following parameters need to be measured and monitored:

- Volume flow rate of feed and filtrate
- Solids concentration of feed, centrate (one or two liquid phases), filtrate, and cake
- Particle-size distribution of feed, centrate, filtrate, and cake
- Impurities concentration present in feed, and wash liquid cake
- Torque
- Electrical current in amperes
- Electrical power in kilowatts
- Bearing temperature
- Vibration levels (displacement, velocity, acceleration) along horizontal and vertical directions in appropriate locations, as suggested by manufacturer
- Rotational speed in rpm

As a reference, Table 12.2 shows a typical decanter worksheet for testing. Similar worksheets can be designed for screen bowls and screen scrolls, with an additional pertinent section for screen drainage. The screen-size opening and screen condition should also be noted. Likewise a similar form can be designed for pusher centrifuges, but without the flocculant addition. Cake height, distance between adjacent cake rings, stroke frequency, and stroke length need to be noted, from which the conveyance efficiency of the cake and the bulk capacity of the pusher can be determined. For a basket centrifuge the cycle times should be added, and feed time, wash time, dry time, and any overhead time (acceleration, deceleration, cake discharge time) need to be recorded. For a disk centrifuge, the cake and effluent discharge rates can be monitored in certain setups in addition to the other standard measurements, similar to the decanter.

Production Testing

Production testing is very much like a pilot test. It is important to determine what is the objective of the test—evaluation of the process or optimization. What are the variables that can be changed—speed or G force, pools, or differential in the case of a decanter? The variables are the back-pressure valve on the centrate stream and the speed in the case of a disk machine. As with the pilot tests, the feed condition needs to be representative. There is generally not much flexibility with production tests. Testing depends on idle time or downtime, when production is not in high gear, and when the flow rate can be changed to either lower or higher than the production flow rate in order to determine the process results under these off-spec conditions. Also, if

TABLE 12.2 Typical Decanter Worksheet

Date					
Run number					
Feed					
Temperature °F (°C)					
gal/min (L/min)					
Specific gravity					
% suspended solids					
lb/min [kg/h (dry solids)]					
Cake					
% total solids					
lb/min (kg/h) wet					
lb/h (kg/h)					
lb/ft^3 (kg/m^3) wet					
ft^3/min (m^3/h) wet					
Effluent					
% suspended solids					
lb/min (kg/h)					
Specific gravity					
gal/min (L/min)					
Flocculant					
Addition					
%					
gal/min (L/min)					
lb/min (kg/h)					
lb/ton (kg/t)					
lb/ton active (kg/t)					
$/ton					
% recovery					
Machine					
rpm					
G/g					
Pool depth					
Gear ratio					
Δ					
Amperes					
Volts					

NOTES: (1) Analysis on suspended solids by filtration and drying for feed, screen and effluent. Cake analysis on total solids by drying. (Drying temperature 103–105°C.)

(2) This data format would be similar for a screen bowl, except that an additional set of data for screen drainage would be added. This set would be identical to the *effluent* data. Screen size should be noted on the data sheet.

(3) This same form can be used for a pusher centrifuge. Flocculant would not normally be used. The amount of cake left on the basket(s) should be noted after testing (pounds per each stage). If possible, cake height on each basket should be recorded.

(4) For a basket centrifuge the cycle times would be added—feed time, dry time, wash time, and final dry time. The flocculant section would normally be eliminated.

(5) For a disk centrifuge the data would be based on flows measured for the discharge streams.

the resulting cake or centrate does not meet specifications, it would be wise to arrange for these products to be isolated and perhaps recycled back to the feed so as not to upset the downstream equipment (for example, by sending off-spec wet cake to the dryer).

Saturation Measurement for Continuous Centrifuges

Unlike the bucket tests, it is impractical to determine the bulk cake volume in both pilot and production tests for continuous filtering centrifuges. The cake sample can be obtained through a sample container with accurately known volume, and by measuring the mass of the wet sample, one can determine the bulk cake density ρ_c. The cake properties can be determined from measuring ρ_c, wet cake weight M_{wet} of a sample, and dry cake weight of the same sample M_{dry}, from which W_s and W_m can be determined via Eq. (12.14c). The values of ε_s and S can be found by the following relationships:

$$\varepsilon_s = W_s \frac{\rho_c}{\rho_s} \qquad (12.14d)$$

$$S = \frac{\rho_s}{\rho} \frac{1/W_s - 1}{\rho_s/W_s\rho_c - 1} \qquad (12.14e)$$

Example. Given: $M_{wet} = 30$ g, $M_{dry} = 16$ g, $\rho_s = 2.5$ g/cm³, $\rho = 1.2$ g/cm³, and $\rho_c = 1.2$ g/cm³, the following values are derived based on Eqs. (12.14c) to (12.14e) with $W_s = 53.3\%$ and $W_m = 46.7\%$: $\varepsilon_s = 0.26$, $\varepsilon = 0.74$, and $S = 0.63$.

Cake Washing

Cake washing by displacing the mother liquor containing contaminants with a compatible wash liquid can be investigated using a bucket centrifuge or a basket centrifuge for batch tests, and using a pusher, a screen scroll, or a screen bowl for continuous flow tests. Wash liquid should be applied prior to desaturating the cake (liquid level falls below cake surface), otherwise air bubbles get trapped inside the cake matrix, rendering a poor sweep by the wash liquid and resulting in poor wash efficiency. In all cases the contaminants (such as undesirable sulfate or chloride) and methods of measuring these contaminants are also identified. Subsequent tests are carried out to monitor the contaminant level as a function of the wash ratio WR (see Chap. 2 for some common definitions of WR). Figure 7.13 shows a chart of percent contaminant versus WR with G force, washing time, nozzle, and screen design as the operating and design parameters for the curves. The behavior is self-evident.

Conclusions

It is important to carry out laboratory tests to determine, even on an approximate or crude basis, the centrifugal effect on separation and filtration. Useful process information can be obtained from simple bench-scale tests, which require only a small amount of test sample, to be followed, if necessary, by a moderate-scale pilot test, which needs more test materials. The pilot tests can be conducted in-house or in the field. There is more flexibility in running pilot tests as they represent a demonstration on a larger scale to investigate variables which cannot be explored under bench-scale testing. This pertains especially to the dynamics associated with continuous-feed centrifuges. Indeed, pilot tests are used to simulate mini production runs. Optimization with the operating variables is often feasible. It would be useful to quantify the effects of the variables on the separation results from which the production machine should be sized. Process and mechanical testing is conducted on production centrifuges on a regular basis to ensure the quality of separation from the centrifuges and the integrity of the mechanical components.

Chapter

13

Centrifuge Selection and Sizing

In this chapter the selection and the sizing of the centrifuge are discussed. Selection is based on two major concerns—technical needs and economics—and sizing is related to the minimum required criteria from a design and operation standpoint to attain the process objectives.

Process Definition

It is of paramount importance to define the solid-liquid separation task. Unfortunately this is also the very step that the centrifuge users fail most frequently. Because the user is uncertain about what a centrifuge can do, he or she often fails to define the process requirements upstream or downstream of the separation step. The following guidelines could be valuable in assisting a user to define the separation requirements.

1. As shown in Fig. 13.1, draw a box or boundary around the steps in the process flow sheet that involve solid-liquid separation. Centrifuges may or may not have already been employed in this box with other separation and peripheral equipment. The feed stream is shown entering the box from the left and four possible streams are exiting the box to the right. These exiting streams may consist of a concentrated heavy cake solid (that is, all the solids that sink in the liquid phases, including heavy plastics), a light cake solid (such as lighter plastic or fat globules which float on liquid), a heavy liquid phase (such as water or solvent), and a light liquid phase (such as liquid fat or oil).

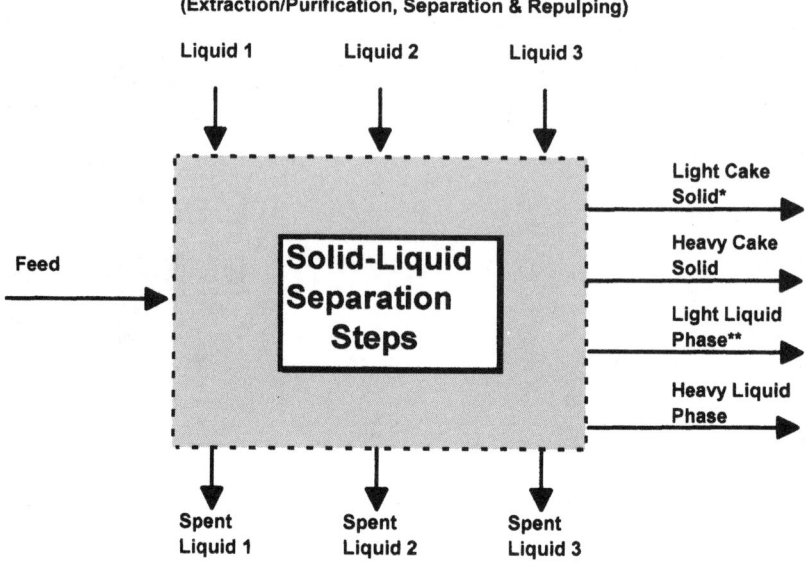

*Provision with two solids separation capability according to their density difference.
**Provision with two liquids separation capability according to their density difference.

Figure 13.1 Solid-liquid separation of a given flow process.

Theoretically this separation can be carried out in a single centrifuge that can separate out two solids and two liquids of different densities, or upto three sequential stages, each employing a centrifuge or centrifuges in combination with other separation equipment. Also, slurries or solvents may be introduced to wash the cake in order to remove impurities, if present. These spent streams most likely need to be further cleaned up before disposal or reuse. They may contain dissolved solids as well as small amounts of suspended fines.

2. Next identify the following variables in all the relevant streams which enter and exit the box:

1. % suspended solids (ss) by weight (or by bulk volume)
2. % dissolved solids (ds) by weight if relevant
3. volumetric throughput in gal/min or L/min
4. solids throughput in lb/h or t/h (dry basis, that is, no liquid)
5. handleability (flowable slurry, granular cake, cake stackable and nonbleeding, paste, hazardous—cannot handle manually, and so on)

3. Match the objectives of the box in Fig. 13.1 with the following possible objectives. (The centrifuge as part of the equipment in the box should also support the overall objective of the separation step—the "black" box.)

1. Solid-liquid (two-phase) separation or liquid-liquid-solid (three-phase) separation (state separation requirements or specifications of each stream?)
2. Clarification (tolerable % solids recovery, or % centrate solids?)
3. Classification by size and by specific gravity (product cut and recovery?)
4. Degritting (tolerable grit %?)
5. Thickening or concentration (thickened cake solids by weight?)
6. Dewatering (cake solids specification? cake compactible, drainable?)
7. Washing within centrifuge (purity needed? wash ratio limit?)
8. Separation and repulping (purity needed? number of stages?)

It is important to define to your best knowledge what is required for the separation process based on these three steps. Upon satisfactory definition of the separation function, some finer points will help to zero in on the selection process.

Further issues. It is useful to know the average particle size in the slurry and the size distribution, especially the fine fraction of a given slurry. The fine fraction refers to the fraction less than -10 μm (0.0004 in) for sedimenting centrifuges and to the fraction less than 45 μm for filtering centrifuges. This is all relative as in some applications fine fraction may refer to the fraction less than 0.5 μm (0.0002 in). The feed solids concentration also dictates the type of centrifuges which suits best for a given application. The capacity for both liquid and solids needs to be taken into account. Would it be more appropriate to go with a batch or with a continuous centrifuge? What would be the expected G force to make the separation? What is the associated horsepower requirement and motor size? Can the material stand handling? What is the minimal cake dryness so that the cake can stack without liquid bleeding out? The centrate discharge may be pressurized to avoid foaming. Does the product require clean-in-place or sanitary-in-place procedures? What are the temperature and pressure required? Is there any special requirement with regard to construction materials for the rotating assembly, temperature and thus thermal stress, corrosion, and chloride attack? The economics and cost-effectiveness should be weighed against the product value and quantity.

Selection Based on Process Function

The process functions of the centrifuges are separation, clarification, classification, degritting, thickening, dewatering, cake washing, and separation and repulping. As shown in Table 13.1, dewatering by drainage and cake washing are most appropriate for filtering centrifuges. This includes the use of vertical baskets, peelers, pushers, and screen scrolls. On the other hand, separation, clarification, classification, degritting, thickening, and separation and repulping are best handled by continuous solid bowls (limited on batch solid bowls) and disks. Screen bowls offer the benefits from both worlds. However, the clarification capacity is somewhat limited.

Selection Based on Size

Materials that are predominantly 45 μm and above and which are relatively incompactible lend themselves to separation by filtration, whereas materials that are finer in size or are compactible lend themselves to separation by sedimentation. Table 13.2 shows these two categories of machines.

Sedimenting centrifuges. For sedimenting centrifuges, decanters and solid-bowl baskets have been used to process particles between 0.02 μm and 5 mm (8×10^{-7} and 0.2 in). The lower end of the size range 0.02–3 μm (8×10^{-7}–1.2×10^{-4} in) may correspond to dewatering of monodispersed fine calcium carbonate, which has high specific gravity, or to classification of kaolin with particles of 0.5–2 μm (2×10^{-5}–8×10^{-5} in). The high end up to 5 mm (0.2 in) corresponds to bulk raw materials. The majority of the applications are in the 1–200-μm (4×10^{-5}–8×10^{-3} in) range, with significant amounts of less than 45 μm (325 mesh).

The disk centrifuges are typically applied to separate particles with sizes ranging from of 0.1 to 10 μm (4×10^{-6}–4×10^{-4} in). Economics override the use of disks when particle sizes exceed 50 μm (2×10^{-3} in) as other types of centrifuges can serve the same function. When particle sizes exceed 100 μm (4×10^{-3} in), pluggage between adjacent disks becomes a concern as the spacing between adjacent disks can be as narrow as 400 μm (0.016 in).

Tubular centrifuges can be used for particle sizes extending down to 0.01 μm (4×10^{-7} in) as they can reach 20,000g. For particles larger than 2–5 μm (8×10^{-5}–2×10^{-4} in) it is uneconomical to use tubular centrifuges, unless there is a close density difference between the solid and liquid phases, rendering separation difficult. Spin tubes are laboratory units and are used to test materials of various sizes using Gs from 200 to 20,000g. The ultracentrifuge is a special type of spin tube,

TABLE 13.1 Classifying Centrifuges by Process Functions

Types	Vertical	Peeler (siphon)	Pusher, single and multiple stages	Screen, vibratory	Screen scroll	Screen bowl	Disk	Solid bowl	Solid bowl, batch
Clarification						×	×	×	×
Classification						×	×	×	×
Degritting								×	×
Thickening							×	×	×
Dewatering, compaction							×*	×	×
Dewatering, drainage	×	×	×	×	×	×		×	
Cake washing	×	×	×		×	×		×*	

*Limited

TABLE 13.2 Classifying Centrifuges by Feed Solids

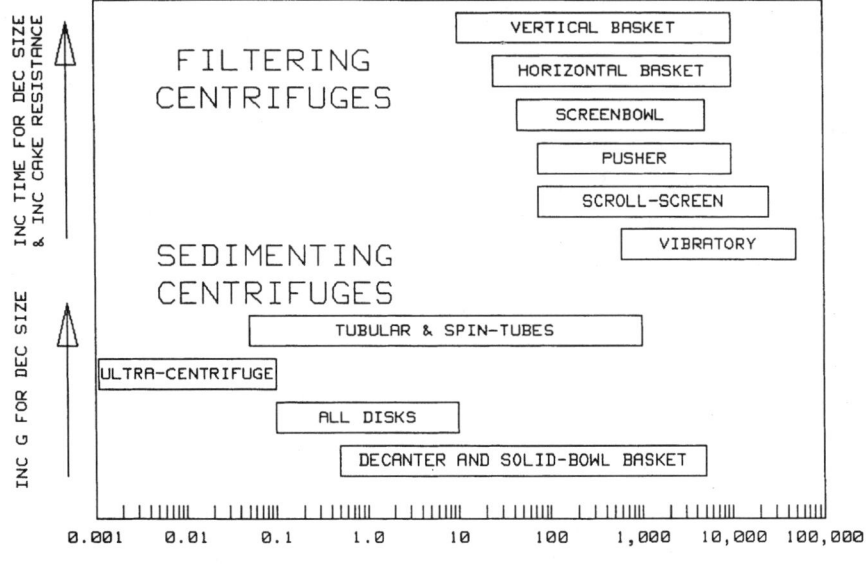

which develops between 200,000 and 500,000g. The particle size to be separated out can be as small as 0.001 μm (4×10^{-8} in).

Filtering centrifuges. Most filtering centrifuges, especially continuous-feed types, are tailored to separate solids in the slurry with particle sizes predominantly greater than 45 μm (325 mesh). The vibrating types dewater coarse raw materials such as sugar cane, molasses, and coal in preparation plants of sizes up to ½ in (13 mm). The lower end stops at 600 μm, or approximately 28 mesh. Particles smaller than this would pass through the screen openings. Screen scrolls and pushers dewater crystalline materials with sizes of greater than 75 μm (200 mesh). The smallest slot size of a wedge wire basket for a pusher is about 0.004 in (100 μm), whereas for conical screens it is about 0.012 in (300 μm). This is roughly two to three times the smallest particle size of slurry using this equipment. This is because particles in clusters bridge over screen openings, thereby allowing the use of more open screens, with slots two to three times larger than the particle geometry. Recent laser-cut screens can provide perforation hole sizes down to 20–40 μm (0.0008–0.0016 in) for screen scrolls, where one can dewater 10–20-μm (0.0004–0.0008 in) particle-size slurry. However, given that the percent open area is reduced, the filtration rate and hence the solids throughput per filter area are also reduced. The percent open area for a typical wedge wire screen is about 9–10% and can be as low as 5%.

Both horizontal and vertical baskets use filter medium on a perforate screen support. Depending on the cloth openings, particles down to 5–10 μm (0.0002–0.0004 in) have been separated using these types of centrifuges with a thin filter cake. It takes longer cycle times to drain cakes with small particle sizes and low permeability. Frequently there is significant filter medium resistance when a fine cloth with a low percent open area is used to capture and retain the fine solids.

Selection by Feed Solids Concentration

The various types of sedimenting and filtering centrifuges are selected based on the feed solids concentration. This is summarized in Table 13.3 based on feed solids by bulk volume.

Sedimenting centrifuges. The solid bowls can take feed solids by weight from a fraction of 0.5% for thickening applications to 70% w/w for degritting. The typical range is between 10 and 50% w/w solids. On the other hand, screen bowls usually handle 15 to 50+% w/w solids. For dilute slurry, the cake formed contains a significant amount of liquid, which when it drains through the screen, also carries the fines with it. On the other hand, screen bowls are not used as classifiers and therefore, unlike solid bowls, seldom would be fed with solids concentrations greater than 50% w/w.

TABLE 13.3 Classifying Centrifuges by Feed Solids Bulk Volume

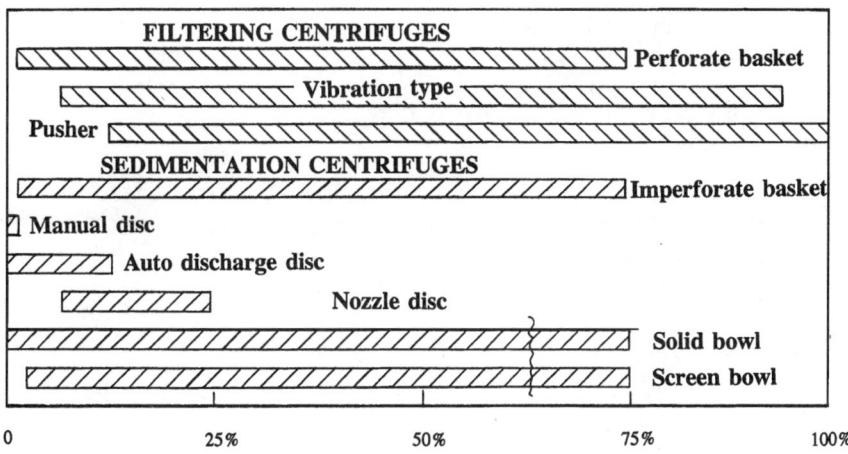

SOLIDS IN BULK VOLUME PERCENT

Disks are used for low feed solids separation. In increasing order for higher solids applications there are the manual disk, intermittent disk, and finally the nozzle continuous-discharge disk. The manual disk handles feed streams with no more than 1% w/w solids, the intermittent disk 5% w/w, and the nozzle-discharge disk 5–10%. Jamming, plugging, wear and tear, and solids discharge are concerns with high-solids applications.

Filtering centrifuges. The pusher handles feed solids in the range of 40–60% w/w. In this feed solids range, the slurry is pumped into the machine from a thickener such as a hydrocyclone. For higher solids concentrations of between 60 and 80%, it is screw-fed into the first-stage basket. The screen-scroll centrifuges behave in a similar manner. With higher solids concentrations, such as for dewatering coal with feed solids at 70–85% w/w, the feed at high tonnage is fed through a head with a feed pipe diameter as large as 10 in (254 mm).

The feed solids for the batch baskets vary from 15 to 50% w/w. Feed materials should be readily flowable, otherwise the machine may get out of balance if the feed solids are not distributed uniformly inside the basket.

It is important to select the machine based on the process objectives, the size and feed solids concentration, the G, and the capacity it can handle. Table 13.4 shows sizes, feed solids concentrations, G, capacities, solids recovery, cake height, cake solids, washing characteristics, retention times, bowl or basket diameters, cake discharge mechanisms, and typical applications. This table includes filtering centrifuges encompassing the vertical baskets, peelers, screen bowls, pushers, screen scrolls, vibrating centrifuges, and the industrial sedimenting types, including disks, decanters, and solid-bowl baskets. It also includes screen bowls for comparison. Additional comparative charts can be found elsewhere.[1]

Sizing

For sedimenting centrifuges where either clarification, classification, or degritting is the objective, consideration should be given to equate pilot and production machines based on one of the following criteria:

1. Leung number $\mathcal{L}e$, as discussed extensively in Chap. 6, with $\mathcal{L}e_1 = \mathcal{L}e_2$
2. Sigma factor Q/Σ, with $(Q/\Sigma)_1 = (Q/\Sigma)_2$
3. Pool volume Q/V_p, with $(Q/V_p)_1 = (Q/V_p)_2$
4. G multiplied by the surface area, with $[Q/(GA)_p]_1 = [Q/(GA)_p]_2$

TABLE 13.4 Comparison of Various Centrifuges

Types	Vertical	Peeler (siphon)	Screen bowl	Pusher, multistage (1-stage)	Screen scroll	Screen, vibratory	Disk	Solid bowl	Solid bowl, batch
Operation	Batch filtration	Batch filtration	Continuous sedimentation and filtration	Continuous filtration	Continuous filtration	Continuous filtration	Continuous sedimentation	Continuous sedimentation	Batch sedimentation
Size, μm	10–10,000	20–10,000	40–5000	80–10,000	80–20,000	500–50,000	0.1–10	0.8–5000	0.8–5000
Feed solids, % w/w	15	15	10	40	35	50	0.5–5	0.2–75	1
G/g	500–2000	1000–2000	500–2000	300–1200	200–700	50–300	To 15,000	To 10,000	To 2000
Capacity, ton/h	To 6–10	25	30–150	To 90	To 100	250	To 400 (gal/min)	To 100	To 4
Centrate, % solids recovery	To 99	To 99	90–99	To 99	96–99	90–99	To 99	To 99	To 99
Screen filtrate	Good	Good	Fair to good	Good	Reasonable to good	Reasonable	Good	Good	Good
Cake height, mm (in)	To 250 (10)	To 230 (9)	Depends on rate	15–80 (15–125)	To 50 (2)	To 30 (1.2)	5–25 (0.2–1)	To 100+ (4+)	Above 100 (4+)
Cake solids, % w/w	To 95	To 95	To 95	To 98	To 95	To 95	To 50	To 80–90	To 80–90
Washing	Very good	Very good	Good	Good	Moderate to good	Minimum	NA	Minimum	NA
Retention time	To 30 min	To 30 min	To 5–10 s	10–60 s	To 30 s	To 30 s	To 10 s	To 10 s	To 60 s

TABLE 13.4 (Continued)

Types	Vertical	Peeler (siphon)	Screen bowl	Pusher, multistage (1-stage)	Screen scroll	Screen, vibratory	Disk	Solid bowl	Solid bowl, batch
Bowl basket diameter, mm (in)	250–2000 (10–80)	250–2000 (10–80)	15–1350 (to 54)	300–1250 (12–50)	250–1100 (10–44)	To 1300 (52)	To 500 (20)	15–1350 (to 54)	To 500 (20)
Filtering medium	Fabric or metal cloth	Fabric or metal cloth	Perforation/ slots/TC ligaments	Wedge wire screen	Wedge wire/ perforate screen	Wedge wire/ perforate screen	NA	NA	NA
Cake discharge	Peel at low speed or remove filter bags	Peel at speed on non-breakable crystals	Scroll conveyor	Pushing	Scroll conveyor	Vibration	Manual/ dropping bottom/ continuous nozzles	Scroll conveyor	Bag
Applications	Pharmaceutical and specialty minerals, need flexibility and low cake moisture with good wash	Products with extreme demand on residual moisture and wash requirement	Chemicals, minerals, salts	Chemicals, salts	Minerals, chemicals, food, industrial	Coarse minerals	Fine solids, food processing, chemicals	Chemicals, biological, minerals, waste food processing	Biological waste, food processing

Subscripts 1 and 2 represent pilot and production machines, respectively, and subscript p indicates condition evaluated at the pool. Based on expensive experience, it has been found that $\mathscr{L}_{\varepsilon}$ provides the best of these criteria for clarification, classification, as well as degritting because it represents the most approriate physical phenomena occurring in centrifugal separation. However, as with other indexes, $\mathscr{L}_{\varepsilon}$ does not account for the entrainment of solids by liquid stream, which usually occurs at very shallow pools or very deep cake piles when the cake is close to the pool surface. The latter occurs with decanters at very low differential speed. Also with decanters, entrainment due to very high differential speed as a result of agitation from the secondary flow from the conveyor blades is not accounted for. Therefore discretion should be exercised when these extreme conditions are encountered.

For dewatering up at the dry beach or for filtering applications where the liquid from the cake can drain under centrifugal gravity, the dewatering time t_d should be used, as discussed in Chaps. 7, 9, and 12. In essence, pilot and production machines should have the same t_d, that is, $t_{d1} = t_{d2}$ where 1 and 2 represent pilot and production machines, respectively.

For dewatering by expression of compactible cake, it will shown in Chaps. 16 and 17 that the compaction stress $p_s \propto (\rho_s - \rho)Gh$ needs to be high and the retention time $t \propto 1/Q$ needs to be long, which translates to a lower throughput to allow time for liquid to percolate from the cake interior to the surface.

For washing, wash ratios in combination with other operating parameters such as G, time t, and cake height h are used for sizing. In addition, practical experience should be used to complement these criteria if necessary.

Costs

Besides the technical decision, the cost of the centrifuge also needs to be factored into consideration. While the cost might change with time, the 1996 figures suggested in this book provide a relative base for comparison when buyers obtain exact price quotes from the centrifuge manufacturers or vendors.

Neither the investment cost nor the operating cost of a centrifuge can be directly correlated with any single characteristic of a given type of centrifuge. The costs depend on the features of the centrifuge tailored toward the physical and chemical nature of the materials to be separated, the degree and difficulty of separation, the flexibility and capability of the centrifuge and its ancillary equipment, the environment in which the centrifuge is located, and many other nontechnical factors, including market competition. The cost figures presented here only represent centrifuges for use in the process industries as of 1996. For a particular installation there may be discrepancies between the costs suggested here and the bids obtained from the manufacturer.

A useful parameter for value analysis is the installed cost of the number of centrifuges required to produce the demanded separative effect (end product) at the specified capacity of the plant. The possible benefits of adjustments in the upstream and downstream components of the plant and the process should be carefully examined in order to minimize the total overall plant costs. The systems approach should be used.

Purchase price

Typical purchase prices, including drive motors, of tubular and disk sedimenting centrifuges are given in Table 13.5. Prices will vary upward with the use of more exotic construction materials, the need for

TABLE 13.5 Comparison of Tubular and Disk Centrifuges

Type	Bowl diameter, in (mm)	Approximate Σ value, 10^4 ft^2 (10^3 m^2)	Designation	Purchase price,* 1996 $
Tubular	4 (102)	2.7 (2.5)	Oil purifier	60,000–80,000
	4 (102)	2.7 (2.5)	Chemical separation	60,000–80,000
	5 (127)	4.2 (3.9)	Blood fractionation	100,000–140,000
Manual discharge disk	13.5 (343)	21 (20)	Hermetic	100,000–130,000
	24 (610)	95 (88)	Centripetal pump	150,000–300,000
Continuous nozzle-discharge disk	12 (305)	12 (11)	Clarifier	100,000–130,000
	18 (457)	25 (23)	Separator	150,000–200,000
	30 (762)	100 (93)	Recycle clarifier	270,000–300,000
Intermittent-discharge disk	14 (356)	13 (12)	Centripetal pump	130,000–150,000
	18 (457)	22 (20)	Centripetal pump	170,000–200,000
	24 (610)	38 (35)	Centripetal pump	250,000–300,000

*All prices are for stainless-steel construction, with the exception of the oil purifier.

explosion-proof electrical gear, special enclosures required for vapor containment, and the degree of portability. This is true for all types of centrifuges.

The average 1996 purchase prices of continuous-feed solid-bowl centrifuges for both 316 stainless steel and steel are shown in Fig. 13.2. On the other hand average purchase prices of continuous-feed filtering centrifuges are given in Fig. 13.3. The latter chart compares prices for screen bowls, pushers, screen scrolls, and oscillating conical baskets. On average, the screen bowl is approximately 10% higher in price than the solid bowl of the same diameter and length. This incremental cost results from the added complexity of the screen section, bowl configuration, and casing differences. The typical prices of both solid bowl and filtering centrifuges do not include the drive motor, which will add another 5–25% to the cost. The higher end of this range represents a variable-speed drive. If a variable-speed back drive is used instead of the gear unit, the additional cost is about another 10–15%, depending on capability.

The average prices of the batch centrifuge are shown in Fig. 13.4. All models include drive motor and control. The inverting filter, horizontal peeler, and advanced vertical peeler are the premium baskets, which are used for specialty chemicals and pharmaceuticals in partic-

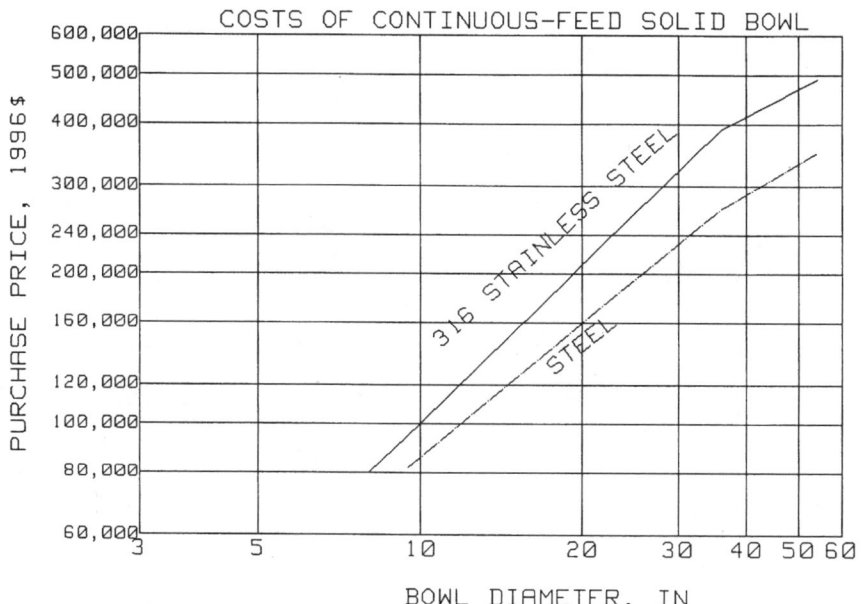

Figure 13.2 Costs of continuous-feed solid bowls.

314 Chapter Thirteen

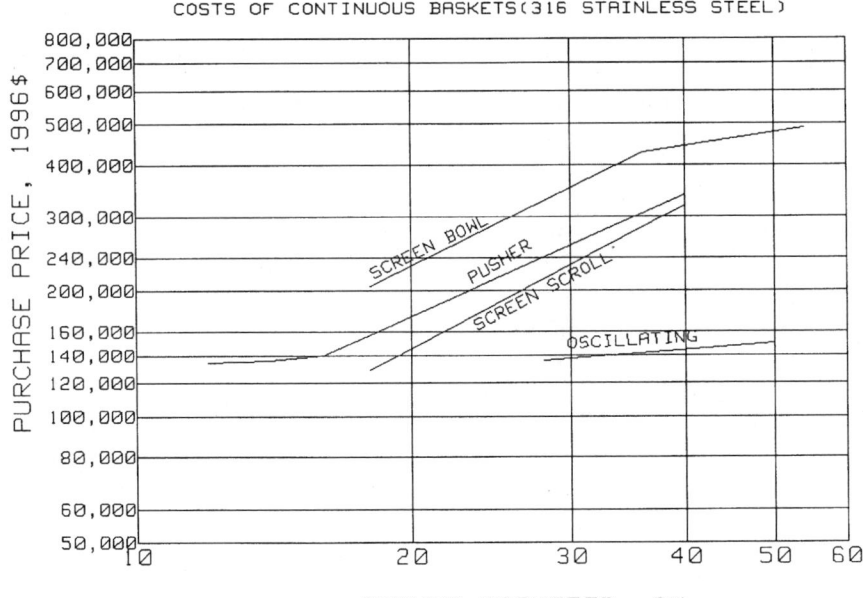

Figure 13.3 Costs of continuous filtering centrifuges (316 stainless steel).

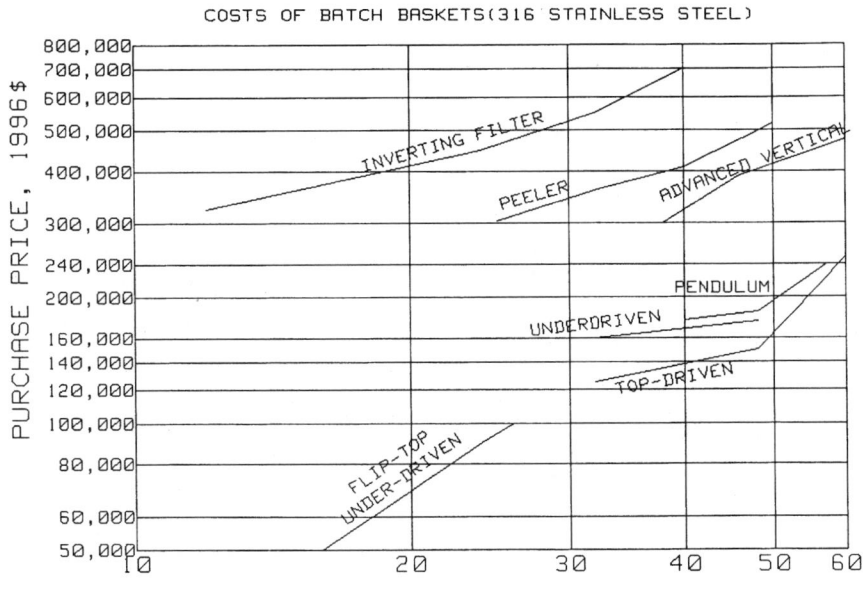

Figure 13.4 Costs of batch baskets (316 stainless steel).

ular. Versatility in control with the use of programmable logic control (PLC), personal computer (pc) interface or digital control systems (DCS), automation, and cake heel removal are the key features responsible for higher prices. The underdriven, topdriven, and pendulum baskets are less expensive with lesser features.

Installation costs

Installation costs of centrifuges vary over an extremely wide range, depending on the type of centrifuge, the area and kind of structure in which it is installed, and the details of installation. Some centrifuges, such as portable tubular and disk oil purifiers, are shipped as package units and require no foundation and a minimum of connecting piping and electrical wiring. Others, such as large batch automatic and continuous scroll-type centrifuges, may require substantial foundations and even building reinforcement, extensive interconnecting piping with certain flexibility, auxiliary feed and discharge tanks, pumps and other facilities, and elaborate electrical and process-control equipment. Minimum installation costs, covering a simple foundation and minimum piping and wiring, are about 5–10% of the purchase price for tubular and disk centrifuges; 10–25% for bottom-drive, batch automatic, and continuous scroll centrifuges; and up to 30% for top-suspended basket centrifuges. If the costs of all ancillaries—special foundations, tanks, pumps, conveyors, electrical and control equipment, and so on—are included, the installation cost may well reach one to three times the purchase price of the centrifuge itself.

Maintenance costs

Because of the care with which centrifuges are designed and built, their maintenance costs are in line with those of other slower-speed separation equipment, averaging about 5–10% per year of the purchase price for centrifuges in light to moderate duty. For centrifuges in severe service and for highly corrosive fluids, the maintenance cost may be several times this value. Maintenance costs are likely to vary from year to year, with lower costs for general maintenance and periodic large expenses for major overhaul. Centrifuges are subject to erosion from abrasive solids such as sand, minerals, and grits. When these solids are present in the feed, the centrifuge components subject to wear, such as feed and solids discharge ports, unloader knives, and helical scroll blade tips, should be protected with replaceable wear-resistant materials. Prescreening of the centrifuge feed to remove the grit should be considered as an alternative option to reduce high maintenance costs on erosion of the mechanical components. Excessive out-of-balance forces contribute substantially to maintenance requirements and should be avoided.

Operating labor

Centrifuges run the gamut from completely manual control to fully automated operation. For the former, one operator can run several centrifuges, depending on the type and application. Fully automatic centrifuges usually require little direct attention during operation.

References

1. W. W. F. Leung, "Centrifuges," in *Chemical Engineering Handbook,* Perry and Green (eds.), 7th ed. McGraw-Hill, New York, 1997.

Chapter 14

Optimization and Troubleshooting

It is important to optimize the operation of a centrifuge so as to obtain the best process performance. Centrifuges, like any other mechanical equipment in the process industry, have frequently been neglected. When the machine fails to meet process specifications and requires troubleshooting, the issue of machine optimization often resurfaces. Interestingly, both optimization and troubleshooting encompass the same operating process variables. In this chapter solid-bowl decanter optimization is discussed in detail as there are more interrelated variables, and thus more complications in optimization as compared to other centrifugal equipment. Therefore, this is a good model for demonstrating how optimization is done best. Subsequently a general approach will be discussed from the standpoint of process function. Troubleshooting will be discussed last.

Optimization of Solid-Bowl Decanter

There are a number of variables that dictate the performance of the solid-bowl decanter, including feed (volumetric) rate Q, feed solids concentration by weight W_f, polymer dosage D_p, differential speed Δ between bowl and conveyor screw, pool depth h_p, bowl speed Ω or centrifugal gravity G, and operating temperature T. The operation of the solid bowl will greatly benefit from optimizing these variables, which influence the process results, that is, centrate clarity and cake solids (or dryness). Table 14.1 summarizes the effects of these seven variables, each affecting the process based on the feed variables (volumetric feed rate, % feed solids, and % fine solids) and the desired process results (%

TABLE 14.1 Compensating Operating Variables, Process Requirements

	Feed condition and process requirement	Feed volumetric rate Q	Feed solids W_f, % w/w	Polymer dosage D_p	Δ	Pool depth h_p	G	T
Feed	Feed volumetric rate increase	NA	Decrease*	Increase to reduce turbulence and to provide constant performance	Increase to convey higher solids rate	Increase‡ to maintain centrate clarity	Increase§ to maintain centrate clarity	Increase
	Feed solids increase, % w/w	Decrease*	NA	Increase*	Increase†	Increase‡	Increase§	Increase
	Fines (−10μm solids) increase	Decrease*	Decrease	Increase*	NA	Increase‡	Increase§	Decrease
Separation quality	Centrate clarity increase	Decrease	Decrease	Increase	Increase†	Increase‡	Increase§	Decrease
	Cake solids increase	Decrease	Moderate increase	Decrease for inorganic and increase for organic	Decrease	Increase for compactible cake and decrease for incompactible cake	Increase§	Increase

*For constant performance (same centrate clarity and cake solids).
†Provided Δ is not too high.
‡Provided pool is not too deep leading to washout.
§Provided torque capacity of drive or gear box (if applicable) is not exceeded.

solids recovery %Rec and cake solids W_s). Mathematically the latter two criteria can be written as functions of these seven variables,

$$\%\text{Rec} = \text{Fn}_1(Q, W_f, D_p, h_p, \Delta, G, T) \qquad (14.1)$$

$$W_s = \text{Fn}_2(Q, W_f, D_p, h_p, \Delta, G, T) \qquad (14.2)$$

where Fn_1 and Fn_2 are two different functions defining the solids recovery %Rec and the cake solids, respectively, for a given operating condition.

Each of these two functions is represented by "abstract surfaces" in an eight-dimensional space. In more tangible terms, once a specification is made, namely, $\%\text{Rec} \geq \%\text{Rec}_{\text{spec}}$ and $W_s \geq (W_s)_{\text{spec}}$, one can solve these inequalities to meet the specifications, that is,

$$\%\text{Rec}_{\text{spec}} \leq \text{Fn}_1(Q, W_f, D_p, h_p, \Delta, G, T) \qquad (14.3)$$

$$(W_s)_{\text{spec}} \leq W_s = \text{Fn}_2(Q, W_f, D_p, h_p, \Delta, G, T) \qquad (14.4)$$

We now have two equations in seven unknowns. Obviously, the solution is not unique. Practical considerations with regard to operation and economics also enter into the problem, frequently by adding constraints such as, but not limited to,

1. Constraints on feed and polymer dosage:
   - $Q_{\text{min}} < Q < Q_{\text{max}}$ or $Q = $ constant
   - $(W_f)_{\text{min}} < W_f < (W_f)_{\text{max}}$ or $W_f = $ constant
   - $0 < D_p < (D_p)_{\text{max}}$
2. Constraints on operating conditions:
   - Δ fixed by gear-box ratio
   - G limited by maximum speed of rotating assembly

The differential speed Δ may be limited due to the availability of gear ratios, or the G values may be fixed by the available sheave sizes. In addition, for gear-box arrangements with fixed locked pinion, Δ, bowl speed, and gear ratio are all interrelated by the relationship $\Delta = \Omega_b/\text{gr}$. Given $G = \Omega_b^2 R$, therefore $G = R(\Delta \text{gr})^2$. Some solidbowl decanters have been equipped with variable-frequency drives (VFD) where the speed, and thus the G force, can change to suit the process need during operation. Also, in some applications hydraulic and electric back drives on the conveyor further allow the operator to change Δ infinitesimally.

Depending on the nature of the constraints, the operator might be able to meet these specifications only for one or more of the operating conditions, where a given condition corresponds to a specific value or range of values for (Q, W_f, D_p, h_p, Δ, G, T). Whichever is the most economical and convenient solution logically dictates the choice.

Because of fluctuations in the upstream conditions in certain applications, it becomes necessary to use some of the operating variables, such as volumetric rate, feed solids concentration, polymer dosage, Δ, pool, speed (and thus G), or operating temperature, to compensate for the change in process conditions and requirements due to changes in volumetric feed rate, solids concentration, polymer dosage, percent of fines in feed, centrate clarity, or cake dryness.

Process variables

It is beneficial to examine the effects of these variables on the process results—centrate and cake—individually.

Volumetric rate Q. As the volumetric flow rate Q increases, the retention time is reduced and there is less chance for the particles to separate. Also, it further enhances the possibility of entrainment of the already settled solids in the thicker cake bed by the fast effluent stream, which has even higher velocity at higher rate. Fine settled solids can easily be entrained and resuspended as compared to the coarser ones. A decrease in %Rec (that is, the fraction of solids recovered by centrifugation) is predicted with increasing Q (see Fig. 2.9b).

As the rate increases, the amount of solids also increases, resulting in a thicker cake. It takes longer for the liquid to drain or to express through a thicker cake bed with greater flow resistance. Any liquid moisture not yet removed from the cake results in reduced cake dryness. This effect has been demonstrated in Fig. 2.11.

Feed solids concentration W_f. When the feed solids concentration is too high, hindered settling sets in, where the separation of a particle is influenced hydrodynamically by its neighbor. Consequently the entire slurry settles at one velocity independent of the particle size. Richardson and Zaki[1] correlated the settling velocity as a function of the feed solids concentration in accordance with a power law, which can be best correlated as $V/V_{Stokes} = (1 - \phi_f)^{4.6}$, where ϕ_f is the feed solids concentration by volume (see Fig. 2.6). This would suggest that %Rec would decrease with increasing W_f. If the cake can form and consolidate readily in a solid bowl given that some of the liquid has been removed, it can dewater more effectively. The cake dryness is expected to increase with increasing feed solids concentration.

Polymer dosage D_p. For fine fluffy materials with small particles it is advantageous to agglomerate these particles into a larger entity for sedimentation, in accordance to Stokes' law, which suggests that the separation velocity increases as the second power of the particle size. Therefore it is expected that %Rec would increase with increasing polymer dosage. After a critical dosage has been reached, %Rec may stay constant, independent of the dosage. This is shown in Fig. 2.10a. The centrate would also turn foamy, suggesting that the unconsumed polymer is wasted. While increasing the polymer dose would increase cake solids for organic solids such as municipal sewage and biosolids, it has a detrimental effect on inorganic solids, where the polymer particle creates a flocculated solid structure that traps moisture and is difficult to dewater. Consequently, cake dryness generally decreases with increasing polymer dosage (see Fig. 2.10b).

Differential speed Δ. As the feed rate increases, Δ should increase accordingly. Figure 14.1 shows the reason. Too low a Δ value leads to a thick cake for the same solids throughput. As the surface of the cake is close to the centrate liquid stream, fine solids deposited on the cake surface can easily be lifted and resuspended by the fast-flowing liquid stream and carried out with the centrate to discharge. On the other hand, when the Δ value is too high, this intensifies the secondary flow

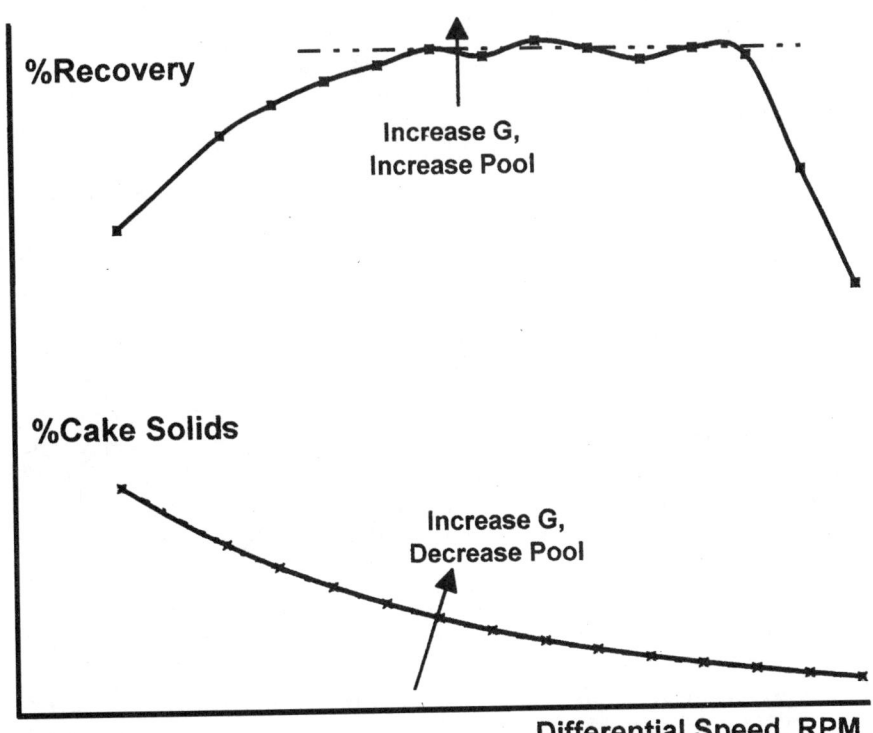

Figure 14.1 % Recovery and % cake solids versus differential speed for decanter.

with vortices formed adjacent to the conveyor blade surface, which tends to disturb the already formed cake. These two behaviors give rise to the shape of the upper operating curve in Fig. 14.1.

The cake solids decrease with increasing Δ due to a reduction in retention time on the dry beach for either incompactible and drainable cake, or compactible and nondrainable cake. The behavior is depicted by the lower curve in Fig. 14.1.

Pool depth h_p. Among all the operating variables, this should be the most readily available variable to adjust for optimal results. Unlike speed or Δ change, which may require additional hardware (such as sheaves or parts for gear ratio change) to implement the change, weir plates are more readily available, and changing the pool setting is a matter of an hour, or sometimes it can be done while the machine is in operation when an adjustable skimmer is in use. To achieve better centrate clarity, the pool should be increased moderately so as to increase the pool volume to keep the pool surface further away from the sediment and avoid entrainment of the settled solids by the incoming feed. Another reason for the pool increase is to get the pool surface closer to the feed accelerator discharge radius to reduce the radial distance between the accelerator discharge and the pool surface. A large radial drop of the feed reduces the accelerator efficiency and thus the separation efficiency. This is shown by the upper curve in Fig. 14.2.

For incompactible cake it is best to operate at a shallow pool with a long dry beach for drainage of liquid from the cake matrix back to the pool. The limitation is a higher torque and possibly higher chatter for concentrated materials that exhibit the stick-and-slip phenomenon. On the other hand, a deep pool is required for highly compactible flowable materials. Cake dewatering is solely by compaction of solids. For example, in dewatering biological sludge, the pool level should be set as high as possible to allow a thick cake layer to build up between the bowl wall and the conveyor hub. This in conjunction with a high G force generates a high compaction stress $p_s = (\rho_s - \rho_L)\varepsilon_s Gh$ acting on the cake adjacent to the bowl wall. With sufficient retention time, liquid can be squeezed or expressed out of the cake, yielding dry cake from the decanter. The two different needs, namely, a shallow pool for incompactible and drainable cake, and a deep pool for compactible and nondrainable cake, are depicted in the lower part of Fig. 14.2.

Centrifugal gravity G. Increasing speed, and thus increasing G, which is proportional to the quadratic power of speed, has a positive impact on both recovery and cake solids. This is demonstrated in Fig. 14.3. For sedimentation, higher G helps to settle out the fine particles. The recovery curve increases with increasing G. Also, increasing the G

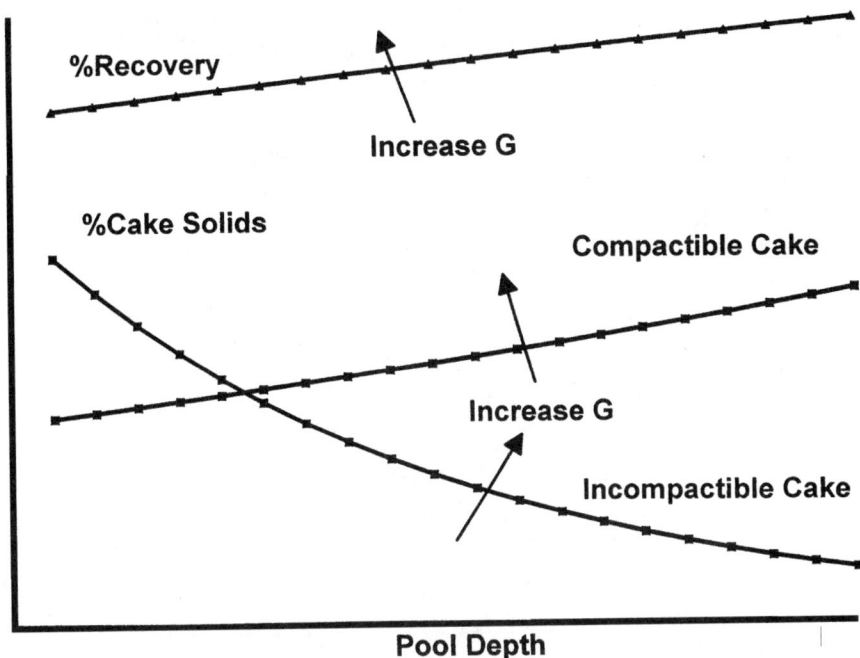

Figure 14.2 % recovery and % cake solids versus pool depth for decanter.

force helps in both types of dewatering mechanisms—compaction and drainage. However, the operator should monitor the torque of the drive and gear unit to ensure that it is within the mechanical limit during the speed increase. For abrasive material such as coal and silica, the G force is always kept below $1000g$ to avoid abrasion. In fact, the wear and tear increase as the bowl speed to the third power, or $G^{1.5}$, given that it is related to the product of G and Δ.

Figures 2.9a,b 2.10a,b, 2.11, 14.1, 14.2, and 14.3 represent the process characteristics of the solid bowl. The seven operating variables can be selected individually or in combinations for optimization to achieve the desired process results, depending on the variables to which the operator has the most access.

It is important to realize that there is conflicting influence of the different variables on the centrate clarity as well as on cake dryness. This is exemplified by Fig. 2.10a,b with regard to inorganic solids, Fig. 14.1 for Δ, and Fig. 14.2 for pool depth. Consequently optimization is in order on all the available operating variables so as to offer the best operating condition in view of meeting the process specifications. Mechanical consideration is important to ensure that the machine can operate under these conditions.

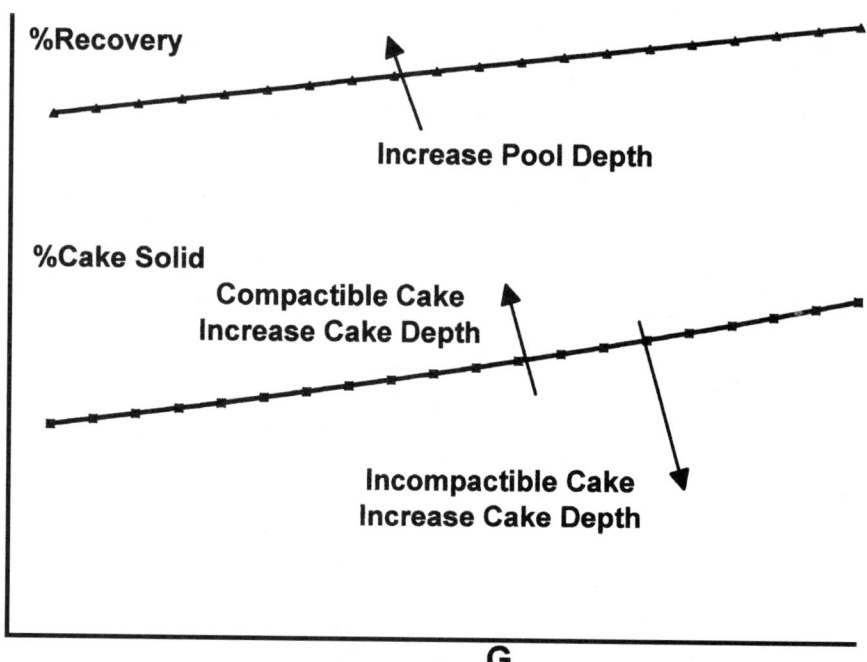

Figure 14.3 % recovery and % cake solids versus centrifugal gravity G for decanter.

Temperature T. The process temperature can be changed in certain process applications. An increasing temperature reduces the liquid viscosity and thereby increases the settling rate. This holds especially for viscous slurries and for three-phase liquid-liquid-solid separation, which more than likely also involves an emulsion coexisting between the two liquid phases. In addition, increasing the temperature helps the liquid in the cake to flow out much more readily during cake expression and drainage. Furthermore, for most liquids a small decrease of interfacial tension results from an increase in temperature. This enhances cake dewatering by reducing the capillary force, which holds the liquid in the cake.

Torque and power consideration

Monitoring torque (and chatter if applicable), vibration, and power is very important in optimization. Process benefits would not have been realized if one were limited by mechanical constraints to reach the optimal process operating condition. Torque and power are indicative of the mechanical capacity of the machine. Often torque is not measured in the decanter because of the absence of a torque device (assembly of strain gauges). However, power is determined through the current drawn, given that power is equal to the product of voltage and current.

Rate effect. As the volumetric feed rate or the feed solids increase, the amount of solids that needs to be conveyed increases too. The conveyance torque and the associated power also increase accordingly.

Polymer effect. When the polymer dosage increases, with organic solids, or biosolids, the cake dryness increases, which yields higher frictional torque on the conveyor blade.

Effect of differential speed. As Δ increases, this reduces the cake height and the conveyance torque.

Pool effect. The torque at the pinion and the spline of the gear box can be reduced significantly by increasing the pool. This reduces cake conveyance torque with the enhancement of buoyancy from the liquid pool. Immersed in the pool, the effective density of the cake is $\rho_{eff} = \rho_s - \rho_L$. The torque and power to accelerate the feed stream are also reduced with a deep pool because both quantities are proportional to the pool radius to the second power. The power provided to the conveyor is $P = T_c\Delta + T_{acc}\Omega + T_w + T_f$. The first two components, cake conveyance and feed acceleration, typically take up 60–80% of the total power at high flow rate, and both benefit from a deeper pool.

Speed effect. Torque and power increase linearly with the G force. These trends are summarized in Table 14.2. They should be considered in conjunction with the process benefits obtained when each of the seven operating variables is adjusted individually or in combination.

Design geometric variables

The design in most cases is also tailored to the process requirements. Table 14.3 summarizes some general rules based on the condition of the feed and the process specifications on centrate and cake.

TABLE 14.2 Compensating Operating Variables, Mechanical Requirements

Mechanical requirement	Feed volumetric rate Q	Feed solids W_f, % w/w	Polymer dosage D_p	Δ	Pool depth h_p	G
Torque increase	Increase	Increase	Increase for organic solids with drier cake	Decrease	Decrease	Increase
Power and current increase	Increase	Increase	Increases moderately	Decrease	Decrease	Increase

Note: Temperature has minimal effect on torque and power.

TABLE 14.3 Compensating Design Variables

Changing condition	Machine diameter	Machine length	Conveyor design	Beach angle
Feed volumetric rate increase	Increase	Increase	NA	Shallow angle
Feed solids increase, % w/w	Increase	Increase	NA	Shallow angle
Fines (-10 μm solids) increase	NA	Increase	Single lead with improved feed accelerator	Shallow angle to convey fines
Centrate clarity increase	Increase	Increase	Preferred ribbon in clarifier allowing axial flow with improved feed accelerator	NA
Cake dryness increase	Increase for granular cake	Increase for fluid cake	Single lead for fluid cake & multiple leads for granular cake	Shallow angle for fluid cake and steeper angle for granular cake

A large-diameter solid bowl is used for high volumetric or mass rate output. It also provides clear centrate and dry cake. Often the length of the machine goes hand in hand with the centrifuge diameter, given an aspect ratio of about 2–4. Since separation takes place at the pool surface, one can scale the settling capacity roughly by the ratio of the product of diameter D and length L. A more rigorous approach is to use the Leung number approach discussed in Chap. 6. As regards the conveyor design, to ensure high centrate clarity, a ribbon conveyor allowing axial flow with an improved feed accelerator system (see Chap. 11) is preferred. The adjacent leads should not be too closely spaced where secondary flow from the differential rotation of the blade and the bowl causes resuspension of sediment resulting in fines entrained in the liquid centrate.

To obtain dry cake, multiple leads are used for granular cake with a finite angle of repose to reduce the cake height, especially at the dry beach, whereas single leads should be used for paste or fluidlike cake (with negligible angle of repose).

The beach angle should be kept shallow for hard-to-convey materials, especially for feed containing a high fraction of fines. On the other hand, if the objective is cake dryness, the beach angle should be steeper for granular materials in which liquid from the cake can drain off the steep beach back to the pool; and it should be shallower for fluidlike cake in which dewatering is by compaction under a longer retention time otherwise the cake can flow back to the pool under a steep beach angle or underneath the blade tips under G force. Therefore with fluidlike cake, close clearance (<0.030 in or <0.76 mm) between the blade tips and the bowl wall is required.

General Process Functions

The optimization of a solid bowl has been detailed in the foregoing discussion. Fortunately there are less variables with other types of centrifuges, either batch or continuous types. Regardless of the centrifuge design, all the centrifugal separators are aimed at meeting one or more of the following eight process functions:

1. Two-phase solid-liquid or three-phase solid-liquid-liquid separation
2. Clarification
3. Classification
4. Degritting
5. Thickening
6. Dewatering (compaction and drainage)
7. Washing
8. Separation and repulping

Tables 14.4–14.12 provide summaries of each of these functions with respect to

- Process function and objectives
- Types of centrifuges
- Variables monitored
- Upsets or perturbations
- Compensating factors

The upsets or perturbations represent realistic factors, which tend to throw a tuned machine off. This can be, for example, a higher solids rate as the result of surge or upstream production; higher solids from the crystallizer; lower operating temperature incurring higher liquid viscosity; finer solids from the mill, crystallizer, or reactor upstream; or polymer aging. As such the pertinent operating variables are adjusted to compensate for the upsets. These will not be discussed as they are to a large extent self-explanatory in Tables 14.4–14.12.

TABLE 14.4 Three-Phase Separation

Process objectives	Both liquid phases free from impurities (other liquid and solid)
Equipment	Sedimenting centrifuges (e.g., solid bowl, disk centrifuges,...)
Variables monitored	Suspended solids and PSD of centrate and feed, light liquid, heavy liquid, cake constituents
Upsets	1. Higher volumetric feed rate 2. Higher feed solids concentration 3. Finer feed PSD 4. Ratio of light-heavy liquids in feed
Compensating factors	1. Change discharge radius of light-heavy phase 2. Increase Δ for solid bowl 3. Increase G 4. Increase dilution of feed slurry to reduce hindered settling

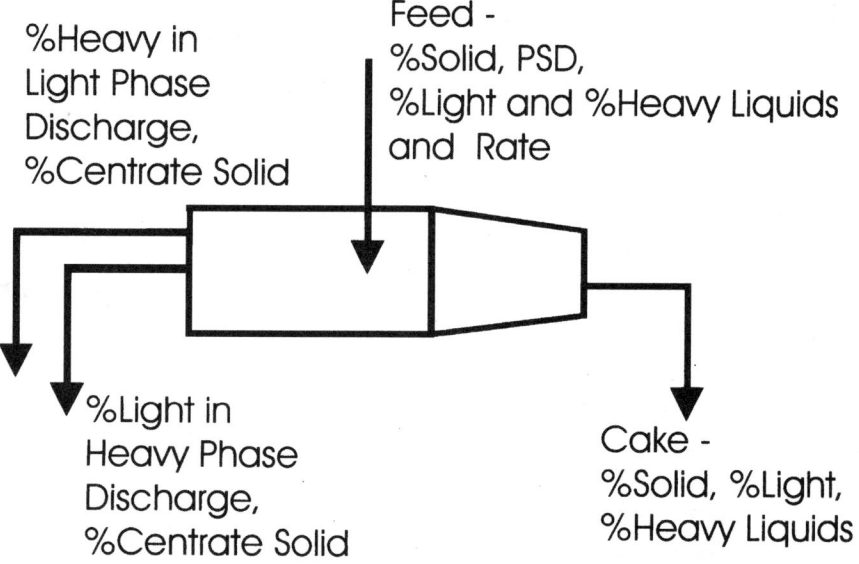

Figure 14.4 Three-phase separation.

TABLE 14.5 Two-Phase Separation and Clarification

Process objectives	Clear centrate with minimal solids
Equipment	Sedimenting centrifuges (e.g., solid bowl, screen bowl, disk centrifuges,...)
Variables monitored	Suspended solids and PSD of centrate and feed

Upsets	1. Volumetric feed rate
	2. Feed solids concentration
	3. Feed PSD
	4. Floc size degradation when polymer is used due to:
	  ■ change in nature of feed (biological)
	  ■ change in nature of polymer (organic)

Compensating factors	1. Increase pool 2. Increase Δ 3. Increase G 4. Better polymer 5. Increased polymer dosage 6. Increase dilution of feed slurry to reduce hindered settling (keeping feed solids throughput [dry basis] constant) 7. Dilute polymer concentration keeping constant or higher polymer dosage

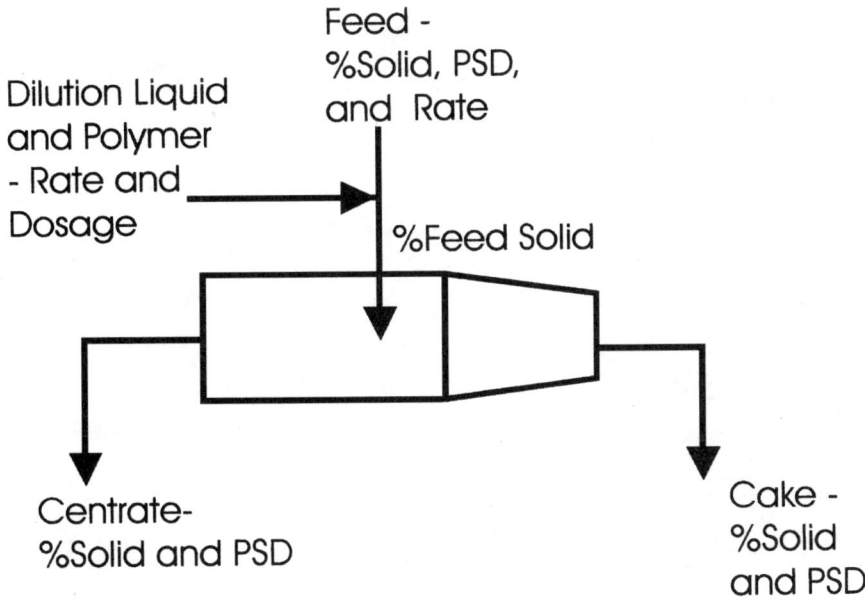

Figure 14.5 Two-phase separation and clarification.

TABLE 14.6 Classification

Process objectives	PSD requirement on centrate and high product recovery in centrate
Equipment	Sedimenting centrifuges (e.g., solid bowl, disk centrifuges, multibowl,...)
Variables monitored	Suspended solids and PSD of centrate and feed
Upsets	1. High volumetric feed rate 2. Higher feed solids concentration 3. Finer feed PSD 4. Change in nature of feed from milling or mining
Compensating factors	1. Increase pool 2. Increase Δ 3. Increase G 4. Increase dilution of feed slurry to reduce hindered settling

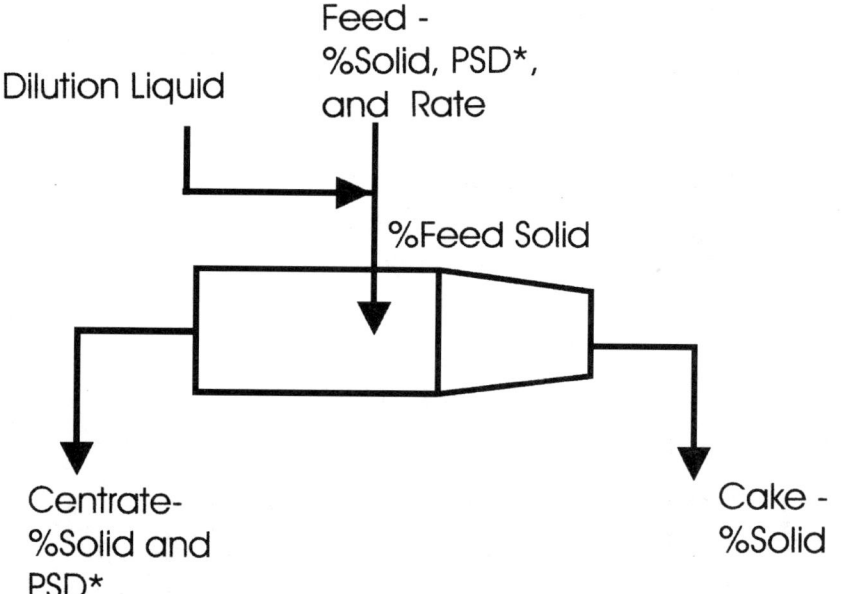

Figure 14.6 Classification.

TABLE 14.7 Degritting

Process objectives	Remove oversized particles from centrate product while keeping high recovery of fine product in centrate
Equipment	Sedimenting centrifuges (e.g., solid bowl, solid bowl basket,...)
Variables monitored	Suspended solids and PSD of centrate and feed
Upsets	1. Higher volumetric feed rate 2. Higher feed solids concentration 3. Finer feed PSD 4. Change in nature of feed from milling or mining 5. Increased level of grit in feed 6. Grit size gets finer
Compensating factors	1. Increase pool 2. Increase Δ 3. Increase G

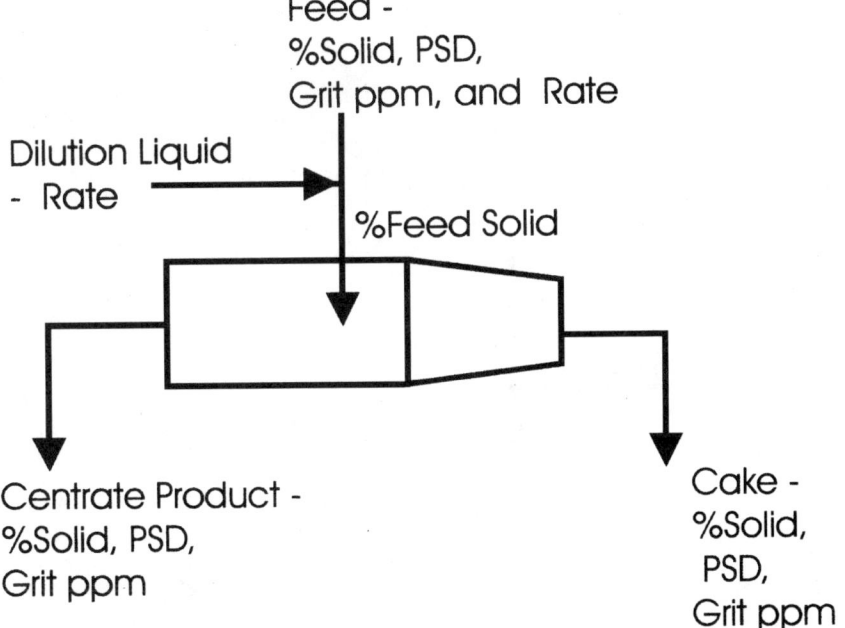

Figure 14.7 Degritting.

TABLE 14.8 Thickening

Process objectives	Thickened cake with clear centrate and minimal power
Equipment	Sedimenting centrifuges (e.g., solid bowl, disk centrifuges,...)
Variables monitored	Suspended solids and PSD of centrate and feed, cake solids, power consumed, kWh/m^3

Upsets	1. Higher volumetric feed rate
	2. Lower or higher feed solids concentration
	3. Finer feed PSD
	4. Floc size degradation when polymer is used due to
	   ■ change in nature of feed (biological)
	   ■ change in nature of polymer (organic)

Compensating factors	1. Increase pool moderately 2. Increase Δ moderately 3. Increase G 4. Better polymer 5. Increase polymer dosage 6. Increase dilution of feed slurry to reduce hindered settling 7. Reduce polymer concentration keeping dosage constant or higher

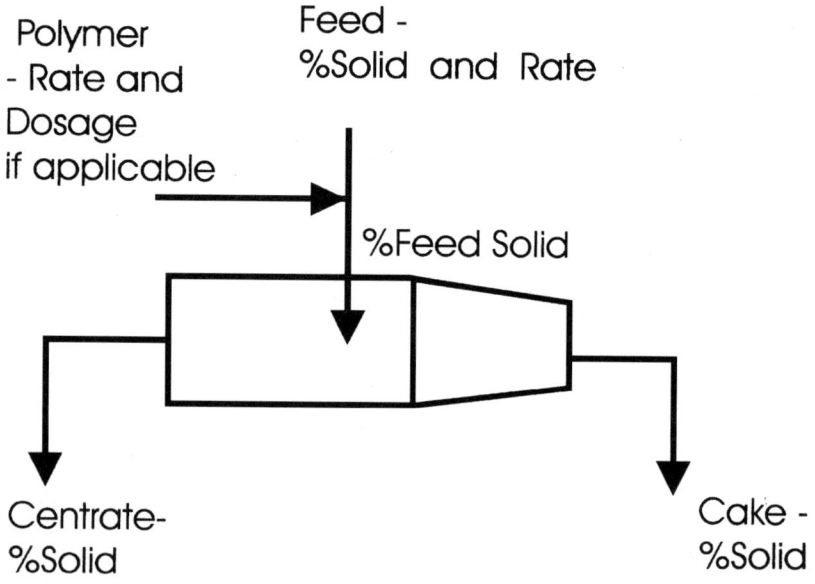

Figure 14.8 Thickening.

TABLE 14.9 Dewatering with Sedimenting Centrifuges

Process objectives	Driest cake or "handleable" (stackable) with clear centrate with minimal solids
Equipment	Sedimenting centrifuges (e.g., solid bowl, disk centrifuges, …)
Variables monitored	Suspended solids and PSD of centrate and feed, cake solids, torque, and power

Upsets	1. Higher volumetric feed rate
	2. Higher feed solids concentration
	3. Feed PSD
	4. Floc size degradation when polymer is used due to
	  ■ change in nature of feed (biological)
	  ■ change in nature of polymer (organic)

Compensating factors	1. Increase pool 2. Increase Δ 3. Increase G 4. Better polymer 5. Increase polymer dosage 6. Increase dilution of feed slurry to reduce hindered settling

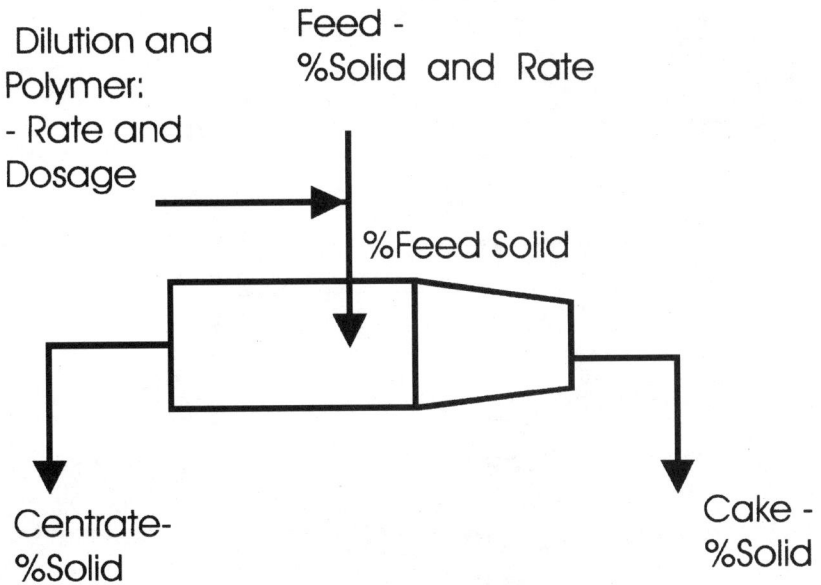

Figure 14.9 Dewatering with sedimenting centrifuge.

334 Chapter Fourteen

TABLE 14.10 Dewatering with Filtering Centrifuges

Process objectives	Driest cake with clear filtrate with minimal solids
Equipment	Filtering centrifuges (e.g., pusher, screen scroll, screen bowl, vibratory screen, basket,...)
Variables monitored	Suspended solids in cake, filtrate, feed, and PSD of feed and filtrate, torque, cake height in pusher...
Upsets	1. Higher volumetric feed rate or lower feed solids concentration both imply increasing liquid to be filtered 2. Finer feed PSD
Compensating factors	1. Decrease cake conveyance speed 2. Increase G 3. Increase dewatering time 4. Thinner cake.

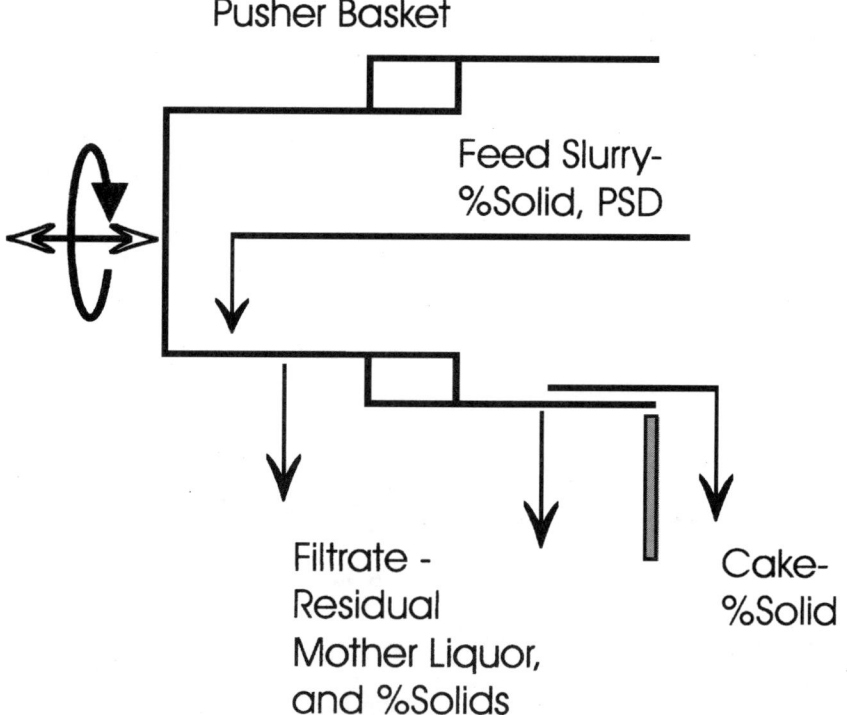

Figure 14.10 Dewatering with filtering centrifuge.

TABLE 14.11 Washing

Process objectives	Cake solids free from impurities
Equipment	Filtering centrifuges (e.g., pusher, screen bowl, screen scroll, basket,...)
Variables monitored	Feed and cake impurities, cake solids, feed rate, wash rate, wash liquid impurities,...
Upsets	1. Higher volumetric feed rate 2. Higher feed solids concentration 3. Finer feed PSD 4. Higher feed impurities in ppm
Compensating factors	1. Increase wash ratio 2. Decrease conveyance speed of cake in conjunction with reducing cake height 3. Increase G 4. Increase dewatering time

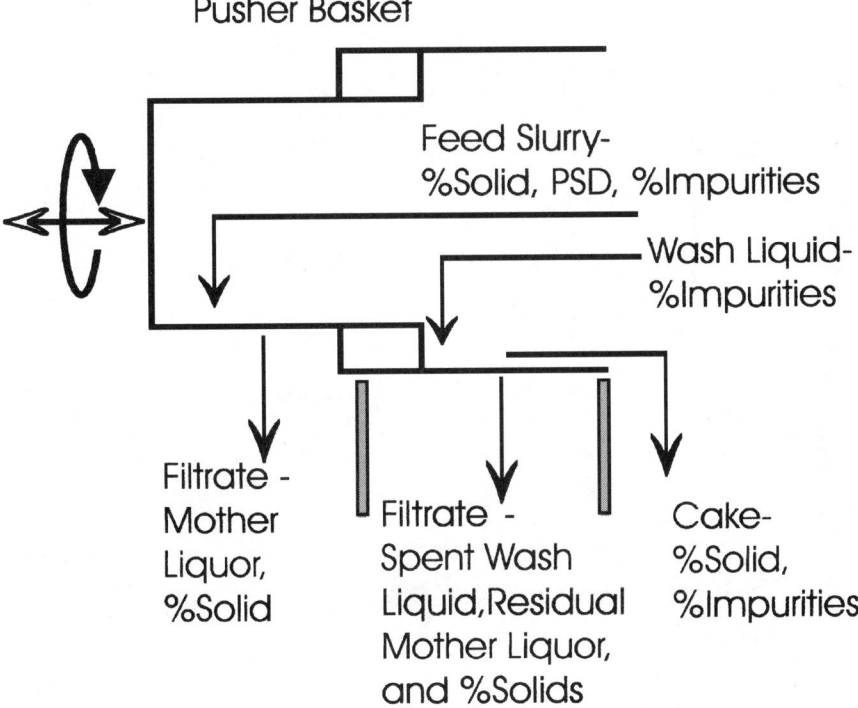

Figure 14.11 Cake washing.

TABLE 14.12 Separation and Repulping

Process objectives	Cake free from impurities
Equipment	Stages of sedimenting centrifuges (e.g., solid bowl, disk centrifuges,...)
Variables monitored	Feed, centrate, and cake solids and impurities of all stages, repulp rate, and impurities in makeup for repulp
Upsets	1. Volumetric feed rate 2. Feed solids concentration 3. Feed PSD 4. Higher impurities in original feed
Compensating factors	1. Increase rate of repulp liquid at each stage 2. Reduce impurities of makeup for repulp 3. Increase pool 4. Increase Δ 5. Increase G 6. Increase dilution of feed slurry to reduce hindered settling

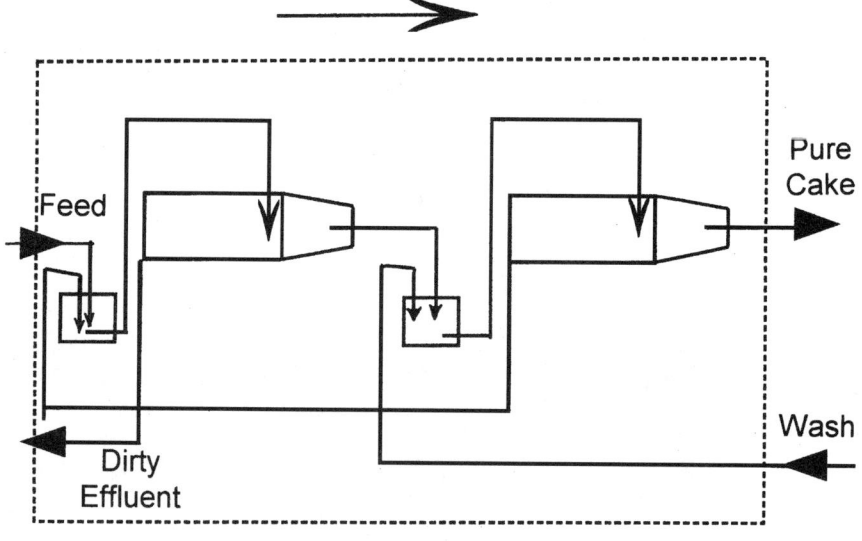

Figure 14.12 Separation and repulping.

Troubleshooting

Frequently, centrifuges are not performing, which requires operators or consultants to troubleshoot. Or the plant is expanding in throughput and the existing equipment is used to absorb this increase in capacity while holding onto the performance specifications such as cake dryness or purity. It is often a prelude to process optimization. In Appendix A a set of guidelines is included for troubleshooting pusher, solid-bowl decanter, screen bowl, disk, and basket centrifuges.

Troubleshooting and optimization are very important to keep a centrifuge operating at its optimal condition. This often extends the longevity of the machine components.

References

1. J. F. Richardson and W. N. Zaki, "The Sedimentation of a Suspension of Uniform Spheres under Conditions of Viscous Flow," *Chem. Eng. Sci.*, pp. 65–73, 1954.

Chapter 15

Kaolin Processing

Centrifuges have been used in kaolin processing since the first solid-bowl decanter made by Bird Machine Company was installed in 1935[1] to classify kaolin for the Georgia Pacific Paper Company. After having been in service for more than 60 years, the solid bowl is still in operation today as a reliable piece of equipment. Centrifuges, primarily solid-bowl and disk centrifuges, are being used to classify, degrit, and, to a very limited extent, dewater and deslime kaolin clay. The majority of the centrifuges presently in use are employed for classification and degritting.

Flow Sheet

Clay is an important ingredient in many industries. The paper industry is an example. The majority of the clay mined in the United States is located in the Georgian basin. It is divided roughly into two categories according to their location—Middle Georgian clay, which is somewhat coarser and therefore requires to be classified before downstream processing, and East Georgian clay, which is finer and is directly suitable for downstream processing after degritting.

Figure 15.1 shows a typical clay processing circuit. After the clay has been mined by dragline, it is shoveled into a blunger and then through a drag classifier before being put into storage. From storage it is taken to either a vibrating screen or a solid bowl for coarse particle classification. The product is sent to a magnetic separator (some use a superconductive electromagnet) to remove the ferrous and ferric compounds and titanium dioxide. The outflow is sent to a high-speed centrifuge, either a disk or a solid bowl, where the clay slurry undergoes finer classification. The product centrate is taken to a flocculator and bleacher, after

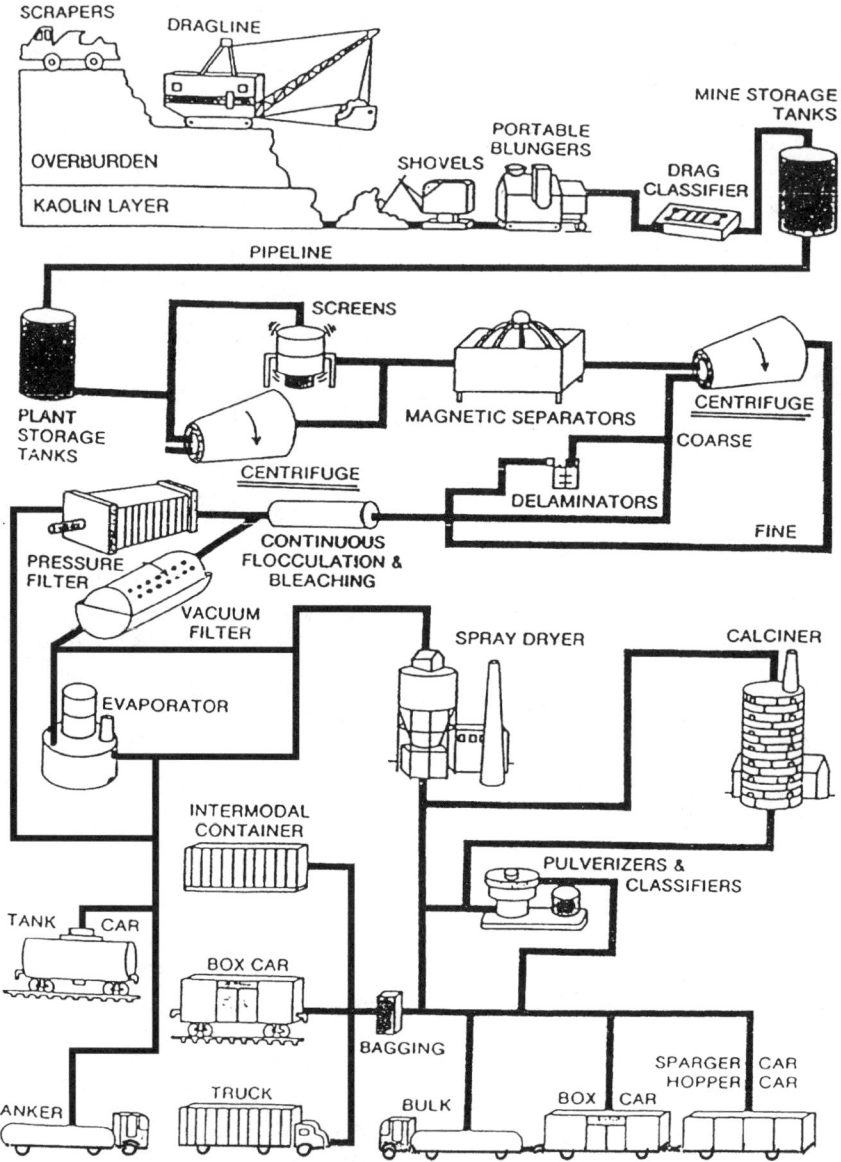

Figure 15.1 Clay processing circuit.

which the purified clay is dewatered, typically using a large vacuum drum filter that rotates at a very slow speed of 0.3–0.5 rpm, and dewaters a very thin cake, 6 mm ($\frac{1}{4}$ in). The cake discharging off the drum contains up to 65% w/w solids. It is forwarded to an evaporator that will dry off the remaining moisture prior to spray drying. Upon exiting the dryer, the clay is ready to be packed and shipped. Alternatively it is taken to a calciner and subsequently pulverized and classified to produce a very high-quality fine clay product.

Product Types

When the clay stream is presented to the centrifuge in the coarse classification, it has a typical particle-size distribution (PSD) as follows:

70–90% under 5 μm

60–80% under 2 μm

50–65% under 1 μm

20–40% under 0.5 μm

Note that the percentage is only quoted for the four sizes of 0.5, 1, 2, and 5 μm. The first three sizes, 0.5, 1, and 2 μm, are the characteristic sizes for fine kaolin processing. A clay "product" needs to meet a certain specification with regard to the cumulative percentage of these three sizes in order to make a high-quality fine coating. Therefore the product coming out of the centrifuge in the liquid centrate should be enriched with 0.5–2-μm clay particles. The particles of 2 μm and larger are removed as cake or reject by centrifugation, whereas particles less than 0.5 μm are considered slime and are not in readily useful form. Table 15.1 shows the different types of product according to the fraction under 2 μm. The higher this percentage, the finer and more valuable is the clay product. Size recovery or yield [see Eq. (2.17)] refers to the percentage of a given size, say, under 2 μm, in the feed which is recovered in the centrate. In general this should be in the range of 60–90%. For size recovery less than this range, this represents a loss of the valuable 2-μm product and is generally not acceptable.

TABLE 15.1 Clay Types

Types	Feed to centrifuge, % under 2 μm	Product leaving centrifuge, % under 2 μm	Size recovery, % under 2 μm
Filler clay	50–65	55–75	70–95
Delaminated clay	45–60	70–75	80–90
Intermediate coatings	50–72	80–83	60
Fine coatings	55–75	90–93	60
Ultrafine coatings*	72–78 <2 μm (68–78 <1 μm)	97–100 <2 μm (92–94 <1 μm)	65–70

*Typical feed to centrifuge 35–60% w/w solids, all others 25–35% w/w solids.

Classification

The size cut of each product has its unique percentage for the respective size distribution, that is, percent under 0.5, 1, or 2 μm. It represents the "signature" of the product. Finer clay has a finer size cut and thus its unique signature, with a high percentage biased toward the under (that is, less than) 2 μm.

Clay classification separates clay particles with different sizes by their respective settling velocities under G. The modified Stokes' law in the G field, as discussed in Chap. 2, reveals that the separation velocity is

$$V_s = \frac{1}{18}\left(\frac{\rho_s - \rho}{\mu_{sl}}\right) R(\Omega d)^2 H(\phi_f) \quad (15.1)$$

where ρ_s is the density of the clay, = 2.58 g/cm³, ρ the density of the water, = 1 g/cm³, μ_{sl} the viscosity of the slurry, Ω the rotational speed of the centrifuge, R the radius, d the equivalent spherical diameter of a clay particle (which is known to be more like platelet or disk shaped), $H(\phi_f)$ the hindered settling factor, and ϕ_f the solids volume fraction in the feed slurry. For concentrated slurry with $\phi_f > 0.1$, the hindered settling factor $H(\phi_f) = (1 - \phi_f)^{4.6}$ from Richardson and Zaki[2] should be used. It can be seen that the separation velocity is sensitive to both particle size and the rotational speed of the centrifuge. Given the separation velocity is tied to the distance (that is, the pool depth) and the time (the inverse of the flow rate for a given pool volume) in which the particles separate, the size cut from centrifugation should depend on the flow rate and the pool depth. In addition, the G force also influences the separation velocity and therefore should have a significant effect in classification.

If the objective is that particles larger than 2 μm settle out whereas particles less than this size overflow in the centrate stream, the centrifuge needs to be "tuned" to operate at this condition by adjusting flow rate, speed, or G force and pool depth, either individually or in combination. The differential speed Δ between the conveyor and the bowl in a solid-bowl decanter can influence the end result.

Cut size

Assume the feed PSD is given by Fig. 6.3a. If the centrifuge can be tuned such that there is a unique size cut, which is the maximum size escaping unsettled in the product effluent, or the minimum size settled in the cake reject, then it is easy to construct the PSD for product and reject (see Fig. 6.3a). In actuality, the measured PSD in the centrate would have a maximum size exceeding that of the predicted cut size. Likewise, the measured PSD in the cake would have a minimum size much less than the predicted cut size. This overlap shows that when the fine product escapes to the centrate, some of the particles settle out with the reject because of their proximity to the bowl wall and the fact that they get trapped and entrained by the sedimenting larger particles. Similarly, some oversize particles which should have settled escape the machine with the centrate. This is shown in Fig. 6.3b.

Operating variables affecting product cut

The best means of representing the product is by its signature. Figure 15.2a (compare Fig. 6.8b the dimensionless counterpart) shows a graph of centrate product % cumulative-under-size (that is, PSD) versus feed rate. The three curves correspond to the size of interest, namely, 0.5, 1, and 2 μm. The concave curves all start out parallel to the horizontal at large flow rates and increase to reach 100% at certain smaller flow rates. The 0.5-μm curve is at the bottom, followed successively by 1 and 2 μm. Under a high feed rate, that is, feed rates very much greater than the typical feed rate during operation, a condition occurs whereby the machine is being flushed—the feed enters and leaves the machine without process separation. As such, the PSD approaches that of the feed. In the example illustrated in Fig. 15.2, the

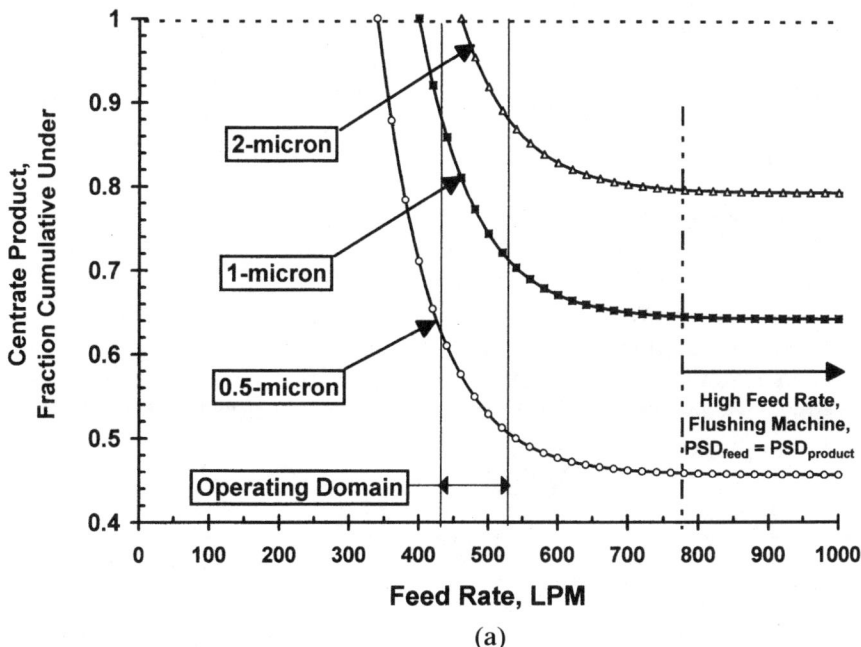

Figure 15.2 Product PSD versus feed rate or inverse speed. (a) Different product sizes for 0.5, 1, and 2 μm at a fixed G.

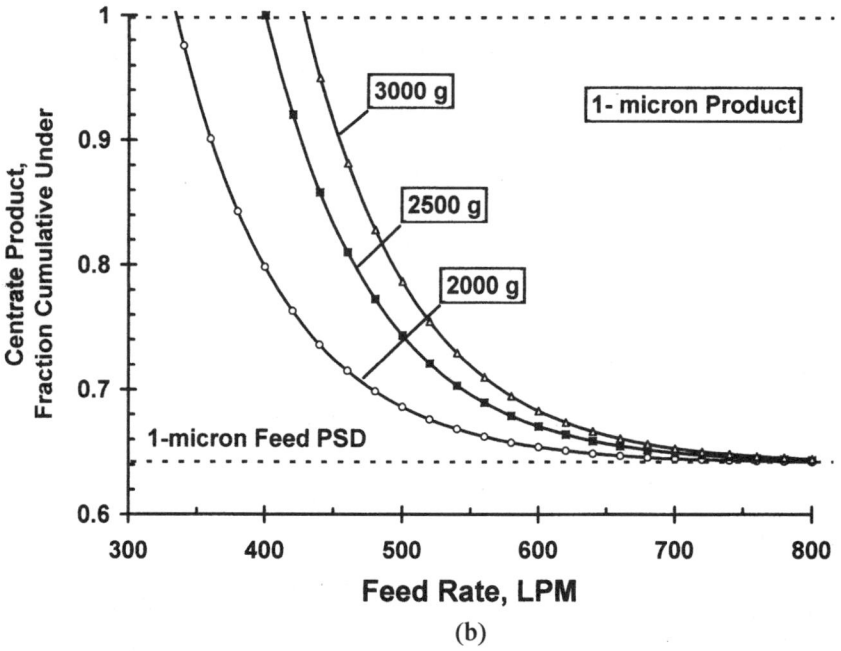

Figure 15.2 Product PSD versus feed rate or inverse speed. (*b*) Different Gs for 1 μm.

fraction cumulative under the 0.5, 1, and 2 μm is, respectively, 0.45 (45%), 0.64 (64%), and 0.78 (78%), respectively. The feed rate for this particular example shows that for rates greater than 780 L/min the machine is flushed, with the product centrate PSD being the same as that of the feed. Below a feed rate of 780 L/min, in this example, classification takes effect and the product starts to enrich in the finer particles. In the machine under consideration, for typical operating feed rates between 430 and 530 L/min the under 2-μm particles in the product can reach between 89 and 100%, under 1 μm between 72 and 88%, and under 0.5 μm between 51 and 64%. The machine can operate at a lower feed rate to yield a product enriched in the 1-μm product. In the example of Fig. 15.2*a*, when the feed rate decreases to below 400 L/min, it is possible to get nearly 100% product less than 1 μm. The operational characteristics depend on a given feed PSD.

In Fig. 15.2*b* the PSD is plotted for the 1-μm product against the feed rate for different Gs. At low G, separation is less effective, resulting in a lower fraction of the 1-μm product and a higher fraction of the coarser particles, 2–5 μm, leaving the centrate. On the other hand, better separation is realized due to the higher G force. This is in agreement with Stokes' law of sedimentation.

Figure 15.3 shows the effect of the differential speed Δ. Except at the extremes of the range, with Δ being either too small or too large,

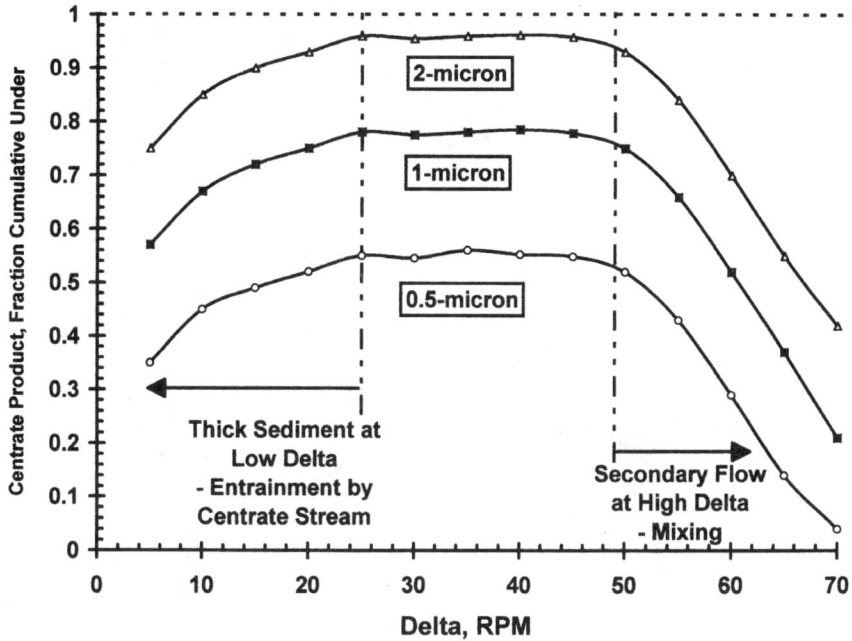

Figure 15.3 Product PSD versus differential speed for decanter.

it has minimal effect on the product quality. At very low Δ the sediment builds to a thick pile, which is likely to be entrained and resuspended by the fast centrate stream flowing by. This is particularly detrimental when operating at a shallow pool. On the other hand, a very high Δ between the conveyor and the bowl, causes a secondary flow pattern whereby fluid flows down toward the bowl wall (either along the blade surfaces or at the midpoint between adjacent blades, depending on whether the conveyor is rotating faster or slower compared to the bowl). This is balanced by another stream of equal amount flowing in the opposite direction, toward the pool surface, so that the net flow is zero. This establishes a vortical flow, which unfortunately causes resuspension of the sediment cake at the bowl wall. The strength of the vortex increases with increasing Δ. As such, it is appropriate to operate at midrange of the differential speed, and this operating Δ should increase with increasing solids throughput.

Figure 15.4 shows that increasing the pool depth is beneficial for solid bowls until the point of washout, where the pool level well exceeds the spillover diameter. At low pool depth the result is dictated by (1) the entrainment of already settled solids by the fast-moving centrate, (2) the localized disturbance from the secondary flow between adjacent blades due to the differential speed Δ, and (3) the acceleration efficiency, which gets worse as the radial drop between the feed

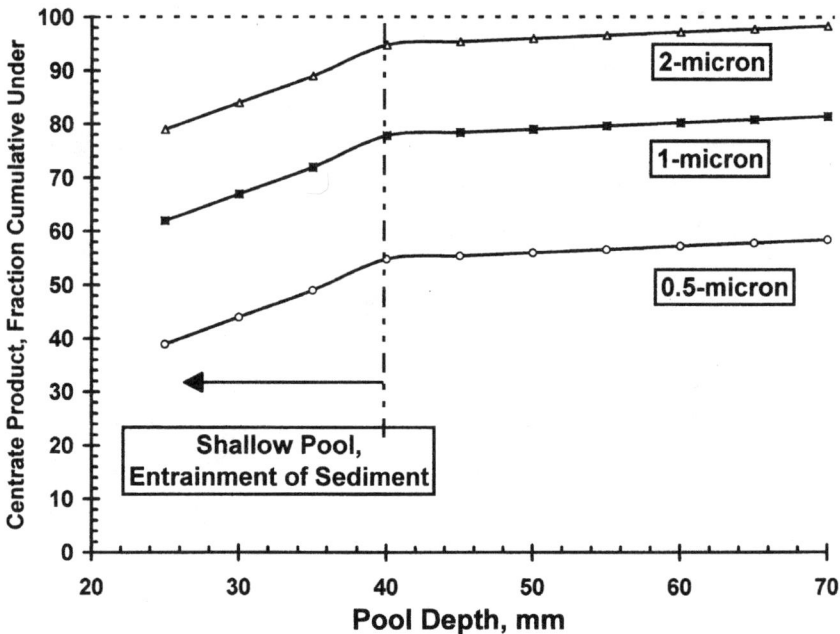

Figure 15.4 Product PSD versus pool depth for decanter.

accelerator discharge and the pool surface increases. The entrainment in (1) and the disturbance from the secondary flow in (2) require some minimal pool volume for resettling of any resuspended solids. Otherwise the solids would be carried out with the centrate. This explains the sensitive increase in the centrate product quality at low pool followed by a very gradual increase at deeper pool. Another complication is that at large radius (that is, shallower pool) the G force driving the separation should be greater provided it has acquired a solid-body rotation through proper acceleration (see Chap. 11).

The feed-rate and G force effects discussed in conjunction with Fig. 15.2 pertain to both the decanter and the disk, whereas the pool effect discussed in conjunction with Fig. 15.3 pertains to the decanter alone. Figure 15.5 applies to disks used in the fine classification circuit (see Fig. 15.1), where the centrate product PSD is plotted against the sludge depth inside the disk. Again, there is minimal influence of the sludge depth, except at very deep sludge depths, where there is entrainment of the oversized solids into the disk stack. The sludge depth is controlled by the rate of recycling a portion of the cake reject, typically at 50–54% w/w solids consistency, back to the solids holding area, bypassing the disk stack. The return tubes need to be wide open and frequently cleaned to avoid solids cloggage. This relieves the loading of the disk stack by the second pass of the feed at a much higher feed solids concentration. A high recycle rate leads to reduced cake output

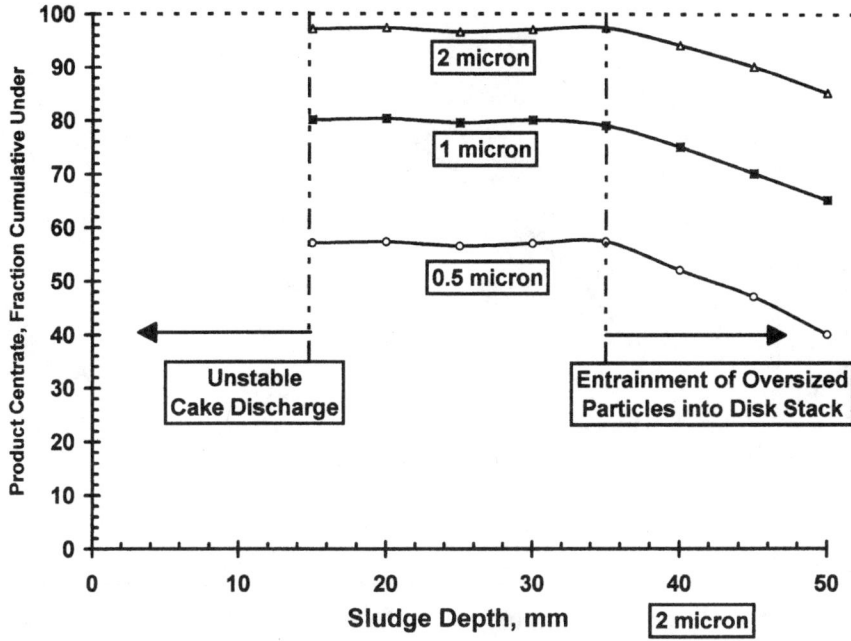

Figure 15.5 Product PSD versus sludge depth for disk centrifuge.

with a deep sludge blanket building in the disk nozzle area, highly concentrated cake, poorer cut, and loss in recovery, all attributed to the cake interfering with the disk stack used exclusively for separation and classification. The recycle rate needs to be adjusted appropriately, and product effluent with under 2-μm particles can reach 95–97%.

Design variables affecting product cut

The clarifier length and improved accelerator efficiency for solid bowls as well as the number of disks used in disk centrifuges all affect performance in a somewhat similar manner. When the clarifier length is too short, or the accelerator efficiency is too low, or disks are inadequate, the separation capacity is greatly reduced, resulting in flushing conditions with the product PSD not too different from that of the feed. For longer clarifier lengths (decanters), higher accelerator efficiency, elevated operating temperatures, and increased number of disks (disk centrifuge), separation becomes more effective, resulting in an enrichment of finer particles and hence superior product quality.

Product recovery

The product recovery in the centrate increases with increasing rate or decreasing G, decreasing pool depth, decreasing clarifier length, de-

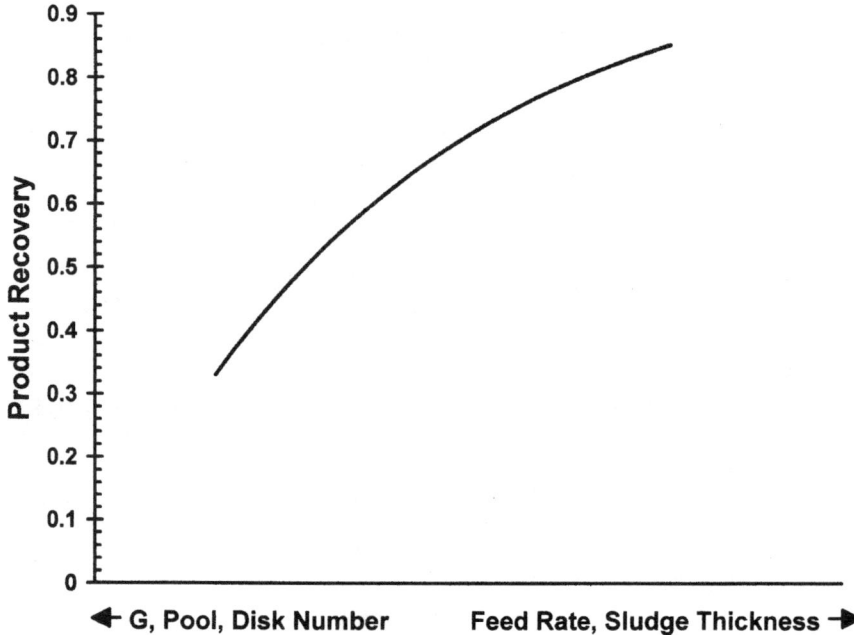

Figure 15.6 Product recovery versus feed various variables.

creasing number of disks, and decreasing accelerator efficiency (Fig. 15.6). All these variables tend to suppress separation and enhance flushing of the centrifuge. When they are adjusted properly, improved product recovery should attain close to 60–80%. Higher recovery, above 80–90%, might result in the recovery of the coarser solids, thus adversely impacting the quality of the product.

Feed condition

As the feed solids become more dilute, sedimentation is more effective due to reduced hindered settling, and the product size cut is improved compared to the case with more concentrated feed slurry. The performance curves of product PSD shown in Fig. 15.2 shift upward for the same feed rate and operating speed. Also, as the feed PSD becomes finer, the performance of the machine improves with respect to size cut and recovery, resulting in higher performance curves.

Example. A pilot-size machine is used to classify clay. The feed slurry has 27% w/w solids with 78% < 1 µm. The 1-µm product from the centrifuge needs to concentrate to 90%. The G force and feed-rate effects are explored. Figure 15.7a shows the 1-µm PSD from the product plotted against feed rate in the range of 110–230 L/min (29–61 gal/min). The pilot centrifuge was tested at 2000, 2500, and 3000g. As shown,

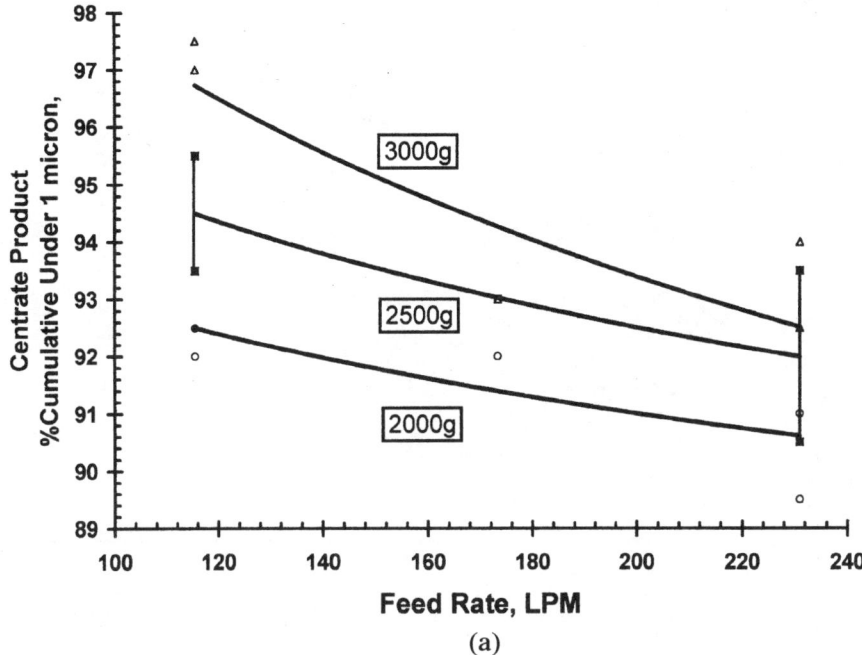

Figure 15.7 1-μm product test results. (a) Product PSD versus flow rate.

the % < 1 μm is between 91 and 97% for the range of flow rates tested. Higher G and lower feed rate result in a higher fraction, as expected. The product recovery is given in Fig. 15.7b. A higher feed rate exaggerates the flushing effect, resulting in higher recovery of fines in the centrate. On the other hand, higher G results in settling of the fine particles and therefore lower product recovery in the centrate.

Degritting

For the degritting process, the feed to the decanter centrifuge contains a bimodal distribution which contains 99.9% clay particles in the product range, that is, less than 2–5 μm, whereas a small percentage, say 0.1% or 1000 ppm, has oversized particles, that is larger than 45 μm (325 mesh) together with a trace amount of mica and other foreign particles from mining. This is similar to Fig. 6.9. The task for the solid-bowl centrifuge is to remove by centrifugal sedimentation these oversized and foreign particles in the reject cake stream from the centrate product.

Two key objectives are noted for degritting. (1) The product grit level should be low, say, less than 10–20 ppm, compared to typically 1000–2000 ppm of grit in the feed. (2) The size recovery of the under

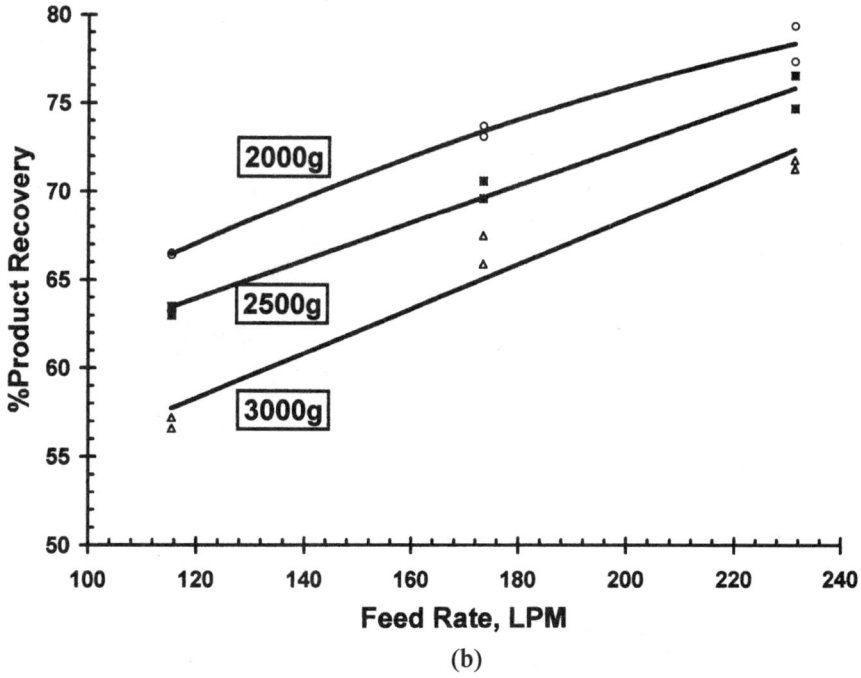

Figure 15.7 1-μm product test results. (b) Product recovery versus feed rate.

2-μm product should be 90% or higher to avoid loss of valuable product during degritting. The operating variables and design parameters for tuning the solid bowl to meet a set of objectives are:

1. Operating
   a. Feed rate, m³/h or gal/min
   b. Centrifugal gravity G
   c. Pool depth h_p
   d. Differential speed Δ
2. Design
   a. Clarifier length L_c
   b. Accelerator efficiency η_a

The clarifier length is a measure of the machine size. Figure 15.8a shows the grit concentration measured in ppm in the centrate product plotted against feed rate for a fixed operating G, h_p, Δ, and a given design with L_c and η_a fixed. As the feed rate increases, the retention time of the feed inside the solid bowl is reduced. This drives up the unsettled grit level in the product centrate. By increasing G and h_p, and with a longer clarifier of length L_c and an improved feed accelerator design, having higher efficiency η_a, the grit concentration decreases for a given feed rate in response to these improvements. The amount

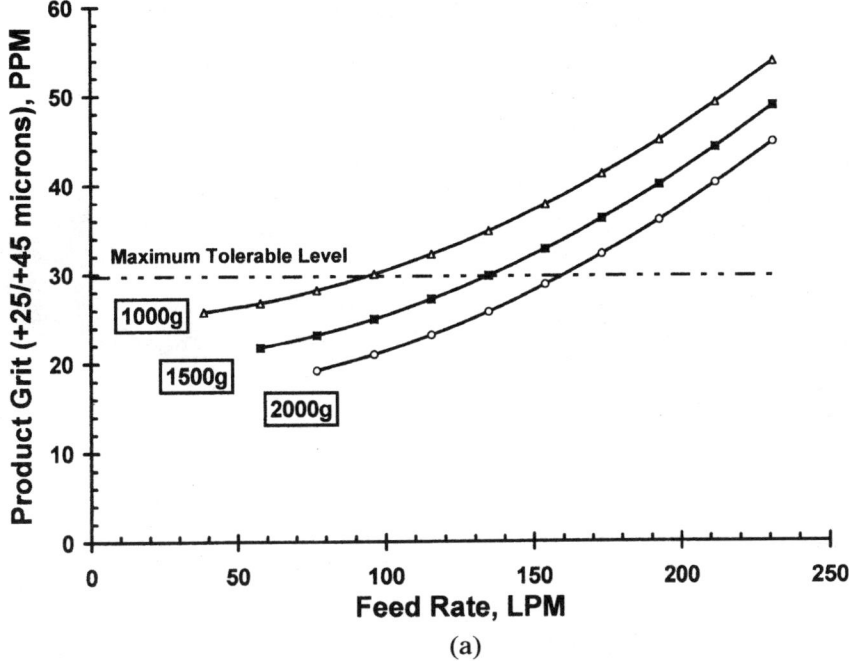

Figure 15.8 Degritting. (a) Product grit versus feed rate.

of decrease in grit level depends on the effectiveness of these variables acting individually or in combination. Therefore the feed rate does not need to be reduced in order to meet a certain grit specification, given other compensating adjustments can be made.

Another measure of the performance is the product recovery. For a fixed condition, as the feed rate increases, so does the recovery due to elevated levels of unsettled grit and fine product being carried out in the centrate stream. This is illustrated in Fig. 15.8b. As improvements are made on grit reduction by increasing G and h_p, and with a longer clarifier and improved feed accelerator design, the product recovery drops because more product settles out. This works opposite to grit removal. Given these two contrasting behaviors, it is best to adopt a high Δ and shallow pool h_p to keep the fine particles in suspension, while a modest G level is used to settle out the grit. In search of the optimal operating condition, a few trials are in order to generate the data shown in Fig. 15.8. Both plots should be examined before determining the optimal operating G, pool, and Δ values.

In all the preceding an increase in slurry viscosity has a negative effect on sedimentation but a positive effect on size recovery. The compensating factors G, h_p, and Δ together with the design variables need to be selected to offset this effect.

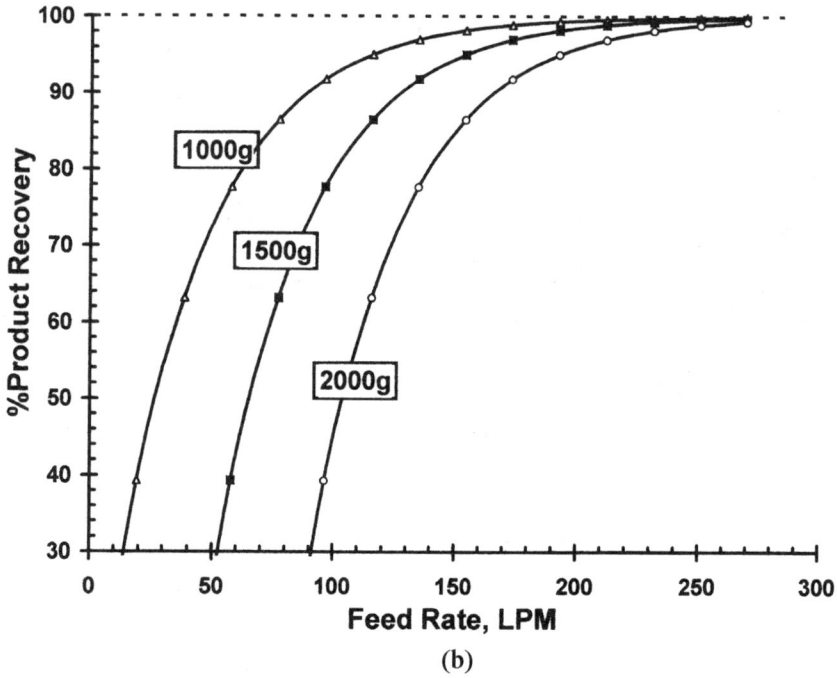

Figure 15.8 Degritting. (b) Product recovery versus feed rate.

Quantity versus Quality

In the foregoing discussion we have seen the behavior of clay classification, specifically where product % cumulative % under-size, grit level, and % product recovery are plotted against feed rate with the G force as a parameter. Similarly, brightness can be plotted against rate with the G force as a parameter. Interestingly, the inverse relationship between quality and quantity is apparent.

Cake Discharge

Up to now we have only focused on classification and degritting in the clarifier section of the solid bowl or the disks. This is under the assumption that what is presented to the beach of the solid bowl or the nozzle of the disk will be discharged. In fact, the mechanism of cake discharge, especially for kaolin cake, is quite interesting yet complicated.

Clay particles are platelet shaped and swell upon being wetted with water. The surface of the clay particle is highly negatively charged. When a thickened material is sheared, the platelets line up into a stiffer structure, resisting further shear stress. The increase in effective viscosity with shear stress is termed shear thickening. This generates higher torque on cake transport.

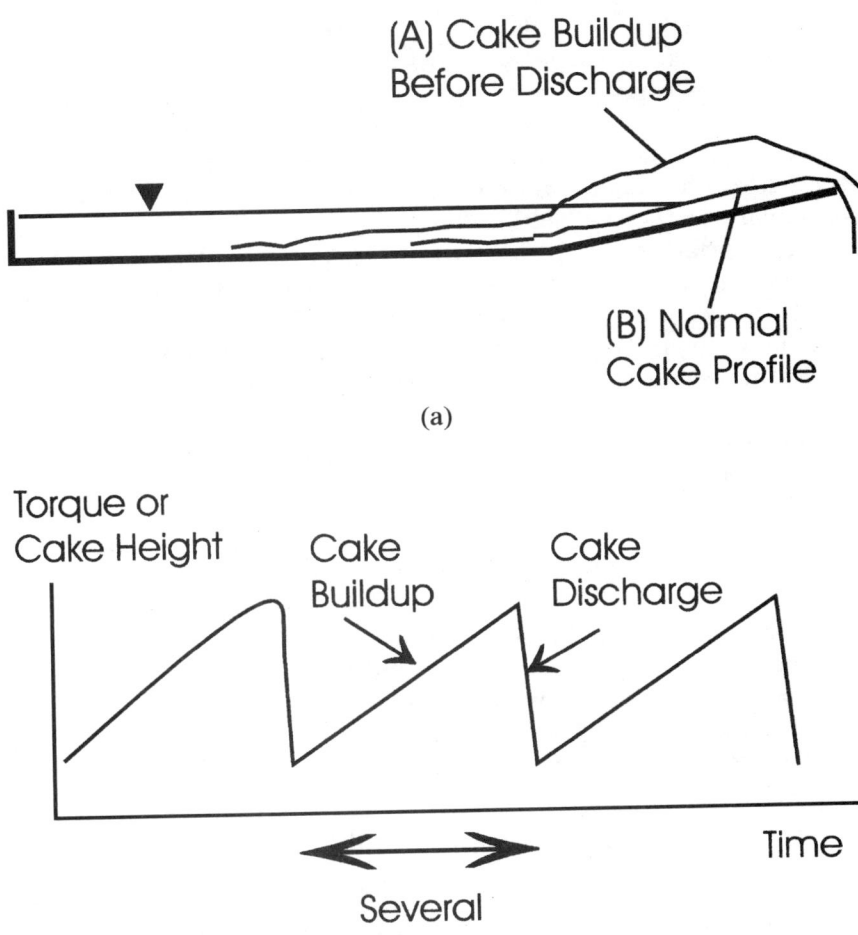

Figure 15.9 Solid-bowl beach discharge. (a) Cake profile. (b) Torque trace with time.

Solid-bowl beach discharge

Figure 15.9a shows a cross section of the solid bowl through the axis of the machine. The conveyor details are deliberately omitted for clarity. As the pool level is set below spillover at the beach, this allows a dry beach for further dewatering. When the cake is below the pool, its effective weight is decreased by the liquid pool buoyancy. After being conveyed out of the pool onto the dry beach, the cake experiences its full weight in the G field. Note that the cake needs to be lifted vertically from the inner bowl wall to the spillover discharge point against the G force. The vertical lift of the cake can be split up into the portion immersed in the pool with effective weight $(\rho_{cake} - \rho)G$ and the remaining portion outside the pool with effective weight $\rho_{cake}G$. If $\rho_{cake} = 1.51$ g/cm^3 (assuming a cake with 55% w/w clay solid at a solids density of $\rho_s = 2.58$ g/cm^3) and $\rho = 1$ g/cm^3, then the effective weight is $0.51G$ below the pool versus $1.51G$ above the pool. The change in resistance from below pool to above pool is as high as threefold. This often causes nonconveyance, especially at the liquid-air interface. Any unconveyed cake solids tend to accumulate on the dry beach. After the cake builds up to a deep pile, with the surface reaching or above the spillover lip ("negative" climb angle in the latter), it falls out under centrifugal gravity and discharges as a block (see the upper cake profile in Fig. 15.9a). The cycle repeats with the cake starting to build up on the dry beach. Each discharge cycle is on the order of several minutes. The torque signature, which is a good representation of the cake thickness inside the machine, reflects this scenario, as evident in Fig. 15.9b. The torque trace is cyclic. In each cycle the torque rises steeply for a few minutes in response to the cake solids gradually building up at the dry beach and subsequently drops precipitously as the cake is unloaded. This intermittent discharge behavior is eased out as the pool is raised. In this case the distance of lifting the cake under $1.51G$ in the dry beach to the spill point is reduced significantly. Concurrently the torque spike is also reduced.

By installing a baffle such as an annular dip weir, as shown in Fig. 4.17a, near the junction between the cylinder and the conical section, the pool level can be set higher compared to the spillover, which further facilitates cake discharge. There is a liquid head difference across the baffle with the upstream (toward the centrate discharge) at a higher level. This difference in liquid level across the dip weir is responsible for driving the flow from the upstream clarifier to the downstream beach. It depends on the restriction, the flow rate, and the rheological properties of the cake. This liquid head can amount to as much as 3–6 mm (0.125–0.25 in) or more. Dip weirs are typically set at a clearance of 13 to 25 mm (0.5 to 1 in) from the bowl wall or bowl strips, depending on whether the cake is more solidlike or fluidlike. A more open gap is used with the solidlike cake, whereas a tighter gap is used with the fluid cake. The torque peaks are further reduced due to this "hydraulic assist" mechanism.

Some additional advantage of the dip weir is that it also serves as an active barrier to stop fine products from entraining into the beach and being conveyed out with the cake. Therefore it improves the recovery of the fines. For example, with an 18-mm (0.75-in) gap the size recovery of 1 μm is typically 88–90%, and with an even tighter gap of 13 mm (0.5 in) the size recovery of the 1-μm product rises to 93–95%.

Figure 15.10 plots the cake throughput against the differential speed Δ. Two cases are shown. In one the pool is low (dashed line) and the cake is conveyed out by the differential speed. In the other case

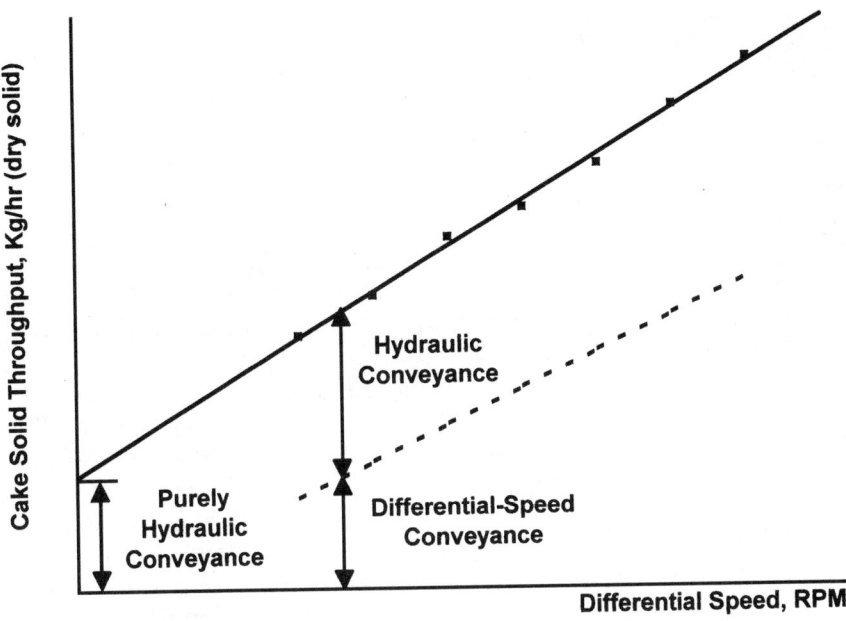

Figure 15.10 Cake solids throughput versus differential speed with hydraulic convergence.

the pool is set above the spillover, with the dip weir or cake baffle using hydraulic assist in addition to mechanical conveyance. In each case the solids rate is linearly proportional to Δ. It is clear that the solids rate is much higher with use of the hydraulic assist. This is due to the fact that (1) buoyancy is used to reduce the effective density of the cake, hence, the lower torque, and (2) hydraulic force due to the liquid head across the dip weir or cake baffle provides additional driving force to discharge the cake. This reduces the cyclic loading of the gear unit because of torque spikes and further reduces the chance of mechanical fatigue. This is especially important for machines designed for high throughput, typically 20–30 ton/h (dry solids) per machine for a 600-mm (24-in)-diameter decanter.

Disk nozzle discharge

The discharge rate of clay through continuous nozzles is determined by the solids concentration of the feed, underflow and overflow as well as the feed rate (Eq. 4.5).

Suppose the nozzles are worn out over a period of time to a larger diameter resulting in higher cake flow rate. For disk centrifuge equipped with cake recycle feature the recycle rate needs to be increased dramatically to increase the resistance to flow at the nozzles. This compensates in part for the high-volume flow rate at the nozzles.

It is perceived that the fines would be lost through the nozzles, resulting in poor recovery. Consequently, the nozzle opening is critical to the operation, and it also becomes a major operating expense for the plant.

Nozzle decanter

Some solid-bowl decanters are equipped with nozzles spaced evenly around the circumference to discharge cake solids. The nozzles are located at the intersection of the beach and the cylindrical clarifier.[3] Again, a problem similar to that in the nozzle disk centrifuge persists where the replacement of nozzles is a major expense, especially with materials that have high grit levels. Another arrangement is possible, where both the nozzles and the beach are used to discharge cake. The nozzles are used to take away materials that have difficulty negotiating up the beach. This reduces the torque requirement of the conveyance since only the materials that can be conveyed up the slope are taken out of the beach. Another innovative advancement[4] which adopts a cocurrent design with feed introduced at one end of the machine and product taken out the other end. The full length of the machine is fully utilized for classification. The cake reject is discharged through the nozzles at the beach-cylinder junction whereas the centrate product is allowed to overflow at the conical beach. Higher capacity, better size cut, and brighter products have been reported even at a slightly lower G.

Dewatering

Dewatering of clay by centrifuges is not common. This step is accomplished primarily by rotary vacuum drum filters with precoating to stop fines from bleeding through the cloth. The filter rotates very slowly, producing very low throughput per unit filter area. Cake solids of 60–65% can be obtained. The technology of using relatively low-speed centrifuges was attempted in the 1970s without success. The primarily reason is low solids recovery, that is, the equipment is not able to capture the fine products of 1–2 μm.

Recently the technology has advanced considerably, which makes dewatering of clay by solid bowl once more viable. To improve solids recovery, a decanter with a long clarifier equipped with a ribbon blade supported by axial vanes (allowing axial flow without disturbing the cake flow along the helix) can be provided. Most important, an improved feed accelerator system permits prompt acceleration to speed of the incoming slurry containing 1–2 μm clay particles, which avoids turbulence at the pool and resuspension of the sediment as obtained from conventional underaccelerated feed. In addition, the latter further takes away the useful clarification length. A dip weir or better still an adjustable cake baffle[5] in the proximity of the beach-clarifier intersection is required to assist in cake discharge. The differential speed needs to be small to increase the retention time for dewatering, and the G force needs to be high, in the range of 3000–4000g. Cake solids of 60–65% can be obtained with a high solids recovery. This should offer an alternative technology for kaolin dewatering.

Desliming

The object is to remove particles less than 0.5 μm. Typically a disk centrifuge with nozzle discharge is used. Feed entering the disk is less then 2 μm in size. The object is to capture particles between 0.5 to 2 μm in the cake, leaving the 0.5 μm in the overflow. Desliming is not a very common operation as it represents another step in processing. Obviously, this is targeted at a more expensive coating product with a narrower range of PSD.

In this chapter the use of continuous industrial sedimenting centrifuges has been exemplified by a number of applications pertaining to classification, degritting, dewatering, and desliming of clay. The same methodology can be applied to classification and degritting of other fine valuable mineral slurries, such as calcium carbonate. A useful analysis can be made on test results, and in subsequent sizing the dimensionless Leung number can be used, as introduced in Chap. 6.

References

1. Bird decanter brochure
2. J. F. Richardson and W. N. Zaki, "The Sedimentation of a Suspension of Uniform Spheres under Conditions of Viscous Flow," *Chem. Eng. Sci.* 3 (1954), S. 65–73.
3. E. Retter, "Solid Bowl Worm Centrifuge with Improved Discharge Openings," U.S. patent 5,252,209, Oct. 12, 1993.
4. F. Muller, J. Kompe, R. Kluge, "Centrifugal Classification with the Centrisizer in a Range of 1 Micron," Sonderdruck, Aufbereitungs Technik, Mineral Processing, 1996.
5 W. W. F. Leung, A. Shapiro, and R. Yarnell, "Decanter Centrifuge with Adjustable Grate Control," U.S. patent 5,643,169, July 1, 1997.

Chapter

16

Dewatering of Compactible Solids

In 1986 high-solids decanter centrifuges were introduced in the waste application market for dewatering municipal and industrial sewage sludges. The new centrifuge can dewater sludges with more than 5–10% higher cake solids than conventional decanter centrifuge design and operation. It immediately gained wide acceptance in the industry as the standard practice.[1–5] The technology can also be used for nonwaste applications such as biosolids dewatering, as found in food-processing and corn-milling operations. What is common with all these applications is that the cake is compactible but not drainable. It follows that solids dewatering is done by compacting and shearing the cake under very high centrifugal body forces as it is being conveyed in the machine, and allowing sufficient time for the liquid to express out of the cake. In this chapter the discussion pertains to waste sludge. However, a similar methodology can also be used for other compactible solids.

Characteristics of High-Solids Decanter

There are four common characteristics of a high-solids decanter: (1) dry cake, (2) clear centrate, (3) operation control, and (4) effective polymer introduction (Table 16.1).

Dry cake

Compaction. An effective means of dewatering compactible and nondrainable cake is to subject it to high compaction stress as well as moderate to high shear stress. This allows the moisture in the void

TABLE 16.1 Characteristics of High-Solids Decanter

1. Dry cake
   High G
   Deep cake bed
   Higher polymer dosage
   Lower feed rate
   Low differential speed
   Long clarifier
2. Clear centrate
   High G
   Higher polymer dosage
   Lower feed rate
   Good feed acceleration
   Long clarifier
3. Stability of operation
   Effective torque control
   Back drive
4. Effective polymer introduction to feed slurry

spaces of the cake matrix to percolate upward away from the bowl wall, toward the upper cake layer, which is less compact and consequently more pervious to flow. The compaction stress is determined from

$$p_s(R) = p_{s0} + (\rho_s - \rho)\Omega^2 \int_{R_c}^{R} \varepsilon_s(R) R \, dR \tag{16.1}$$

where p_{s0} is the solids stress at the cake surface, which is zero; R_c is the radius at the cake surface, Ω the rotational speed, ε_s the solids volume fraction in the cake, and ρ_s and ρ are the densities of the solids and the liquid, respectively. For incompactible cake ε_s = constant and given $p_{s0} = 0$, $p_s(R) = \frac{1}{2}(\rho_s - \rho)\varepsilon_s\Omega^2(R^2 - R_c^2)$. For both compactible and incompactible cake, with small cake thickness (where the cylindrical geometry can be approximated by a planar geometry), a simple yet familiar form of the stress equation becomes

$$p_s = (\rho_s - \rho)\varepsilon_s Gh \tag{16.2}$$

where $G = \Omega^2 R$ and the cake thickness $h \, (= R - R_c) \ll R_c$. The lab test described in the following provides an interesting support of the cake compaction theory.

A series of laboratory compaction tests was conducted to confirm the cake compaction and dewatering mechanism. A fresh sample of 50% primary sludge mixed with 50% waste-activated sludge obtained from a municipal wastewater treatment facility was thickened to a 9% w/w sludge. The prethickened sludge was subject to centrifugal compaction in a large spin tube, 33 mm (1.3 in) in diameter by 157 mm (6.2 in) long, with the tube outer radius at 208 mm (8.2 in) when it was in a horizontal position under full rotation. The test was carried out with different sludge bed thicknesses h = 13–140 mm (0.5–5.6 in) and G

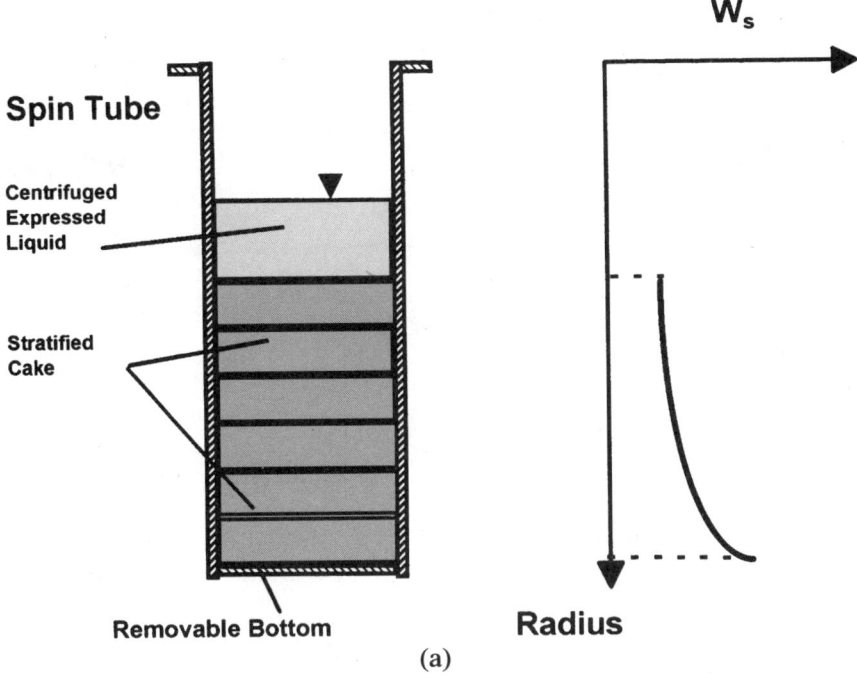

Figure 16.1 (a) Stratification of cake solids in large spin tube.

forces at 1000, 2000, and 3000g to generate a wide range of solids compaction stresses p_s. After the cake has been exposed to the G force for a long time, the compacted cake reaches an equilibrium dryness with all the free liquor expressed out of the cake. This is shown by the schematic in Fig. 16.1a. The liquor above the cake surface was decanted off and the cake was pushed out of the tube by a plunger after the bottom cap is removed. The thick cake was dissected into individual thin segments, each 13 mm (0.5 in) thick, starting from the cake surface to the cake bottom, to analyze the solids content in order to establish a solid profile of the centrifuged cake in the tube (see Fig. 16.1a). The ash content of the cake varied between 12 and 18% with pH = 5.2, and the solids density was determined to be 1.2 g/cm³.

Figure 16.1b shows the percent by weight of solids plotted against the radial location. A small dosage of polymer of 4.0 kg/t (7.9 lb/ton) was used in the test. The curves represent the measured profile obtained from the thick cake (66–94 mm) runs, whereas the data symbols represent the thin cake (12–26 mm) runs. For each cake thickness and rotational speed, the tube was tested with and without lubricant (Teflon spray coat) coating the inner surface of the tube. The result was somewhat sensitive to the frictional characteristics of the tube wall, which is made of polished stainless steel. The thin cake

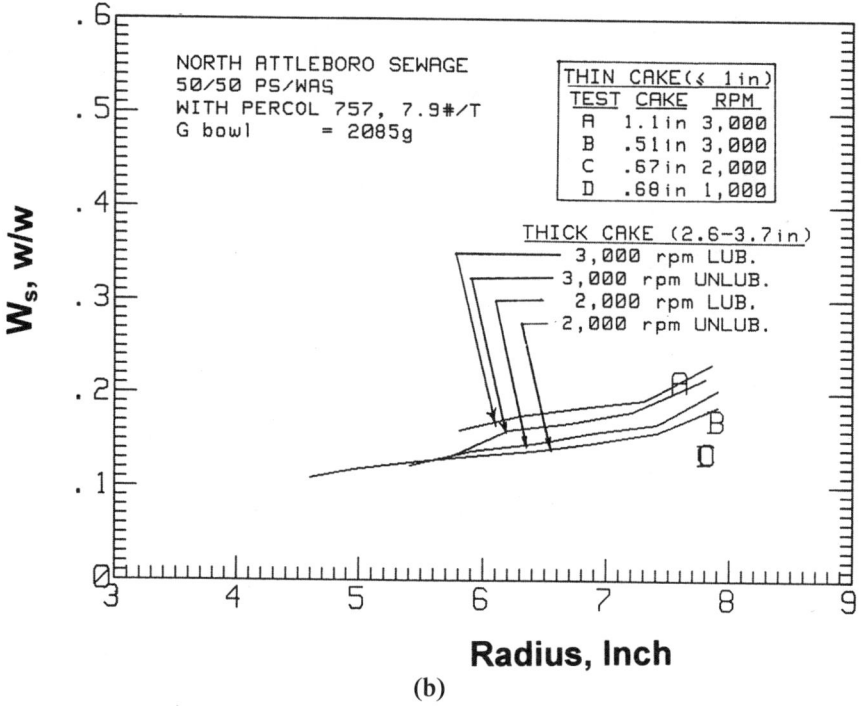

Figure 16.1 (b) Measured cake solids versus radius with polymer addition.

and the thick cake, at the same speed and radial location, do not give the same solids weight fraction. However, a plot of the solids volume fraction ε_s versus the solids compaction stress p_s in Fig. 16.1c reveals that the data of the thin and thick cakes are in accord with each other for identical values of p_s. This supports the fact that the cake solids weight fraction W_s and consequently ε_s [given that the saturation $S = 1$ for compactible and nondrainable cake; see Eq. (2.12e)] are indeed unique functions of both p_s and dewatering time t. In the tests only the equilibrium condition was established at extended spin time without recourse to the transient effect. Under this equilibrium condition, p_s and ε_s are related by the *inverse* relationship given by Eq. (16.1), $p_s = f(\varepsilon_s)$, which further depends on the compaction behavior for a given type of sludge or compactible cake. Using Eq. (16.1) p_s can be determined for the test samples, and the results are graphed in a log-log format in Fig. 16.1d. It is clear that within the range of the test results an approximate *explicit* relationship for $\varepsilon_s = f(p_s)$ in the form of a power law can be used, where

$$\frac{W_s}{W_{s0}} \approx \frac{\varepsilon_s}{\varepsilon_{s0}} = p_s^k \qquad (16.3)$$

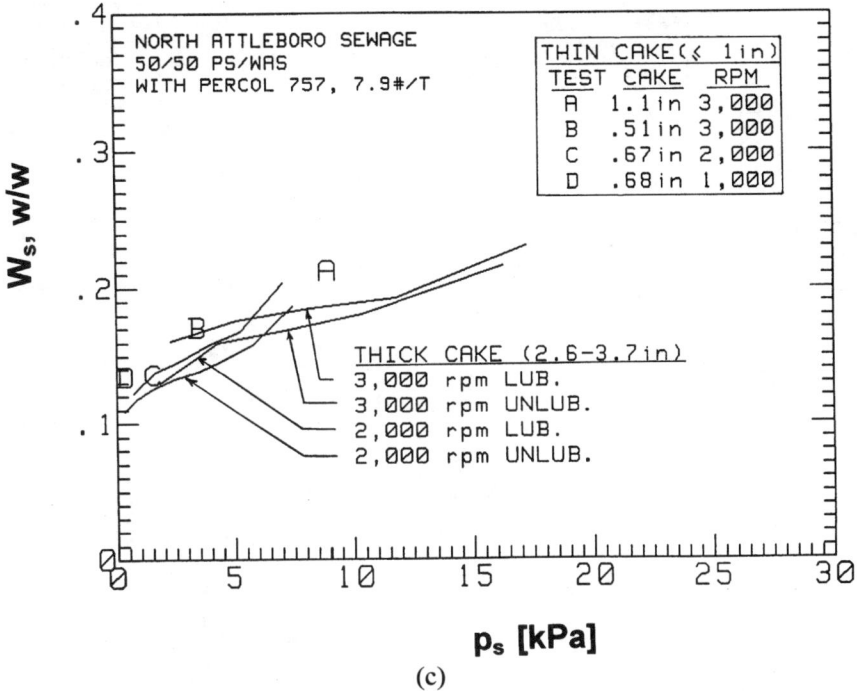

Figure 16.1 (c) Measured cake solids versus compaction stress with polymer.

The data in Fig. 16.1d are best correlated by Eq. (16.3) with $k = 0.23$. Results from tests conducted on the same sludge but without polymer addition are shown in Fig. 16.1e. There the data are best fit with a power law with $k = 0.20$. When comparing Fig. 16.1d and e, W_s is higher for the sludge with the polymer added. In quantitative terms, both k and W_{s0} (and thus ε_{s0}) are greater in magnitude for the case with polymer addition. Also, it is best to lubricate the test tube to reduce the undesirable wall effect which undermine cake compaction from (1) sidewall friction and (2) linear and parallel geometry of the sidewalls of the spin tube, both of which are absent in cylindrically configured bowls. The fact that $k < 1$ both with and without polymer addition suggests that as the cake gets more compacted under increasing stress, the resistance to compaction also increases more than proportionally. This results in a diminishing return when the cake is subject to higher compaction stress.

There are three additional shortcomings in this simplified "lab demo." (1) In the tests the cake is beneath the pool, whereas in actuality the cake solids at the centrifuge discharge are above the pool. Therefore the lab results should yield a conservative estimate on cake dryness. (2) The shear stress exerted on the cake due to the churning motion of the helix during cake conveyance is absent in this batch test. (3) The solids frac-

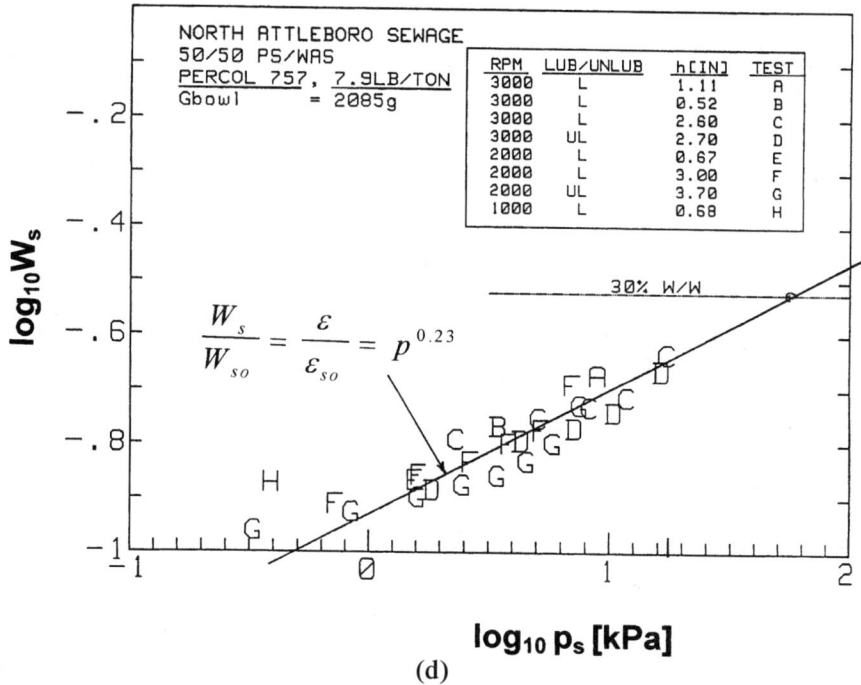

Figure 16.1 (*d*) Measured cake solids versus compaction stress with polymer on log-log plot.

tion reported in the test corresponds to the equilibrium solids concentration after the cake has been subjected to the stress for a long period of time. One could have repeated the same tests for different spin times to simulate the finite residence time in the machine. These three factors together with the side-wall effects provide an overall conservative estimate of the dry solids content for real applications, but they point out the important factor of solids compaction with regard to cake dryness.

Based on the foregoing considerations it is important to maximize the compaction stress p_s acting on the solids structure in the cake, which is proportional to the product Gh. It is interesting to compare the Gh of the high-solids decanter with that of the conventional decanter. The high-solids decanter operates between 3000–4000g, with cake heights h anywhere between 130–180 mm (5–7 in). This should contrast with the conventional decanter, which operates between 2000 and 3000g with h = 25–50 mm (1–2 in). It follows that the Gh product for high-solids decanters is five to seven times that of the conventional decanter (see Table 16.2).

Also, when the cake stacks up inside the pool rather than outside in the bowl, the compaction stress is different as it is affected by liquid buoyancy, which is related to the pool level. Thus the stress experi-

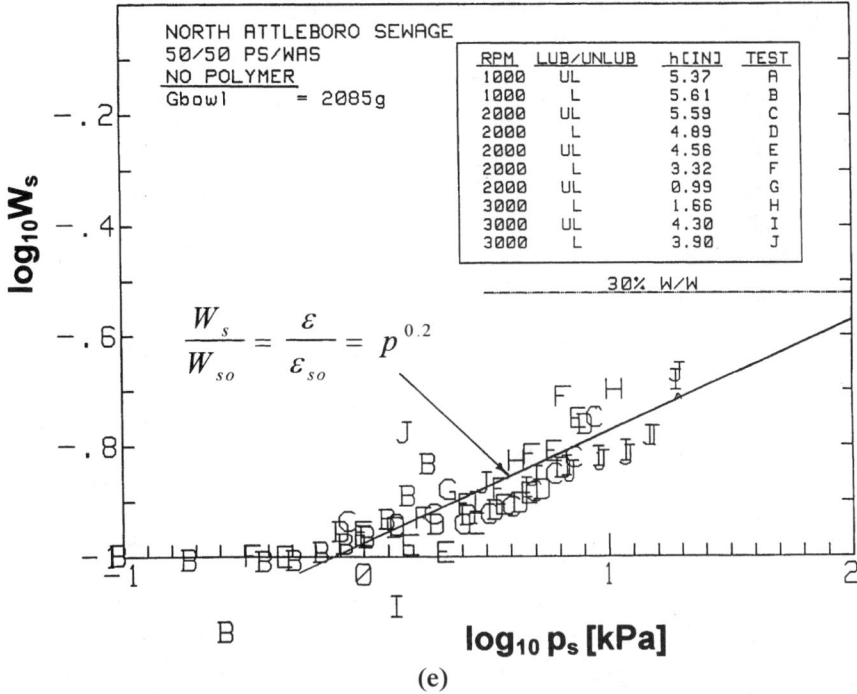

Figure 16.1 (e) Measured cake solids versus compaction stress without polymer on log-log plot.

enced by the cake adjacent to the bowl wall is the sum of the two contributors—the cake column above and below the liquid pool. Applying Eq. (16.1),

$$p_s(R_b) = \rho_s \Omega^2 \int_{R_c}^{R_p} \varepsilon_s(R) R \, dR + (\rho_s - \rho) \Omega^2 \int_{R_p}^{R_b} \varepsilon_s(R) R \, dR \quad (16.4)$$

where R_b, R_p, and R_c denote the radii at the bowl wall, the pool surface, and the cake surface, respectively.

Retention time. Another important factor for obtaining dry cake is that it takes time for the liquid to express out of the cake, especially for a thick cake. For our example, if the cake resides in the decanter for a shorter period of time before all the moisture gets squeezed out under the G field, the cake solids should be less than what they would have been if obtained under equilibrium. It follows that a long retention time is important for cake dewatering. In Table 16.2 the retention times of the cake for the two types of machines are compared. It can be seen that for high-solids decanters the retention time is between 4 and 20 times that of the conventional decanter. This is made

TABLE 16.2 Comparison of Conventional and High-Solids Decanters

	Conventional decanter	High-solids decanter
$\dfrac{G}{g}$	2000–3000	3000–4000+
h	25–50 mm (1–2 in)	130–180 mm (5–7 in)
$\dfrac{Gh \text{ (high solids)}}{Gh \text{ (conventional)}}$	1	5–7
Δ	10–20 rpm	0.5–5 rpm
$\dfrac{L}{\ell}$	8–12	8–12
$u = \dfrac{L_{\text{lead}}}{2\pi}$	0.04 m/s (1.6 in/s)	0.004 m/s (0.16 in/s)
$t = \dfrac{L}{u} = \left(\dfrac{L}{L_{\text{lead}}}\right)\dfrac{60}{\Delta^*}$	0.5–1 min	2–20 min
$\dfrac{t \text{ (high solids)}}{t \text{ (conventional)}}$	1	4–20

*Δ in rpm; L = machine length; L_{lead} = average lead.

possible by having a long clarifier, where a significant portion has been used for cake compaction, and a low differential speed as maintained by a back drive on the conveyor, which typically runs between 0.5 and 5 rpm. Differential speeds as low as 0.1–0.2 rpm can be obtained. However, this is at the expense of a very high conveyance torque, which might limit the solids throughput.

Given these two key ingredients for high-solids decanters—high p_s or high Gh, and long retention time—it is not surprising to find that almost the entire machine is filled with very thick cake solids. Figure 16.2b shows a schematic of this scenario. This is in contrast to the conventional decanter shown in Fig. 16.2a. Figure 16.3 shows a photograph of the conveyor screw of a decanter with the cake solids in place after dewatering biosolids sludge. Note that the solids fill the spaces between adjacent conveyor flights over almost two-thirds of the machine length, from the beach back toward the clarifier. This corresponds to the schematic in Fig. 16.2b.

Polymer. Increasing the polymer dosage also helps increase cake solids. However, both centrate clarity and economics dictate the polymer dosage. Most polymers used are polyelectrolytes. Liquid emulsion

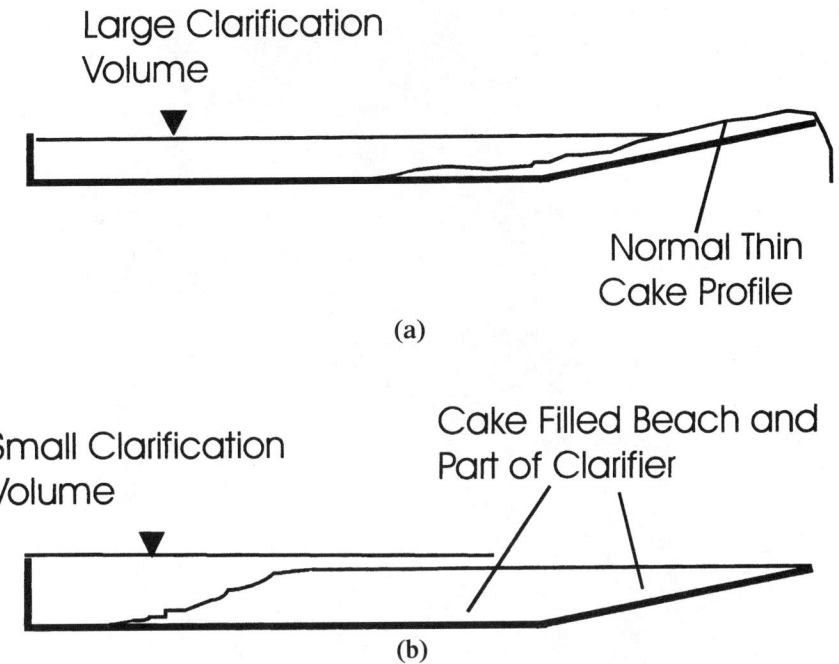

Figure 16.2 Solids distribution. (a) Conventional decanter. (b) High-solids decanter.

polymer and solid polymer are used widely in dewatering sewage sludge some applications require chemical conditioning (lime or ferric chloride) before polymer addition.

Dip weir and baffle. The differential speed in the high-solids decanter is typically about 4 to 10 times less than that used in conventional decanters, yet the solids throughput is reduced only to 15–25% below the nominal capacity of the conventional equipment. The mechanical conveyance using differential speed is assisted by hydraulics (see Fig. 15.10) from deep pool operation with the use of an annular dip weir and baffle. When a dip weir or baffle (such as those shown in Fig. 4.17a–c) is used, the pool level in the clarifier is much higher than the spillover point at the beach. A deeper pool enhances centrate clarity and cake compaction as it allows a thicker cake to form before conveyance to the beach.

Centrate clarity

High-solids dewatering applications typically require over 95% solids capture. The high G force coupled with the agglomeration of the fines into large solids with the use of polymers enhances sedimentation. In addition, it is important to have good feed acceleration as there is

(a)

Figure 16.3 Solids distribution in high-solids decanter. (*a*) Cake solids filling up two-thirds of conveyor toward conical beach.

very little clarification volume in the cylindrical clarifier of the machine. Figure 16.2*b* illustrates that mixing and turbulence caused by underaccelerated feed cannot be tolerated due to the limited clarification volume as compared with Fig. 16.2*a* for conventional operation. It is necessary to increase the polymer dosage in the case where the flocculated solids formed from underaccelerated feed are broken as

(b)

Figure 16.3 Solids distribution in high-solids decanter. (b) Closeup of 4–5-in-thick cake.

the feed enters the rotating pool. A longer clarifier is needed since a portion of it is taken up by cake inventory. Also, the feed rate is reduced by 15–25% of the nominal value as compared to conventional decanters for the same size equipment, given the reduced pool volume for clarification.

Note that in Fig. 16.2b the clarifier is split into two regions, one for cake compaction and the other for clarification. The controlling factor is the differential speed Δ. It is prudent to adjust Δ to get a good balance between centrate clarity and cake dryness. For example, if the centrate becomes cloudy, and given that a high polymer dose has already been used, then it becomes necessary to increase Δ slightly to "clean up" the centrate. This effectively moves the cake solids load toward the beach end, thus freeing up more space for clarification.

Operation control

Back-drive torque. The conveyor is driven by a second drive, either in the form of a hydraulic motor, or by an ac or dc electric motor through a gear box. In the latter case the electric motor drives the pinion shaft directly at speed Ω_p, and the differential speed is determined once the speed of the bowl (driven by the main motor), and the gear ratio gr are known,

$$\Delta = \frac{\Omega_b - \Omega_p}{gr} \qquad (16.5)$$

The conveyor speed follows from

$$\Omega_c = \Omega_b - \Delta \qquad (16.6)$$

[Note that when the pinion speed is greater than the bowl speed with the use of cyclo gears, the subtraction in Eqs. (16.5) and (16.6) reverses.] This arrangement is commonly known as back drive, as opposed to the arrangement where the conveyor of the conventional decanter is driven by the gear box, which in turn is driven by the main motor through the bowl. In the latter case Δ is fixed by the gear ratio of the gear box and the bowl speed, and the pinion is locked (that is, $\Omega_p = 0$) by a stationary bracket.

It is important that the back drive provide adequate torque capacity for cake conveyance. Typically the measured torque at the spline of the conveyor T_s ($= T_p$ gr) is inversely related to Δ, as shown in Fig. 16.4. Torque is an indication of the cake thickness inside the bowl. At low Δ the cake is quite thick and therefore T_s is high. Vice versa, at high Δ the cake thickness is low due to a higher conveyance speed for

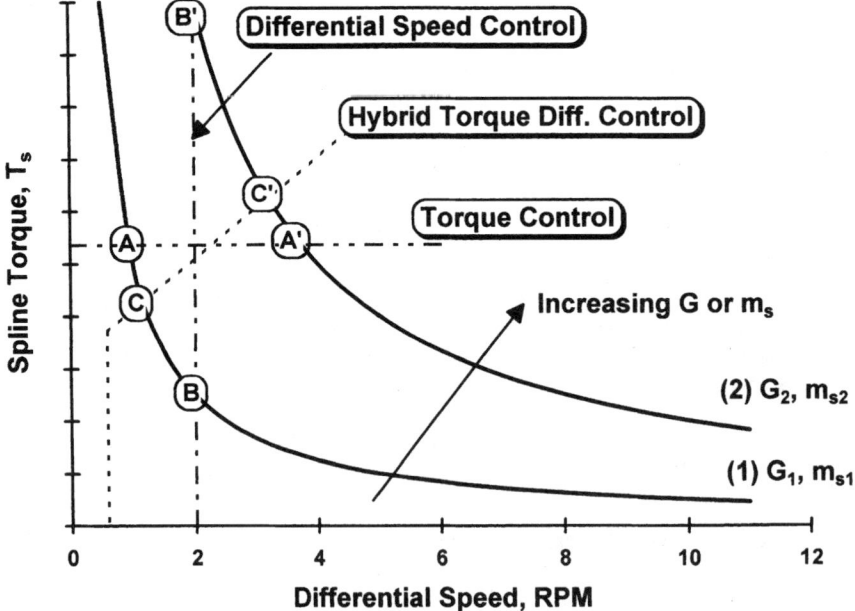

Figure 16.4 Torque versus differential speed plot showing performance curves for different control schemes.

the same solids flow rate, and consequently T_s is low. If the operator needs to run at low Δ, say 1–2 rpm, to obtain drier cake, the spline torque is expected to be very high. Therefore in selecting the back drive one should ensure that there is adequate torque capacity to operate at this low Δ. Obviously, the T_s versus Δ curve is not known a priori, and production or pilot-test experience on the sludge is required to size up the back-drive torque capacity.

In Fig. 16.4 the torque versus differential speed is an inverse relationship. Curve 1 corresponds to a given G cake throughput m_s and curve 2 to a higher G or throughput that has a higher spline torque for a given Δ.

Torque-differential control scheme. Depending on the throughput and the desired cake dryness as discussed in the foregoing, an operator can run at any point in the T_s versus Δ chart in Fig. 16.4. For example, one can operate at a fixed torque T_s = constant and let Δ take on the values as dictated by the operating curves for given G and m_s. This corresponds to a horizontal line (see Fig. 16.4). It is clear that as solids throughput increases, so does Δ in order to maintain a constant torque. This is referred to as torque control. Given that the torque is an indication of the cake solids depth and increased cake compaction, high T_s almost assures dry cake solids. Figure 16.5 is a plot of torque versus cake solids and demonstrates a good correlation between these two variables.

Another common scheme is to set Δ = constant, independent of the spline torque T_s. This ensures a constant retention time of the cake in the machine and further avoids that the machine control will hunt for the Δ, which for some schemes may lead to overshooting and even instability. This is known as Δ control. However, the cake height and thus the torque can increase at the same time in response to fluctuations in m_s. There should be a provision to have a torque relief option to ensure that the torque stays below the maximum.

Another useful control scheme is to allow (Δ, T_s) to take on a prescribed locus, such as the linear Δ-T_s control shown in Fig. 16.4. After T_s and Δ both have exceeded their set thresholds, the linear relationship takes over. In this case Δ can also surge up and down in response to a torque change. With this scheme the parameters are the slope of the Δ-T_s line and the two thresholds. Usually the slope of the line is set slightly less than unity in the plot of T_s versus Δ. As the slope is set to either a very small or a very large value, it approaches either torque control or Δ control.

In all of the preceding the equilibrium operating point will have to satisfy both the process and the control scheme. This translates into the intersection of the two curves—operating curve and control curve.

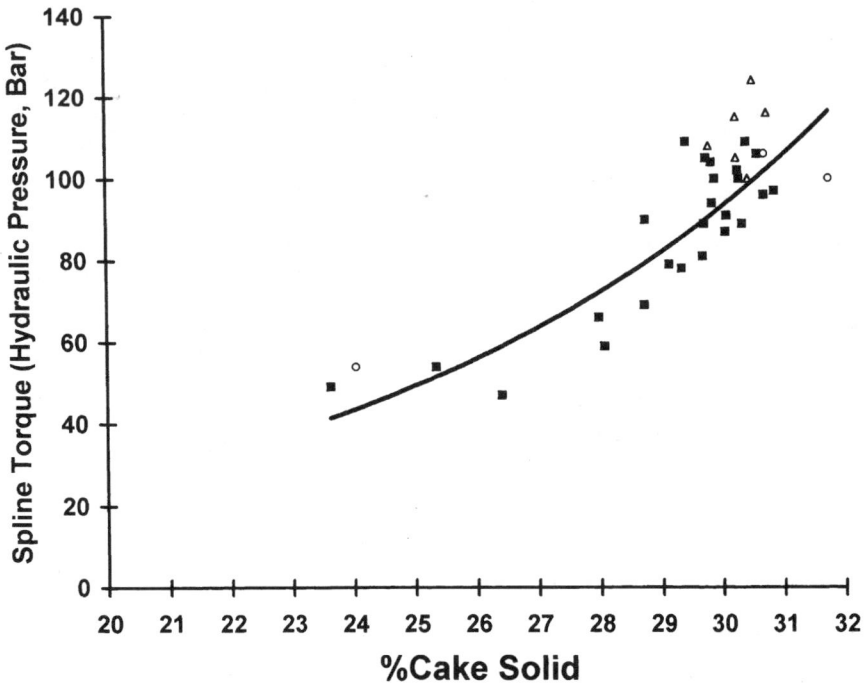

Figure 16.5 Torque versus cake solids concentration.

With the torque-control scheme, as the operating condition switches between curves 1 and 2 due to fluctuations in the solids loading m_s, the operating condition goes back and forth between points A and A'. With the Δ-control scheme, the operating condition fluctuates between points B and B', while with the linear Δ-T_s control, the operating condition fluctuates between C and C' after the thresholds Δ_{th} and T_{sth} are met. Even though the imposed loading can fluctuate or change instantaneously, it takes some time for the process in the centrifuge to reflect this change, such as in the measured torque, centrate solids, and cake solids. This is due to the slow cake solids transport and dewatering inside the centrifuge, especially for a large machine, which has significant delay in responding to changes that occur at an erratic pace. This unavoidably results in a constant catch-up situation, even with the most efficient feedback control. Figure 16.6 shows the torque-versus-time and differential-speed-versus-time plots. Note that Δ control takes place when the torque exceeds a prescribed threshold level. Based on Fig. 16.6 and the foregoing discussion, it can be seen that quasi equilibrium is the best the high-solids centrifuge can attain. Despite this, with a good compensating control and adequate back-drive torque capacity, it can still achieve a near-

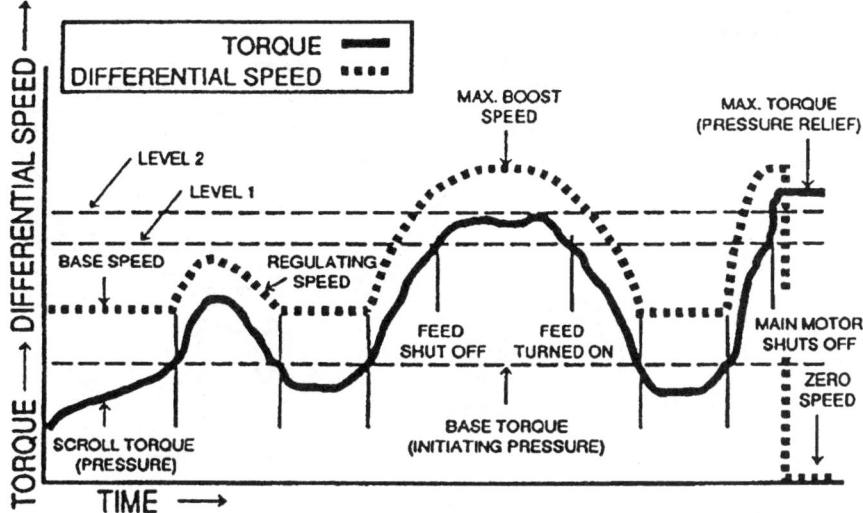

Figure 16.6 Torque and differential speed versus time.

steady-state operating condition without runaway, unstable washout, or solids plugging.

Effective polymer introduction

Polymer may be added through the feed line, say 30–60 m (100–200 ft) before entering the machine, or internally within the machine through separate feed and polymer pipes and compartments. In the former, the polymer has adequate time to contact the feed particles to form a stable flocculated solid (floc in short) and not oversized which could be sheared off is the feed zone of the machine. The required time depends on the feed and the polymer, and jar tests are generally required to establish this condition. In the latter, it relies on the flow dynamics to mix the two streams upon entry into the separation pool. It requires polymer to form a stable floc with the feed particles instantaneously. Regardless of either arrangement, the feed accelerator needs to lay the feed or preflocculated feed gently onto the pool surface to avoid a high shear gradient, which breaks most delicate floc and results in higher polymer consumption.

A number of examples will illustrate the high-solids characteristics discussed in this section.

Examples

1. As shown in Fig. 16.7a, in plant A, a waste treatment facility handling 95% industrial waste and 5% municipal waste, a high-solids decanter is taking a combined prethickened primary (70%) and sec-

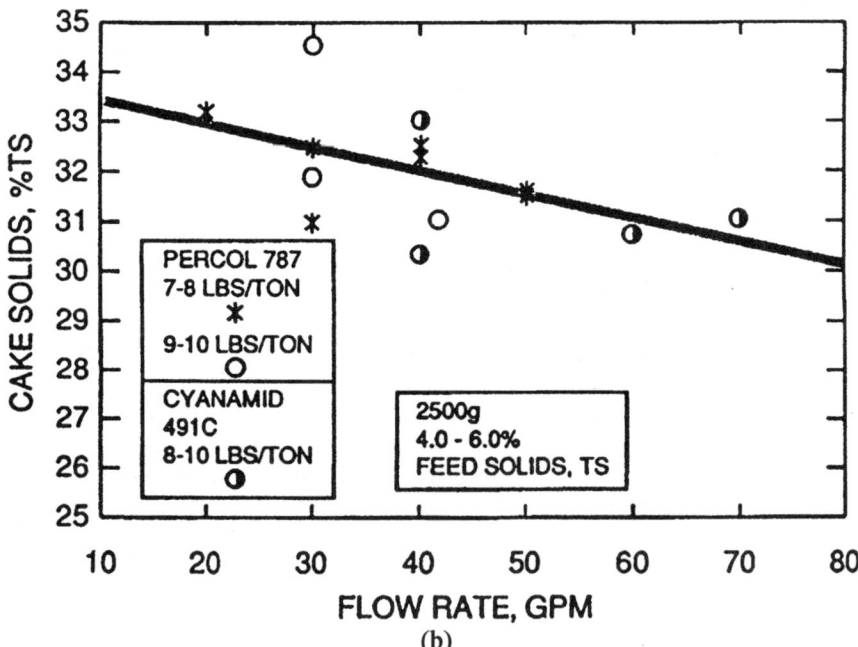

Figure 16.7 Plant A. (*a*) Flow sheet. (*b*) Cake solids versus feed rate.

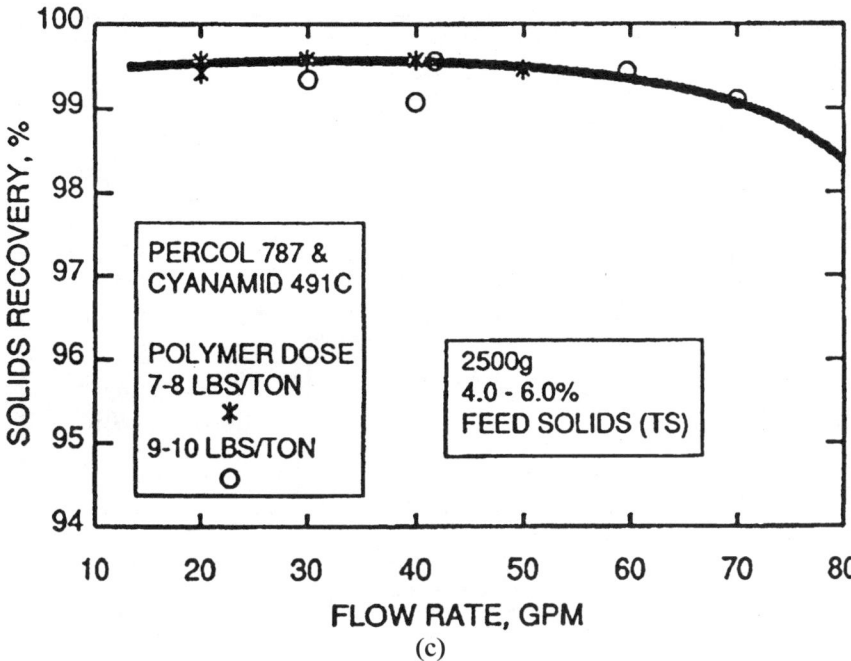

Figure 16.7 Plant A. (c) Solids recovery versus feed rate.

ondary (30%) waste at 5% suspended solids and 60% volatile, and dewatering the feed to above 30% cake. Figure 16.7b and c shows the cake solids and % solids recovery (a measure of centrate clarity) plotted against feed rate. The cake solids reach 33% at 20 gal/min (4.54 m³/h) and decrease to 31% at 70 gal/min (15.9 m³/h). In all cases the recovery is maintained above 99% with a small decrease at increasing rate. The differential speed is 4 rpm or less while the G is 2500–3000g. T_s is registered at 28,600–34,000 in · lb (3230–3840 N-m). A conventional decanter would have produced 20–25% cake under similar feed conditions.

2. In plant B, a 457-mm (18-in)-diameter high-solids decanter is used to dewater mixed sludge. The polymer dosage, illustrated in Fig. 16.8a as a function of flow rate, generally increases with the flow rate. Up to 12 lb/ton (6 kg/t) was used at 40 gal/min (9.1 m³/h). Figure 16.8b shows a plot of the cake solids versus differential speed. Lower Δ increases both the cake thickness and, more importantly, the retention time. This results in drier cake. The limit is the back-drive torque capacity. This is more clearly shown in Fig. 16.8c, where the cake solids on the vertical scale are replaced by the measured torque at the spline from a hydraulic motor (torque %pressure readout). The inverse T_s-Δ behavior, as suggested by the schematic of Fig. 16.4, is clearly demonstrated in Fig. 16.8c. The inverse hyperbola curves cor-

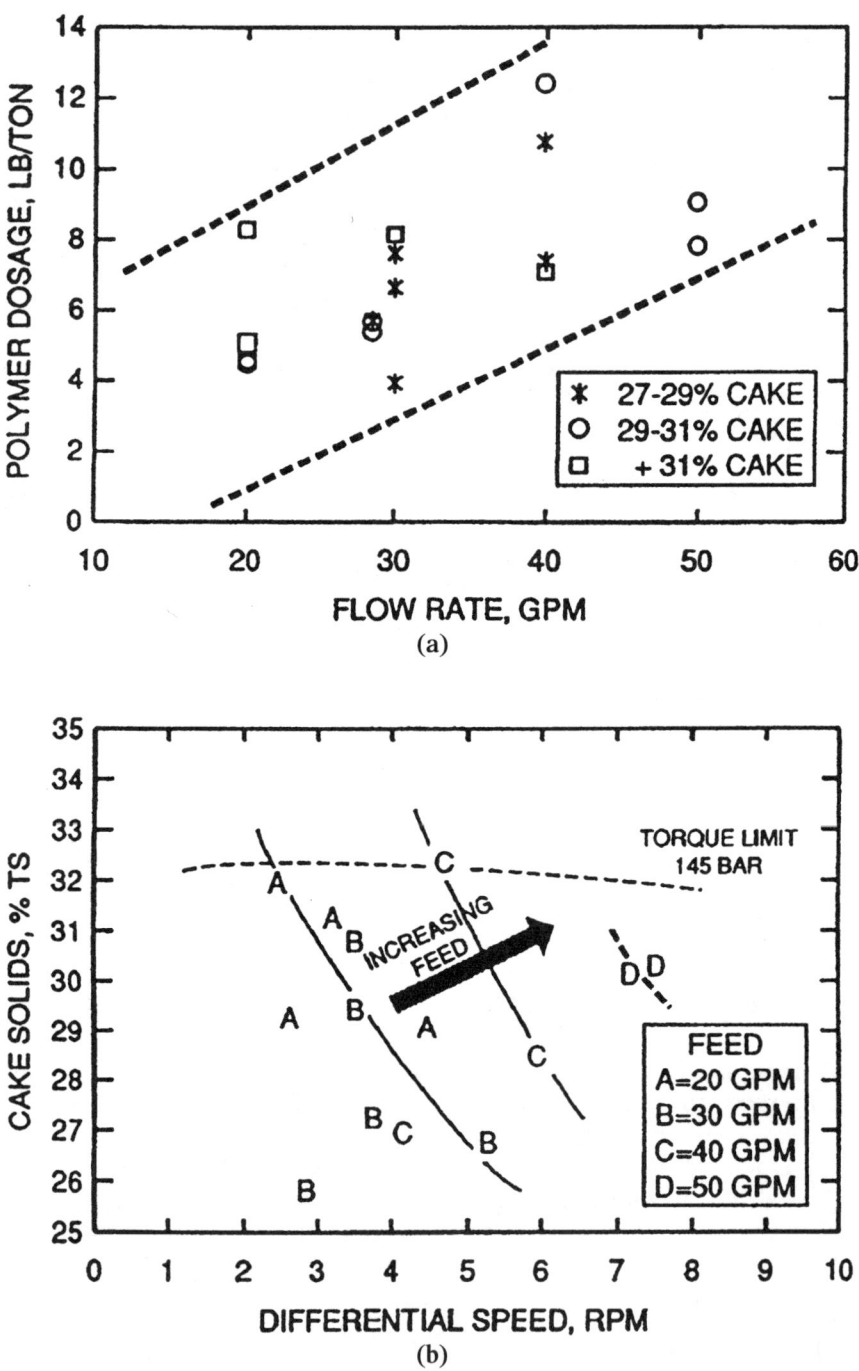

Figure 16.8 Plant B. (*a*) Polymer dosage versus feed rate. (*b*) Cake solids versus differential speed.

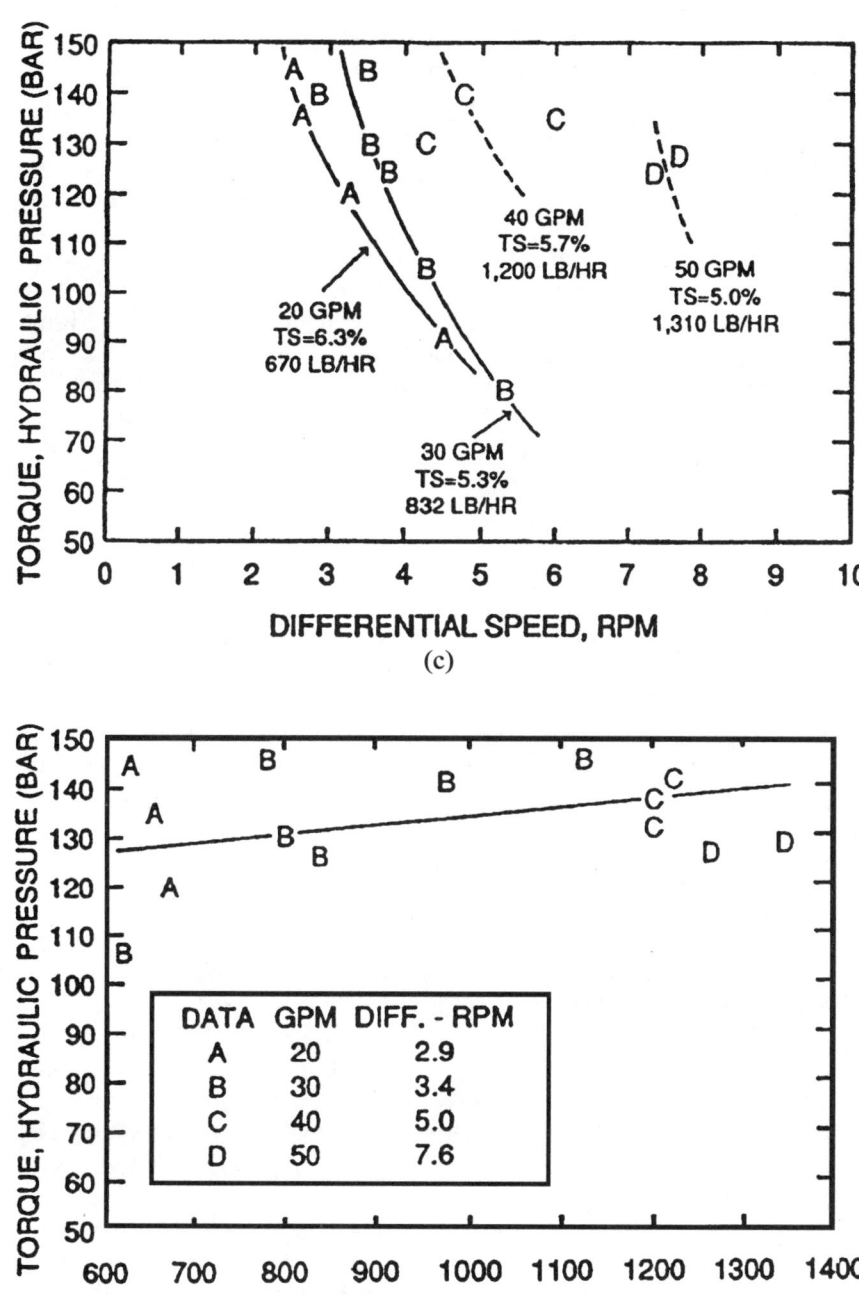

Figure 16.8 Plant B. (c) Torque versus differential speed. (d) Torque versus solids throughput.

respond to constant throughput, with the curves shifting upward at higher throughputs. As discussed in Chap. 2, T_s increases with G and solids throughput. Figure 16.8d further shows this behavior. Here the machine is operating close to the maximum torque capacity to generate the highest cake dryness.

3. In plant C cake solids are tested at different Gs, as shown in Fig. 16.9a. It is seen that the cake solids increase generally with increasing G, supporting the fact that a high compaction stress p_s is required to warrant dry cake. Other than higher G, equally important for higher p_s, is the cake depth, which is controlled by operating at low Δ. Figure 16.9b illustrates this point. As Δ decreases, a thicker cake with a longer retention time for compaction results in drier cake.

4. A 457-mm (18-in) high-solids decanter is processing a raw mixed sludge with 40% primary and 60% waste-activated sludge. The feed has 5% suspended solids and is required to be dewatered to a cake with a minimum of 27% solids before compost. Figure 16.10 shows the cake solids (left scale) and %recovery (right scale) graphed against flow rate. It can be seen that the cake solids are consistently at 30–31% solids, whereas recovery is maintained above 96%. The polymer dosage is between 11 and 15 lb/ton (5.5–7.5 kg/t). Another high-solids decanter of 430-mm (16.9-in) diameter, with a different configuration, operating side by side on the same feed and at the same G, yields 28–29% cake at 48 gal/min (11 m^3/h) while recovery is 94–95%. The same polymer is used at dosages of 12–15 lb/ton (6–7.5 kg/t). This example demonstrates the difference in performance due to different design of the high-solids decanters.

Finally, Table 16.3 compares the various types of waste sludge which are processed by conventional and high-solids decanters. Primary sludge is the easiest to dewater as it has fibrous materials, while the other extreme, waste-activated sludge, is the most difficult to dewater as it contains fine and light solids. It is not practical for a wastewater treatment plant to dewater waste-activated sludge directly. Instead it is combined with primary sludge, and the fibers in the mixed sludge are in ratios such as 40%:60%, 50%:50%, or 60%:40%. The greater the fraction of waste-activated sludge in the mix, the wetter is the cake. The range given in Table 16.3 under raw mixed sludge reflects this ratio.

When the solids are dewatered to their driest consistency, they have the appearance of fine round granules with 6–10 mm in diameter. This represents a high surface-to-volume ratio for the solid and is the driest that can be obtained from centrifugal compaction and expression. However, it is deceiving to estimate the cake solids based on the appearance of small granule sizes, as they can vary from 16% for aerobic digested sludge to 34% for mixed sludge. Despite this, it can be safe to conclude that the cake is as dry as it can get by mechanical

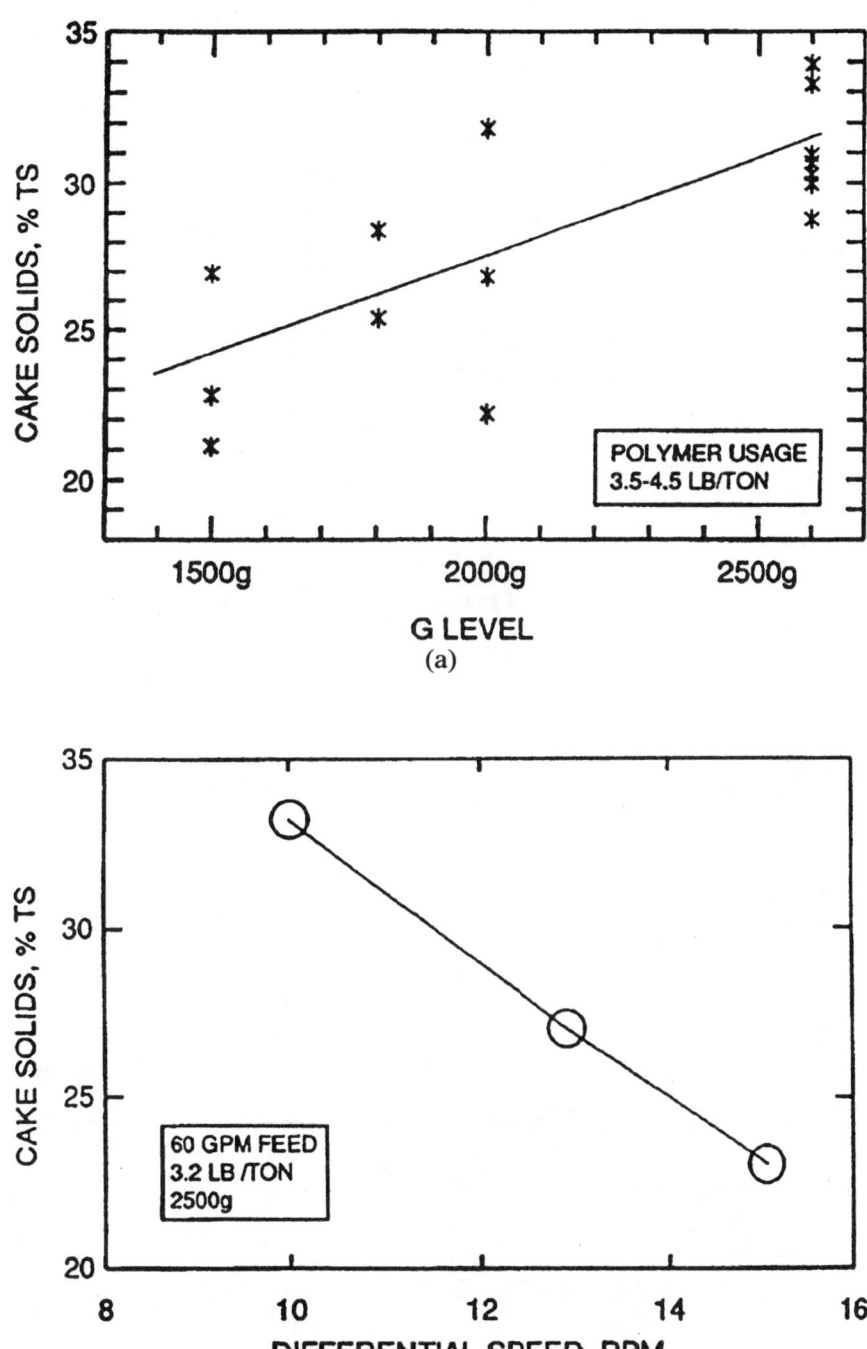

Figure 16.9 Plant C. (a) Cake solids versus G level for anaerobically digested sludge. (b) Cake solids versus differential speed for anaerobically digested sludge.

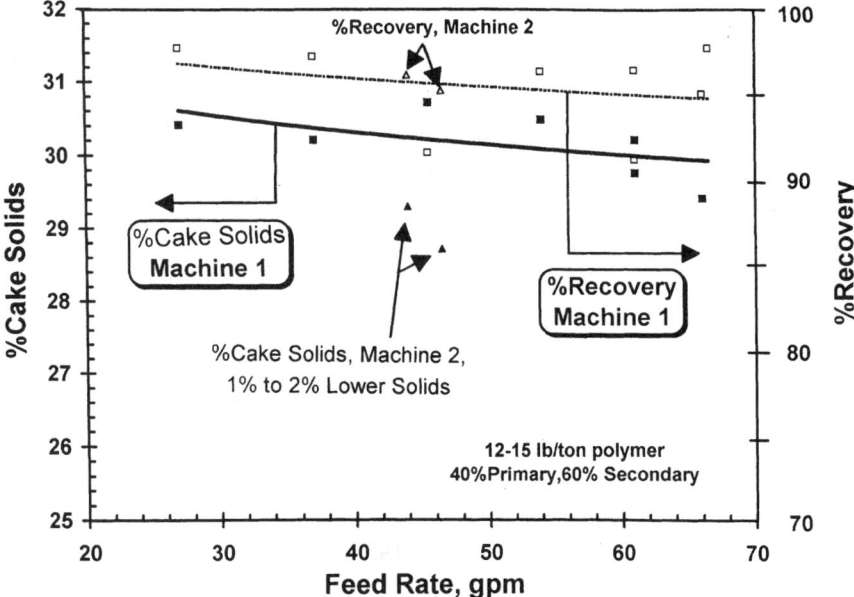

Figure 16.10 Performance of two high-solids machines of comparable size but different designs.

TABLE 16.3 Processing of Different Types of Waste Sludge

Sludge type	Conventional decanter	High-solids decanter
Primary	18–26	30–40
Raw mixed primary and secondary	20–25	28–34
Anaerobic digested	16–19	20–25
Aerobic digested	11–13	17–18
Waste activated sludge	9–11	14–16

compaction. Also the cake is distinctly differentiated from one having a lower solids content, as the latter has a soupy or yogurtlike appearance and slops out on the floor.

Economics and Optimal Operation

As demonstrated in Table 16.3, the high-solids decanter can generate anywhere between 5 and more than 10% drier solids as compared to the conventional decanter. Other than accomplishing technological success, it is important to consider the economics. The incremental benefit of getting drier solids should be weighed against the incremental operating costs—higher polymer dose, higher G, and lower throughput. The costs (capital and operation) and benefits should be expressed per dry ton of solids processed. As to the benefits, every percent increment in cake dryness translates into savings on disposal cost, with the exact amount depending on the downstream handling of the cake solids, such as truck hauling, landfill (or land application), composting, and incineration. The method of disposal may change over time, depending on availability and environmental regulations, and economics can shift either favorably or unfavorably.

As an illustration, consider the case where the sludge has to be hauled to an incineration plant and the key operating expense in the centrifuge is the polymer costs. Let the costs of hauling per ton of wet cake be $\$_H$, the costs for incineration per ton of water (including heating to 100°C (212°F) from room temperature, and latent heat of evaporation) $\$_I$, and the polymer costs per pound of dry polymer $\$_P$. The solids recovery is %Rec and the centrate leaving the centrifuge is fed back to the header of the plant and blended in with the incoming feed. For every ton of dry solids feeding the centrifuge (from plant inflow, not including recycled centrate solids), the operating costs on the polymer becomes $(D_P \$_P / \%\text{Rec})$, where D_p is the polymer dosage in lb/ton or kg/t (polymer to dry cake solids ratio). Likewise, the costs of hauling the wet cake for every ton of dry solids feeding the centrifuge is $(\$_H / W_s)$, where W_s is the cake solids expressed in percent. The costs for incineration per ton of dry solids feeding the centrifuge is therefore $\{(1 - W_s)/W_s\}\$_I$. Thus the total cost $\$_T$ becomes

$$\$_T = \frac{\$_P D_p}{\%\text{Rec}} + \frac{\$_H}{W_s} + \frac{\$_I}{W_s}(1 - W_s) \qquad (16.7)$$

For a given sludge, a given machine design, and a given set of operating conditions, the cake solids depends on the polymer dosage for a given throughput, which depends on the size of the machine. Figure 16.11 shows two curves for cake solids corresponding to a given high-solids decanter for processing the raw mixed sludge and the anaerobically digested sludge. A power law is used to correlate the data,

$$W_s = K D_p^{1/n} \qquad (16.8)$$

Substituting Eq. (16.8) into Eq. (16.7),

$$\$_T = \frac{\$_P (W_s/K)^n}{\%\text{Rec}} + \frac{\$_H}{W_s} + \frac{\$_I}{W_s}(1 - W_s) \qquad (16.7')$$

It can be seen that as W_s increases, the polymer cost also increases, whereas both the hauling and the incineration costs decrease. Therefore it is clear that at either low or high cake solids the total costs due to all three contributions are high with an optimum, or minimum, total cost at moderate cake solids, which hereafter is referred to as "optimal" cake solids from a costs standpoint. The optimal cake solids and optimal polymer dose are determined algebraically from Eq. (16.7'),

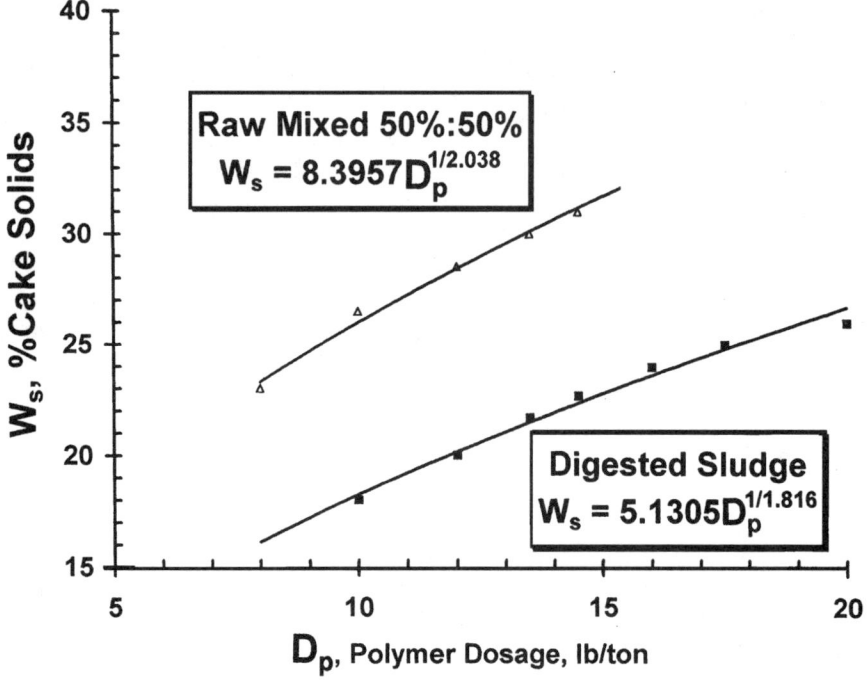

Figure 16.11 Cake solids versus polymer dosage for a 50:50% raw mixed sludge and anaerobically digested sludge.

$$(W_s)_{op} = \left[\frac{(\$_H + \$_I)\%\text{Rec}K^n}{\$_P n} \right]^{\frac{1}{1+n}} \quad (16.9a)$$

$$(D_p)_{op} = \left[\frac{(\$_H + \$_I)\%\text{Rec}}{\$_P K} \right]^{\frac{n}{1+n}} \quad (16.9b)$$

Examples

Raw mixed 50%:50% with cake solids versus polymer dosage given by Fig. 16.11.

1. Given hauling costs of $10/ton of wet cake, incineration costs of $8/ton of water (1400 Btu/lb for heating, evaporation, and thermal inefficiency; $3/million Btu), and polymer costs of $1.40/lb of dry polymer (or liquid polymer with a certain activity). The result is shown in Fig. 16.12a. Here the optimal condition corresponds to 34⁺% cake and not to the driest cake, even if the cake can reach 40–50% by centrifugation with the proper polymer dosage. Under this optimal condition,

Dewatering of Compactible Solids 383

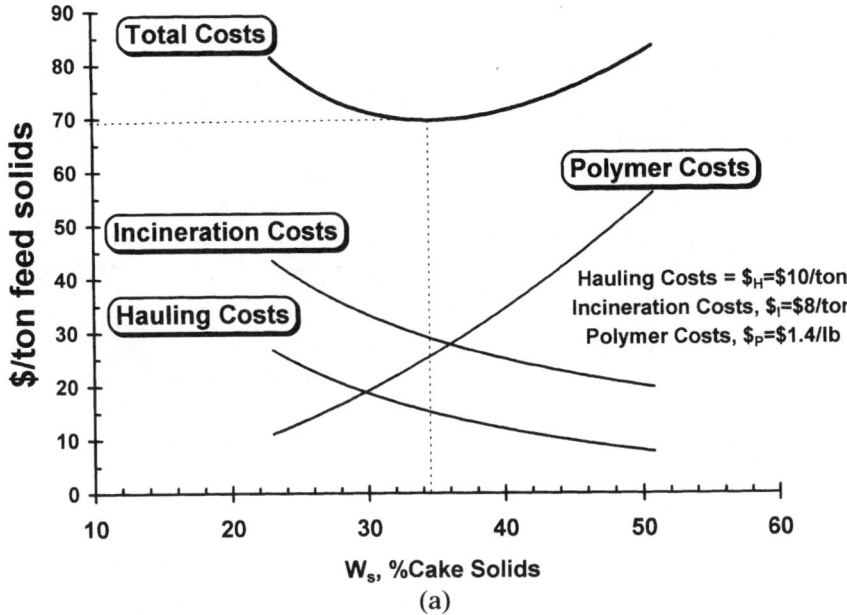

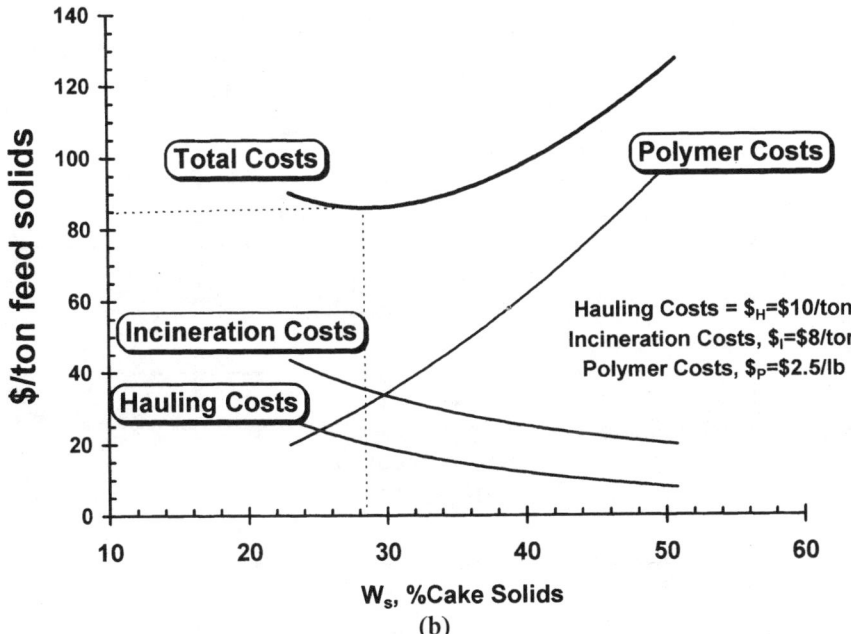

Figure 16.12 Polymer, hauling, incineration, and total costs per ton of feed solids versus cake solids for the raw mixed sludge with characteristic given by Fig. 16.11. (a) $1.40/lb polymer cost. (b) $2.50/lb polymer cost.

polymer costs and the incineration costs are comparable at $27–28/ton of feed solids and the hauling cost is only $15/ton of feed solids, with the total cost slightly less than $70/ton of feed solids.

2. When the polymer cost increases to $2.50/lb dry polymer, the cost of the polymer becomes even more significant. The results are shown in Fig. 16.12b. It is desirable to reduce the polymer dosage with less dry cake at 28+%, even at the expense of increasing the incineration costs. The polymer, incineration, and hauling costs are, respectively, $30/ton, $35/ton, and $20/ton, with a total cost of $85/ton. The higher processing costs reflect the increase in polymer cost.

It is clear from the foregoing that the optimal condition does not necessarily call for the driest cake, as one would intuitively believe. As the economics shift, the most important factor is the cost ratio, which is defined as the ratio of the handling costs (hauling and incineration or hauling and landfill) to the polymer costs, as is evident from Eqs. (16.9a) and (16.9b). Figure 16.13 plots the optimal cake solids and optimal polymer dosage in accordance to the cost ratio for raw mixed sludge and digested sludge with the characteristics given in Fig. 16.11. For the raw mixed sludge, in Example 1, the cost ratio is 12.86 and the optimal cake solids and polymer dosage are 34% and 17 lb/ton (8.5 kg/t), respectively. In Example 2, the cost ratio drops to 7.2 due to the higher poly-

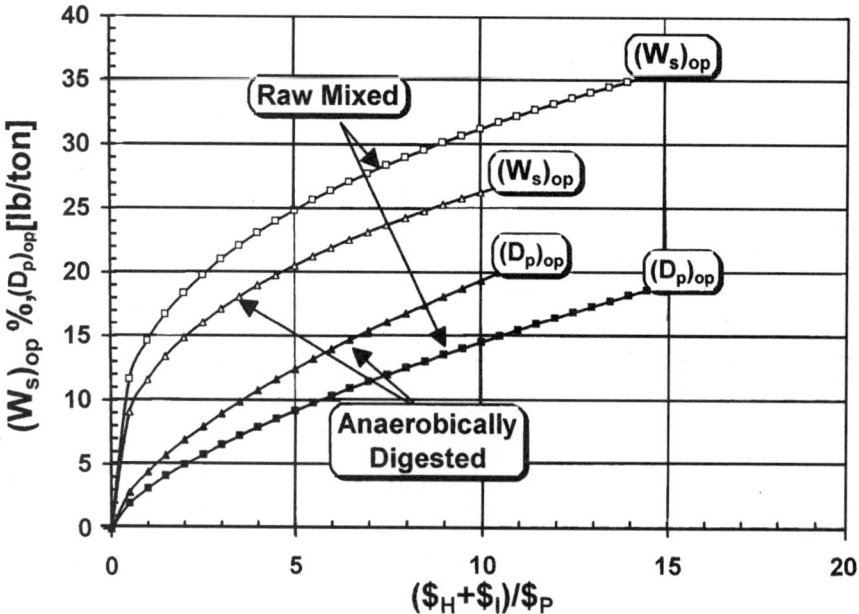

Figure 16.13 Optimal cake solids and polymer dosage versus cost ratio for both raw mixed and digested sludges.

mer costs. The optimal cake solids and polymer dosage are 28% and 12 lb/ton (6 kg/t), respectively. Similar methodology applies to the optimal curves for the digested sludge. Moving forward, polymer research and advances in manufacturing will most likely lead to cheaper and more effective polymer in the future. Also, tighter regulation will lead to enforcing more incineration over landfill and plants may have to be located further away from urban areas, leading to higher hauling costs. This will bring about higher cost ratios, making it economical for centrifuges to operate at higher cake solids with higher polymer dosages.

Combined Centrifugation and Drying

Centrifuge cake obtained from the high-solids decanter can be further dried by passing hot gas through the cake as it is discharged from the centrifuge. In a unique design as shown in Fig. 16.14, both the centrifugation and drying are combined into a single unit whereby the dewatered sludge discharged from the rotor is disintegrated into fine particles which are immediately entrained in the hot drying gas stream in the annular space between the rotor and the stationary casing. Due to the large surface-to-volume ratio of the fine granular solids as discharged from the high-solids decanter, drying rates are extremely fast and within a few seconds the product reaches its final moisture. A 95% cake solids can be produced for a fully dry product whereas a 66% cake solids is produced for a partially dry product. Also, because of the low product temperature and the short residence time in the design as shown in Fig. 16.14, minimal odorous gases are liberated from the sludge.

Drying gas is heated by an adjustable burner which can be fired by a number of different fuels such as fuel oil, natural gas, or digestor gas depending on the availability on-site. Where high-grade waste heat is available this can be utilized by indirect heating to further improve the thermal economy.

Economics of Combined Centrifugation and Drying

As an example, the disposal cost of conventional centrifuge dewatering in Germany in 1995 is about 1400 DM/t (about US $730/ton [short ton]). This includes cost on landfill, polymer, electricity, personnel, maintenance, and amortization with 95% of the total costs on landfill. The disposal costs using the high-solids dewatering is about 1200 DM/t (US$630/ton) which is reduced from that of the conventional; nevertheless it is still very high. On the other hand, using centrifugation and drying all combined into a single unit, for partial drying the specific costs reduces to only 800 DM/t (US$420/ton) and for fully drying to 95% cake solids, the costs further drop to 680 DM/t (US$357/ton). This rep-

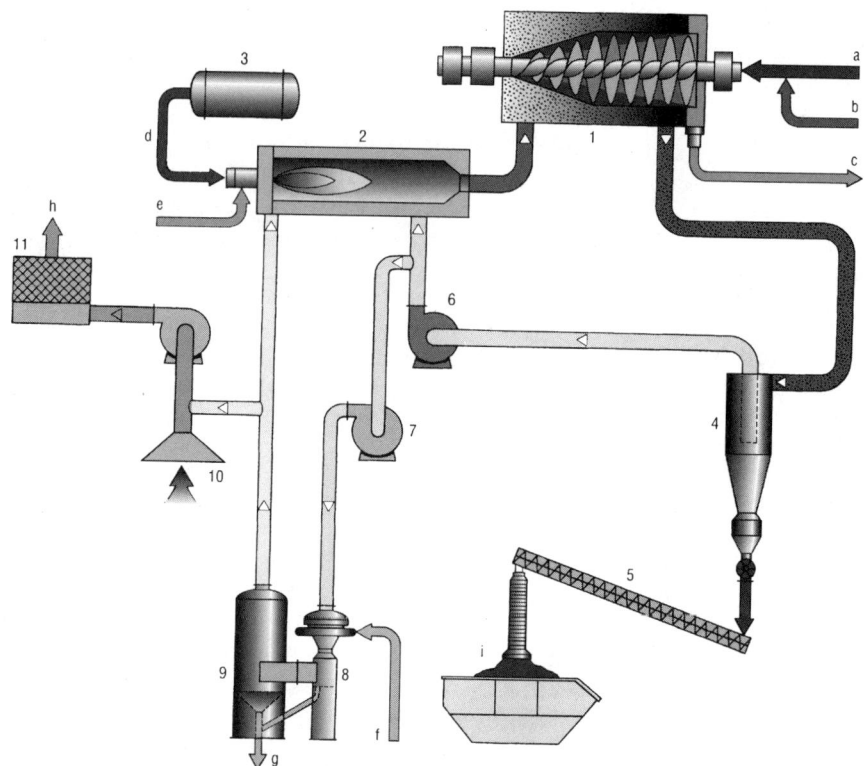

Figure 16.14 CENTRIDRY® 1. Centridry® with sludge dewatering and drying all combined into a single unit, 2. Hot-gas generator, 3. Fuel tank, 4. Product separator, 5. Conveyor screw, 6. Circulating fan, 7. Exhaust fan, 8. Venturi scrubber, 9. Droplet separator, 10. Room ventilation, 11. Biofilter; (a) Sludge, (b) Polymer, (c) Centrate, (d) Digester gas/fuel oil, (e) Air; (f) Washwater, (g) Wastewater, (h) Clean exhaust, and (i) Product. (*Courtesy of Bird Humboldt.*)

resents a significant savings on sludge disposal despite the costs associated with centrifugation and drying include a relatively higher percentage of the total costs on electricity, fuel oil, personnel, and amortization when compared to the first two approaches. This is summarized by the chart shown in Fig. 16.15. (Note in the above, we assume based on the exchange rate $1 US = 1.73 DM). The above merely serves as an example to compare the savings incurred due to a combination of centrifugation and drying versus alternative means. The exact costs figures vary depending on countries and geographical locations. It is expected the cost figures are higher in densely populated affluent areas and less in sparsely populated areas. In the year 2004, some countries are moving toward banning landfill and land application in which case the combined centrifugation and drying becomes an even more attractive approach to handle waste sludge.

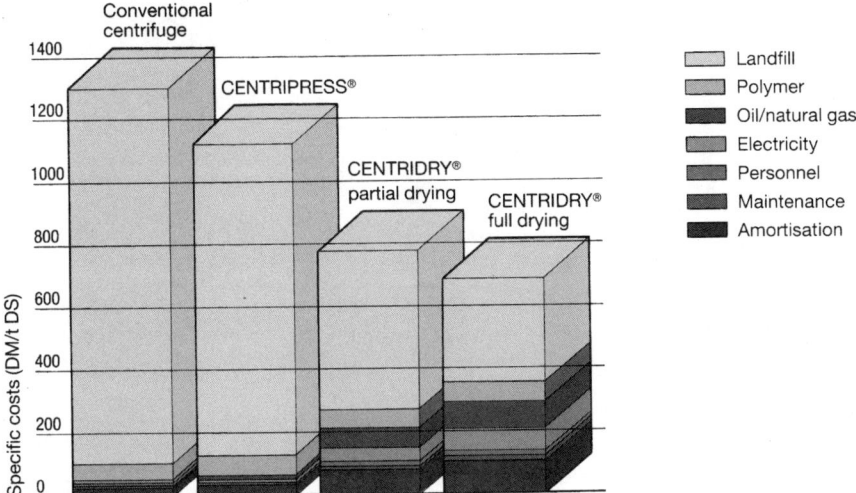

Figure 16.15 Comparison of overall processing and disposal costs for dewatered and dried products for a conventional centrifuge, Centripress®, and Centridry® operating at a throughput of 20 m³/h (88 gal/min) based on cost factors—10 DM/kg, electricity 0.2 DM/kwh, fuel 0.04 DM/kwh, landfill 300 DM/t, amortization 8% in 10 years, and operation 6000 h/year. (*Courtesy of Bird Humboldt.*)

The technology on dry solids dewatering can be used in food processing with the objective of dewatering food waste or other valuable food by-products. In certain applications polymer is not used because of cost or incompatibility with the processed materials. Nevertheless, similar technology, using high-solids decanters, applies. The torque and the differential speed need to be adjusted according to the specific process.

References

1. Bird Humboldt Brochure, #5-376.2e(US) "Centripress® 3000," December 1996.
2. W. W. F. Leung, "High-Solids Decanter Centrifuge," *Fluid/Particle Separation J.*, vol. 5, Mar. 1992.
3. K. Miyano, K. Nishida, and W. W. F. Leung, "Advanced Centrifuge Application on Biological Sludge Treatment in Japan," in *Proc. Am. Filt. Sep. Soc. Ann. Conf.* (Chicago, IL, May 3–6, 1993).
4. J. H. Reynolds et al., "State-of-the-Art Evaluation of High Cake Solids Centrifuge Technology for Municipal Wastewater Solids Dewatering," presented at the Water Environment Federation Annual Conference, Anaheim, CA, Oct. 3–7, 1993.
5. J. S. Berk et al., "Effect of Process Variables on High Torque Dewatering Centrifuge Performance," presented at the Water Environment Federation Annual Conference, Anaheim, CA, Oct. 3–7, 1993.

Chapter 17

Cake Compaction Theory

Cake compaction occurs concurrently with and also immediately following sedimentation, whereby the cake consolidates under centrifugal force. In sedimenting centrifuges liquid expresses radially inward to the cake surface countercurrent to cake compaction, where solids move outward toward the solid bowl wall. Therefore the viscous drag on the cake solids from the expressed liquid retards cake consolidation. In contrast, with filtering centrifuges the liquid from the cake expresses radially outward to the perforate bowl wall in the same direction as the cake solids movement, whereby the viscous drag further enhances cake consolidation. Despite this, for highly compactible materials, where both the permeability and the solids volume fraction are strictly dependent on the compaction pressure, it is prudent to use a sedimenting centrifuge operating with high G. An example is the dewatering of biosolids discussed in Chap. 16. Whereas for moderately compactible to incompactible materials it is best to use filtering centrifuges, which operate at modest G levels, to avoid (1) forming an impermeable skin layer adjacent to the filter medium or (2) infiltration of fine solids in blinding off the filtering medium under high centrifugal gravity. Both factors affect filtration adversely.

In this chapter we examine the theory and model[1,2] of cake compaction with the intent of gaining additional insight into the subject matter.

Model

Consider a rotating centrifuge bowl (having a solid wall for the case of a sedimenting centrifuge and a perforate wall for a filtering centrifuge) with a cake already formed (that is sedimentation is completed) and a pool of liquid that stands above the cake. The liquid saturating the cake (with liquid saturation s = 1 is expressed radially inward for a sedimenting centrifuge and outward for a filtering centrifuge. The flux of the liquid and solid phases is given by the following equations:

1. Liquid phase,

$$q = u\varepsilon = \frac{\tilde{Q}_s}{2\pi R} \tag{17.1}$$

2. Solids phase,

$$q_s = u_s \varepsilon_s = \frac{\tilde{Q}_s}{2\pi R} \tag{17.2}$$

where u is the velocity, ε the volume fraction, q the flux, and $\tilde{Q}$ the flow rate per unit axial width. The last two variables are measured per unit length along the axis of the centrifuge. The subscript s represents the solid phase; variables without subscripts represent the liquid phase.

From continuity, we have for the liquid phase,

$$\frac{1}{R}\frac{\partial}{\partial R}(u\varepsilon R) + \frac{\partial \varepsilon}{\partial t} = 0 \tag{17.3}$$

and for the solid phase,

$$\frac{1}{R}\frac{\partial}{\partial R}(u_s\varepsilon_s R) + \frac{\partial \varepsilon_s}{\partial t} = 0 \tag{17.4}$$

By definition the sum of the solids volume fraction ε_s and the liquid volume fraction ε should add to unity. Hence

$$\varepsilon_s + \varepsilon = 1 \tag{17.5}$$

Adding Eqs. (17.3) and (17.4) and using Eq. (17.1), (17.2), and (17.5),

$$u_s \varepsilon_s R + u\varepsilon R = (q_s + q)R = (q_{sb} + q_b)R_b \tag{17.6}$$

Equation (17.6) implies that the total solid and liquid fluxes are constant, independent of R and depend only on t. This result is set equal to that at the bowl wall, denoted by the subscript b. Note that u_s, ε_s, q_s, u, ε, and q are all functions of R and t.

The Darcy–Shirato equation (accounting for solids movement) written in the rotating reference frame becomes

$$q\left(1 - \frac{u_s}{u}\right) = \frac{K}{\mu}\left(-\frac{dp}{dR} + \rho\Omega^2 R\right) \tag{17.7a}$$

where p is the liquid hydrostatic pressure, K the cake permeability, μ the liquid viscosity, and Ω the angular speed of rotation. This shows that the liquid flux relative to the motion of the moving solids is driven by the pressure gradient and the centrifugal gravity $\rho\Omega^2 R$. When $u_s = 0$, with Eq. (17.1) this further reduces to

$$\frac{dp}{dR} - \rho\Omega^2 R = -\frac{\mu q}{K} = -\frac{\mu \tilde{Q}}{2\pi K R} \tag{17.7b}$$

Force balance under equilibrium. A force balance over a circular segment of the cake, as shown in Fig. 17.1 (similar to Fig. 2.12), reveals that

$$d(pR) + d(p_s R) - (p + k_0 p_s)dR = (\rho\varepsilon + \rho_s\varepsilon_s)\Omega^2 R^2\, dR$$

where k_0 is the ratio of the lateral force (similar to the hoop stress in Fig. 2.12, but compressive rather than tensile) to the radially directed compaction stress. This has an effect similar to the Poisson ratio effect. Usually $k_0 = 0.4$–0.7. The foregoing equation is best written as

$$\frac{dp}{dR} + \frac{dp_s}{dR} + (1 - k_0)\frac{p_s}{R} = (\rho\varepsilon + \rho_s\varepsilon_s)\Omega^2 R \tag{17.8}$$

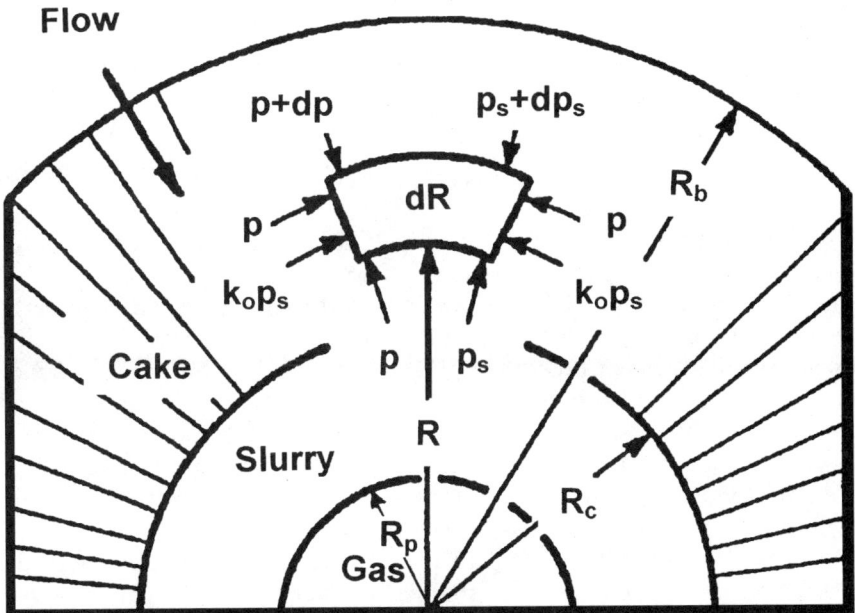

Figure 17.1 Liquid pressure and solids compaction stress acting on an element in a centrifuge cake.

For negligible solids velocity, after combining Eqs. (17.7b) and (17.8) and eliminating dp/dR, the liquid flux becomes

$$q = \varepsilon u = \frac{\tilde{Q}}{2\pi R} = \frac{K}{\mu}\left[\frac{dp_s}{dR} + (1-k_0)\frac{p_s}{R} - \varepsilon_s \Delta\rho \Omega^2 R\right] \quad (17.9)$$

with $\Delta\rho = \rho_s - \rho$.

For compactible cakes it is generally assumed that both ε_s and K are functions of the compaction stress p_s. Indeed, a useful form borne out of fitting experimental data is as follows,[2]

$$\varepsilon_s = \varepsilon_{s0} p_s^a, \quad p_s > p_{s0} \quad (17.10a)$$

$$\varepsilon_s = \varepsilon_{s0}, \quad p_s \leq p_{s0} \quad (17.10b)$$

and

$$K = K_0 p_s^{-b}, \quad p_s > p_{s0} \quad (17.11a)$$

$$K = K_0, \quad p_s \leq p_{s0} \quad (17.11b)$$

where a and b as well as K_0 and ε_{s0} are property constants depending on the cake. These relationships hold for discrete ranges of p_s, and as such the property constants can be specified for different ranges of p_s.

Transient Consolidation

The transient consolidation equation is obtained by combining Eqs. (17.3), (17.5), and (17.9). Thus

$$\left(\frac{\partial p_s}{\partial t}\right)\left(\frac{d\varepsilon_s}{dp_s}\right) = \frac{1}{R}\frac{\partial}{\partial R}\left\{\frac{KR}{\mu}\left[\frac{\partial p_s}{\partial R} + (1-k_0)\frac{p_s}{R} - \varepsilon_s(p_s)\Delta\rho\Omega^2 R\right]\right\} \quad (17.12)$$

where ε_s and $d\varepsilon_s/dp_s$ are functions of p_s. The latter can be obtained, for example, by taking the derivative of Eq. (17.10a), where $d\varepsilon_s/dp_s = a\varepsilon_{s0}p_s^{a-1}$. Equation (17.12) is a nonlinear diffusive equation. The problem is well defined with Eq. (17.12) together with an initial condition at $t = 0$ and two boundary conditions, imposed at the cake surface $R = R_c$, where $p_s = 0$, and at the bowl wall $R = R_b$, where an assumed constant cake consistency ε_{sb} has been reached. (Note that this also defines p_{sb}, given the assumed unique relationship between solids volume fraction and compaction stress.) The resulting nonlinear partial differential equation is best solved numerically.

The result provides a transient cake consolidation behavior with filtration of the mother liquid via the perforate bowl wall or expressed out inwardly from the cake interior to the surface for the solid-wall bowl.

Steady-State Consolidation

Compactible cake

While steady-state consolidation is somewhat simpler, it is still complicated. When the time dependence in Eqs. (17.3) and (17.4) disappears, from Eqs. (17.1) and (17.2) we have

$$qR = u\varepsilon = \frac{\tilde{Q}}{2\pi} = \text{constant}$$

$$q_s R = u_s \varepsilon_s = \frac{\tilde{Q}_s}{2\pi} = \text{constant}$$

This shows that the flow rate of solids $\tilde{Q}_s$ and that of liquid $\tilde{Q}$ are constant, independent of radius R. Using this, Eq. (17.9) can be rewritten as

$$\frac{dp_s}{dR} + (1-k_0)\frac{p_s}{R} - \varepsilon_s(p_s)\Delta\rho\Omega^2 R = \frac{\mu\tilde{Q}}{2\pi K(p_s)R} \qquad (17.13a)$$

The companion equation is obtained by integrating Eq. (17.7b) over R from the pool surface to the interior of the cake. The liquid equation becomes

$$p = \tfrac{1}{2}\rho(R^2 - R_p^2)\Omega^2 - \frac{\mu\tilde{Q}}{2\pi}\int_{R_c}^{R}\frac{dR}{RK(p_s)} \qquad (17.13b)$$

where R_c and R_p are the radii to the surface of the cake and the pool, respectively. Note that in Eq. (17.13b) K is a function of p_s. Equation (17.13a) is a nonlinear first-order ordinary differential equation. It can be solved numerically, such as by a fourth-order Runge–Kutta routine, once the ε_s and K are defined by constitutive relationships such as Eqs. (17.10) and (17.11). Once p_s is calculated, ε_s and K can also be determined. The liquid equation [Eq. (17.13b)] can be determined. The first boundary condition is $p_s = 0$ at $R = R_c$. For a given assumed $\tilde{Q}$, p is determined until the calculated p matches the second boundary condition imposed at the bowl wall. For a filtering centrifuge with negligible filter medium resistance, this corresponds to $p_b = 0$. Iterations are set up on $\tilde{Q}$ until $p_b = 0$ at the basket wall. The following example[3] is used to illustrate the behavior.

Example. Given: $R_p = 0.3$ m, $R_c = 0.4$ m, $R_b = 0.5$ m, $\Omega = 1000$ rpm, $\tilde{Q} = 2.64 \times 10^{-4}$ m²/s (part of the solution, determined from iterations), and $k_0 = 0.5$. Then

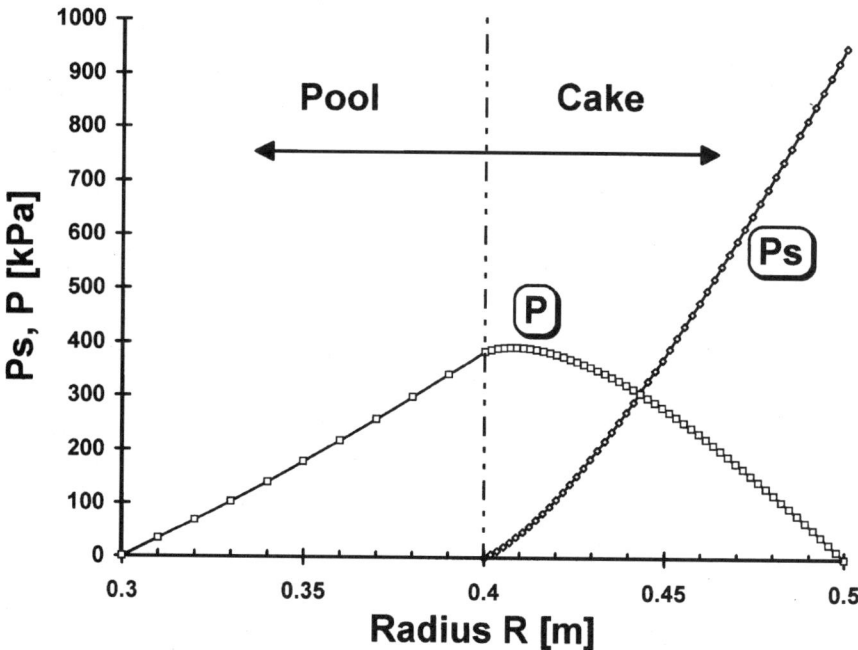

Figure 17.2 Liquid pressure and solids compaction stress distribution in filtering centrifuge.[3]

$\varepsilon_s = 0.135 p_s^{0.059}$, $\quad 1000 \text{ kPa} > p_s > 8 \text{ kPa}$

$\varepsilon_s = 0.23$, $\quad p_s \leq 8 \text{ kPa}$

$K = 1.02 \times 10^{-12} p_s^{-0.36} [\text{m}^2]$, $\quad 1000 \text{ kPa} > p_s > 8 \text{ kPa}$

$K = 4 \times 10^{-14} [\text{m}^2]$, $\quad p_s \leq 8 \text{ kPa}$

Figures 17.2 to 17.4 show the results of compaction stress, liquid pressure, solids volume fraction, and permeability. As can be seen in Fig. 17.2, the compaction stress increases from 0 at the cake surface, $R_c = 0.4$ m, to a maximum at the perforate basket wall, $R_b = 0.5$ m, whereas starting from the pool surface at $R_p = 0.3$ m the liquid pressure first increases following a parabolic law in the region above the cake, reaches a maximum of 384 kPa at the cake surface located at $R_c = 0.4$ m, and decreases from that point to zero at the basket wall. Note that the decrease of liquid pressure within the cake follows a concave downward movement for a compactible material, where both solids volume fraction and permeability are functions of the compaction stress. This contrasts the behavior of an incompactible cake, as shown in Fig. 7.2, where both the solids volume fraction and the permeability are constant, independent of the stress, and where the curve is concave upward. Most inter-

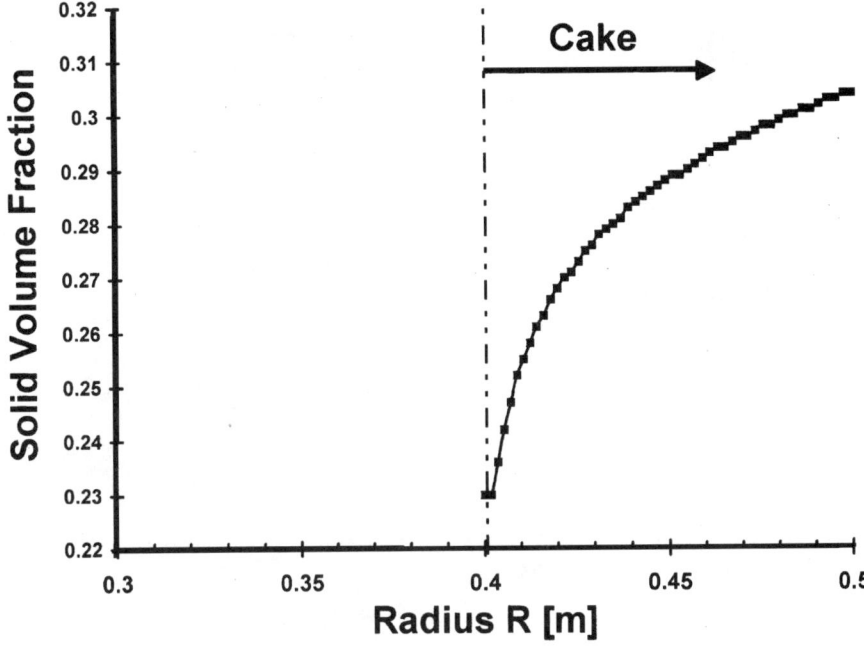

Figure 17.3 Solids volume fraction distribution in filtering centrifuge.[3]

estingly the solids volume fraction increases from 0.23 to over 0.3, as depicted in Fig. 17.3, while the cake permeability decreases from 4×10^{-14} m² to 7.2×10^{-15} m² at the basket wall (Fig. 17.4).

Incompactible cake

When the solids volume fraction and the cake permeability are constant and independent of the compaction stress p_s, Eq. (17.13a) can be integrated to yield

$$p_s = \Delta\rho\Omega^2 R_c^2 \varepsilon_s \frac{(R/R_c)^2 - (R_c/R)^{1-k_0}}{3 - k_0} + \frac{\mu\tilde{Q}}{2\pi K} \frac{1 - (R_c/R)^{1-k_0}}{1 - k_0} \quad (17.14)$$

where $\tilde{Q} > 0$ for filtering centrifuges with liquid draining out toward the perforate bowl wall, and $\tilde{Q} < 0$ for sedimenting centrifuges with liquid from the cake expressing toward the cake surface, away from the solid bowl wall. The latter retards compaction whereas the former enhances compaction. However, a sedimenting centrifuge has lesser limitation on structural strength since a filtering centrifuge operates at much higher speeds Ω and therefore higher G. At the termination of expression, in the case of a sedimenting centrifuge, the flow rate stops, and under equilibrium the stress distribution is

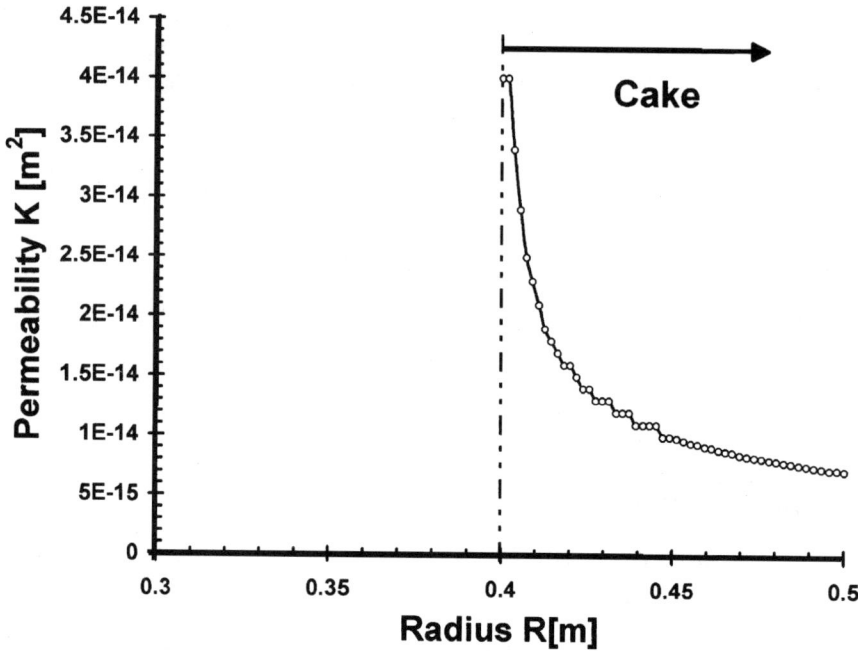

Figure 17.4 Cake permeability distribution in filtering centrifuge.[3]

$$p_s = \Delta\rho\Omega^2 R_c^2 \varepsilon_s \frac{(R/R_c)^2 - (R_c/R)^{1-k_0}}{3 - k_0} \quad (17.15a)$$

When $k_0 = 1$, this corresponds to the isotropic condition for the compaction stress, with the result further simplified to

$$p_s = 1/2 \Delta\rho\Omega^2 R_c^2 \varepsilon_s [(R/R_c)^2 - 1] \quad (17.15b)$$

This relationship has been used in Chap. 16 for high-solids dewatering of sewage sludge.

For incompactible cake, assuming $u_s = 0$, the liquid flow equation [Eq. (17.7b)], when combined with Eq. (17.1) and together with the empirical relationship on pressure drop across a filter medium $p_b = \tilde{Q}\mu r_m/2\pi$, can be integrated to yield

$$\frac{\mu\tilde{Q}}{2\pi R_b} = \frac{\frac{1}{2}\rho(R_b^2 - R_p^2)\Omega^2}{(R_b/k)\ln(R_b/R_c) + r_m} \quad (17.16a)$$

This relationship pertains to a filtering centrifuge and has been established previously in Chap. 7. At the end of the filtration, the pool level decreases and at some point R_p reaches R_c, beyond which drainage inside the cake occurs as discussed in Chap. 7. For sedimenting centrifuges R_b and R_p both remain constant. Also, instead of the pressure drop across a filter medium the boundary condition is the liquid pressure at the bowl wall p_b, which is part of the unknown and still to be determined. Similarly Eq. (17.7b) can be integrated to yield an expression for the sedimenting centrifuge. Thus

$$\frac{\mu \tilde{Q}}{2\pi K} = -\frac{p_b - \frac{1}{2}\rho(R_b^2 - R_p^2)\Omega^2}{\ln(R_b/R_c)} \qquad (17.16b)$$

where $\tilde{Q} < 0$ and $p_b > 1/2\rho\Omega^2(R_b^2 - R_p^2)$.

Moderately compactible cake

For moderately compactible cake, where a small change in solids volume fraction and permeability occurs, a linearized stress-rate calculation can be made by assuming average properties of the solids volume fraction and cake permeability. Equation (17.14) can be written in dimensionless form as

$$p_{sD} = (\varepsilon_s)_{av} \frac{R_D^2 - (R_D)^{k_0-1}}{3 - k_0} + \tilde{Q}_D \frac{1 - (R_D)^{k_0-1}}{1 - k_0} \qquad (17.17)$$

and

$$p_{sD} = \frac{p_s}{\Delta\rho\Omega^2 R_c^2} \qquad (17.18a)$$

$$\tilde{Q}_D = \frac{\mu \tilde{Q}}{2\pi K_{av} \Delta\rho\Omega^2 R_c^2} \qquad (17.18b)$$

$$R_D = \frac{R}{R_c} \qquad (17.18c)$$

In Figs. 17.5 and 17.6 p_{sD} is calculated using Eq. (17.17) for $(\varepsilon_s)_{av} = 0.4$ and 0.6. $\tilde{Q}_D$ is set to positive (liquid flow along increasing radius) up to 4 for filtering perforate-wall centrifuges, and to negative (liquid flow in the direction of decreasing radius) for sedimenting solid-wall centrifuges. When $\tilde{Q}_D = 0$, the expression for sedimenting centrifuges has stopped and an equilibrium has been established, where the centrifugal force due to the cake contents (solids and liquid) is supported by both the hydrostatic pressure and the compaction stress transmitted through the particle-particle contact in the cake solids matrix; that is, the right-hand side of Eq. (17.13a) is zero. Note that for compactible cake, with the compaction stress and the permeability and

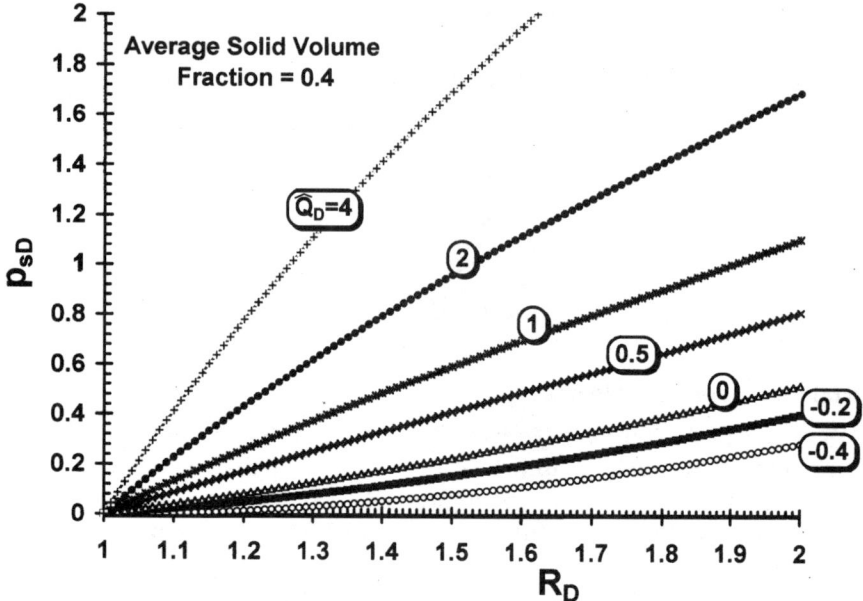

Figure 17.5 Dimensionless solids compaction stress p_{sD} versus dimensionless radius R_D with $\tilde{Q}_D = -0.4$ to 4 and $\varepsilon_s = 0.4$, $k_0 = 0.5$.

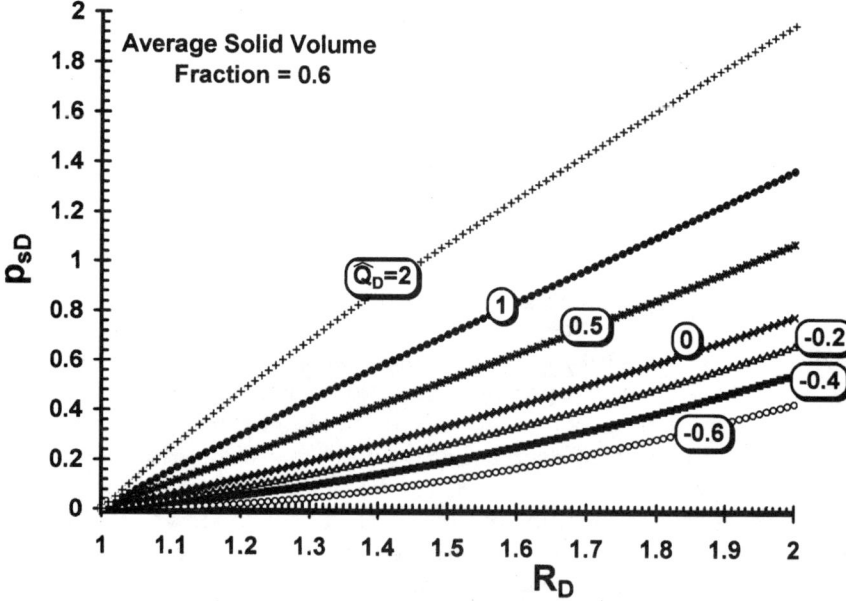

Figure 17.6 Dimensionless solids compaction stress p_{sD} versus dimensionless radius R_D with $\tilde{Q}_D = -0.6$ to 2 and $\varepsilon_s = 0.6$ and $k_0 = 0.5$.

solids volume fraction interacting, the compaction stress p_{sD} increases much more rapidly with increasing radius. Unlike in Figs. 17.5 and 17.6, the curve goes concave upward (see Fig. 17.2).

It appears that the dimensionless compaction stress p_{sD} is much higher for filtering centrifuges, where $\tilde{Q}_D > 0$, than for sedimenting centrifuges, where $\tilde{Q}_D < 0$. This is correct when the two machines are operating at the same G. However, this interpretation is not appropriate given that both p_s and $\tilde{Q}$ are normalized with respect to Ω^2, and sedimenting centrifuges typically operate at between two and more than four times higher gs than their filtering counterparts. Under equilibrium with $\tilde{Q}_D = 0$, the viscous drag is absent, and ultimately higher compaction stress results. Based on the foregoing discussion, filtering centrifuges are excellent for incompactible to moderately compactible cake with relatively coarse solids where a lower G force is adequate. For highly compactible cake, filtering centrifuges are not suitable because of the formation of the impermeable "skin" cake layer adjacent to the filter medium and possible blinding of the medium by infiltration of the fines. Significant pressure drops occur at the skin layer and the filtration rate is reduced precipitously. In this case, sedimenting centrifuges should be employed with higher G forces.

This chapter provides a brief theoretical treatment of cake compaction. The subject is difficult due to the strong interaction between the constitutive properties of the material and the compaction stress exerted within the cake matrix, which renders the problem highly nonlinear, not to speak of complexities such as time-dependent creep, which may be present.

References

1. F. Tiller, "A Comparison of Pressure and Centrifugal Filtration," in *Proc. Am. Filt. Sep. Soc. Ann. Conf.* (Freeport, TX, Sept. 15, 1989).
2. F. Tiller and N. B. Hysung, "Comparison of Compacted Cakes in Sedimenting and Filtering Centrifuges," in *Proc. Am. Filt. Sep. Soc. Ann. Conf.* (Mar. 19–24, 1990), vol. 2.
3. F. Tiller, personal communication.

Appendix A

Troubleshooting Industrial Centrifuges

In this appendix our discussion is restricted to troubleshooting industrial centrifuges that are not operated correctly, which will lead to process inefficiency and consequential mechanical problems, or to strictly mechanical problems.

The following are general guidelines for some selected industrial centrifuges such as pusher, solid-bowl decanter, screen-bowl, disk, and batch basket centrifuges. Symptoms, diagnostics, and possible solutions to both mechanical and process problems are presented.

Troubleshooting Pusher Centrifuges

1. Cake too wet
   - Screen openings blocked with solid particles. Clean screen or use tighter screen with aperture that opens up toward larger radius for self-cleaning (such as wedge wire screen).
   - Screen opening too small, resulting in less drainage area (that is, percent open area) on screen deck. Use larger screen opening, especially near cake discharge end, but smaller screen opening near feed zone.
   - Retention time too short. Reduce push frequency to increase retention time of cake on screen.
   - G too low to dewater cake on screen deck. Increase G force accordingly; watch for high filtrate solids concentration.

2. High filtrate solids
   - Screen too open, especially basket area near feed zone. Tighten screen opening.
   - Product cake lost through large gap between push plates and basket inner diameter, leaking into filtration compartment. Typical gap between push plate and basket diameter should be 0.010–0.015 in (0.25–0.375 mm) metal to metal and close clearance for Teflon ring and metal basket.
3. High vibration
   - Insufficient stroke frequency, resulting in thick cake interfering with accelerator. Increase stroke frequency.
   - Feed unevenly distributed (circumferentially) onto first-stage basket, causing unbalance. Install better feed distributor and accelerator system (see cone and double-disk accelerators discussed in Chap. 11).
   - Cake does not form on first-stage basket, too wet. Feed too dilute or rate too high due to:

     a. Feed slurry too dilute in solids, that is, high liquid content, which needs to filter out by first-stage basket. Increase feed solids concentration by prethickening feed prior to pusher centrifuge such as using hydrocyclone, where underflow of hydrocyclone, typically operated at 40–60% w/w of suspended solids, is fed to a pusher.

     b. Feed rate too high. Better feed acceleration system, such as double disk with curve vanes and smoothener.
   - Basket speed too high, cake "wedges" into screen slots. Reduce basket speed and/or slot size.
   - Rubbing of basket on product buildup behind basket heads or outside basket, manifest in spikes in torque and current. Determine cause of buildup and fix cause.
   - Loose front bushing. Tighten bushing or replace if defective.
   - Lost plows from basket or broken bolts causing unbalance. Replace plows and bolts.
   - Defective isolators or foundation structure. Replace isolators and reinforce foundation structure where appropriate.

Decanter Process Troubleshooting

1. Effluent solids too high

   a. Without polymer
      - Feed solids concentration increased. Reduce feed solids or reduce rate.

- Mass and/or volumetric rate increased. Reduce rate.
- Pool too low. Increase pool marginally so as not to upset cake dryness by reducing dry beach.
- Δ too low or too high. Adjust Δ appropriately by changing gear ratio, hydraulic, or electric back drive.
- Process temperature dropped resulting in higher viscosity of suspension. Restore process temperature.
- G or bowl speed too low. Increase bowl speed by sheave change or adjust variable-speed drive.
- Finer particle-size distribution in feed slurry (for nonbiological solids). Reduce feed rate or take appropriate remedial action on upstream reactor, crystallizer, or mill,....
- Worn blade resulting in very thick unconveyed heel layer left inside bowl; rough operation and vibration also follow. Reblade conveyor.

b. With polymer
- Polymer dosage dropped. Restore polymer dosage.
- Polymer solution too concentrated to be effective. Dilute polymer keeping the polymer dosage rate (that is higher volumetric rate and lower polymer concentration).
- Polymer degraded due to age. Use more polymer to consume remaining batch or change to fresh polymer.
- Feed characteristics changed, such as finer solids. Increase polymer dosage and use torque or centrate to control polymer consumption or monitor manually.

2. Cake solids too wet
- Solids rate too high. Reduce rate appropriately.
- Δ too high with inadequate dewatering time. Reduce Δ appropriately, but not too low to a point causing higher centrate solids.
- Pool too deep (for cake that requires dry beach for dewatering). Reduce pool depth to a certain extent in order not to compromise centrate quality.
- Speed or G too low. Adjust bowl speed appropriately, watch out for higher torque.
- Finer particle-size distribution in feed slurry (for nonbiological solids). Reduce feed rate or take appropriate remedial action on upstream reactor, crystallizer, or mill,....

3. Cake not conveying up conical beach
- Cake too fluid. Feed behavior changed, take appropriate actions.
- G too high. Reduce G; higher centrate solids are expected.

- Feed solids have high specific gravity. Increase Δ and/or increase pool depth.
- Clearance between blade tip and bowl increased. Reblade conveyor with close clearance or install false beach liner such as gypsum (provided it is compactible with process) to close up clearance.
- Worn bowl strips. Replace strips and ensure that clearance between strip and blade tip is within specifications.
- Glazed solids layer left on bowl wall causing cake to orbit without conveying. Remove residual cake heel left on bowl wall; reblade conveyor to tighter clearance with respect to bowl wall or strips lining bowl wall.
- Feed containing fines. Correct upstream process; otherwise increase Δ, increase pool, and/or reduce G.
- Operating pool too shallow. Increase pool marginally until cake conveyance. Too deep a pool might lead to wetter cake as dry beach section is reduced.
- Δ too small. Increase Δ appropriately but not too high, reducing retention time and compromising cake dryness.
- Climb angle too steep for cake. Reduce lead, thus reducing climb angle.

4. High torque with frequent breaking of shear pin or tripping. When a shear pin is used as a protection device, it should be sized appropriately.
    - Worn blade tip with excessive blade-tip torque. Reblade conveyor tip with sharp tip and relief in back of blade. Tip should have a measured clearance from bowl or strip of 0.03–0.06 in (0.75–1.5 mm) or less.
    - Cake not conveying up beach. See all factors discussed in item 3.
    - Pool too shallow. Increase pool to reduce torque. For biological sludge cake, pool increase should be minimal.
    - Rate too high. Torque increases with solids throughput; reduce rate appropriately.
    - Δ too low. Conveyance torque and differential speed vary inversely with each other. Low Δ results in high torque due to thicker cake solids and vice versa. Increase Δ judiciously not to cause wet cake.
    - Speed too high. Part of the total torque is due to solids conveyance, which is proportional to G or bowl speed squared. Increase back-drive torque capacity or simply increase Δ or pool.
    - Drag from solids buildup external to rotating bowl and convey-

or. Reduce solids buildup (usually with torque spikes and vibrations).
- Thrust bearing worn and/or faulty spline. Replace thrust bearing and/or fix spline shaft.
- Misalignment or problem with gear unit. Realign or fix gear unit problem.
- Plugged solids hopper problem or generally experience "binding." Fix binding problem.
- Foreign materials in machine. Remove material and remove conveyor from rotating assembly if necessary.

5. Feed leakage back along pipe and trunnion
    - Feed rate too high. Better feed accelerator.
    - Solids settle inside feed compartment plugging up discharge outlet. Better feed accelerator.
    - Venting problem. Check to determine overpressure zone; incorporate adequate venting on solid and liquid discharge hoppers.
    - Poor feed distribution and accelerator. Serious drooping of feed as it is discharged to a target at a distance from exit plane of feed pipe. Feed in thin film splashing off from intersection or discontinuity in surfaces. Incorporate better feed distributor and accelerator system.
    - Large annular opening between feed pipe and rotating trunnion. Close up gap with baffle.

6. Vibration. Read out vibrations using a vibration meter at locations as recommended in Fig. A.1.
    a. No process load

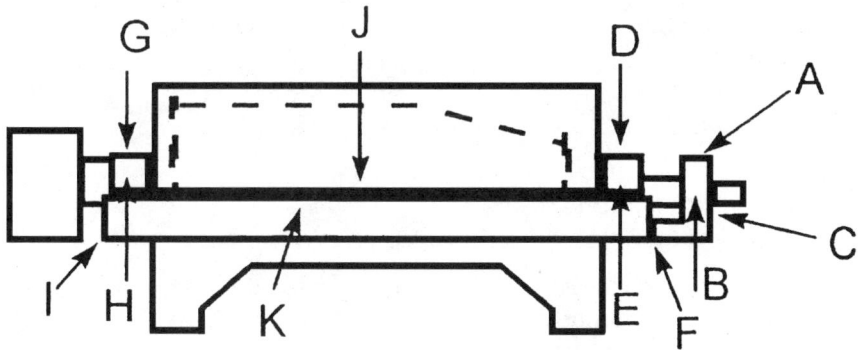

Figure A.1 Locations of vibration meters.

- Rotating assembly (bowl and conveyor) unbalanced. Balance rotating assembly.
- Misalignment in rotating assembly, either bowl or conveyor, or both. Realign rotating assembly.
- Improper lubrication. Check and correct lubrication.
- Improper adjustment of vibration isolators. Adjust isolators.
- Leftover drier solids from previous runs. Flush at low bowl rpm (about 100 rpm) and high differential speed (if possible). Automate this cleaning procedure as part of machine shut-off procedure.
- Worn bearings. Replace bearings.
- Gear box improperly aligned. Align gear box.
- Pillow block bearing damage. Inspect and replace bearings.
- Loose parts or discharge chute in contacting rotating machine. Tighten parts and remove stationary parts from contacting rotating assembly.
- Uneven wear of screw conveyor, including missing wear protective tiles. Reblade and rebalance.

b. With process load
- High torque and possible chatter. Reduce torque (see above), which also reduces chatter; implement chatter torque dampener device to take energy out of system.
- Poor cake conveyance. Increase Δ and pool to increase conveyance.
- Thrust bearing problem. Fix bearing.
- Misalignment. Realign.
- Increased heavier medium or coarser particle size. Better control upstream process.
- Portion of conveyor flights may be blocked by solids bridging across flights. Determine cause and flush machine.
- Drag on rotating assembly from undischarged accumulated solids in hopper; bowl outer surface would be polished. Fix solids discharge mechanism to avoid buildup.

Screen-Bowl Centrifuge Process Troubleshooting

The troubleshooting strategy of the solid bowl is directly applicable to the solid-bowl portion of the screen bowl. Therefore only the strategy pertaining to the screen is summarized here.

1. Cake too wet

- Screen openings blocked with solid particles. Clean screen or use tighter screen.
- Cake heel layer buildup blinding off screen due to increased blade-tip clearance from screen surface. Reblade conveyor.
- Screen opening too small, resulting in less drainage area (that is, percent open area) on screen deck. Use larger screen opening, especially near cake discharge end.
- Pool too deep. Cake has no time to prethicken prior to screen; reduce pool appropriately below spillover.
- Retention time too short. Increase gear ratio to increase retention time; expect higher torque due to thicker cake on screen surface.
- G too low for cake dewatering on screen deck. Increase G force accordingly. For chemical applications without erosion, G can be up to $2000g$, but for mineral applications G is usually under 500–$700g$ to reduce erosion.

2. High filtrate solids
- Screen too open, especially section toward beach. Tighten screen opening.
- Pool set too close to spill over screen, leading to wet cake and high filtrate flow taking fine solids with filtrate liquid through screen opening. Reduce pool to get a small dry beach where cake can consolidate before presenting to screen; this reduces loss of fines through screen.

Disk Centrifuge Process Troubleshooting

1. Effluent solids too high
- Speed too low. Increase speed.
- Solids blanket getting inside disk stack, taking up clarifying volume. Adjust liquid-solid interface toward large diameter of disk stack to fully utilize disk stack for clarifying and settling finer solids.
- Feed rate too high. Reduce rate appropriately.
- Feed solids concentration too high. Reduce feed slurry volumetric rate or adjust feed solids concentration.
- Solids discharge mechanism problematic. Manual discharge—too high feed solids concentration; intermittent discharge—increase frequency to solids discharge; nozzle discharge—nozzle opening too small or too large.
- Finer feed solids. Reduce rate or rectify problems upstream (separation stage, crystallizer, mill, reactor,....)

2. Cake too wet
   - Solids feed rate too high or too low. Overload unit, resulting in high discharge rate; underfeed unit, where compaction force on cake is minimal because of thin cake.
   - Speed too low. Increase speed and G.
   - Fines in feed increased. Reduce rate or fix upstream fines.
3. Light-liquid phase contaminated by heavy-liquid phase
   - Interface radius too small. Increase interface radius by adjusting ring dam discharge diameter or adjusting back-pressure valve for discharge via pressurized centripetal pump.
   - Solids plugging up heavy-phase discharge (opening to top disk). Clean up plugged area to allow a continuous discharge of heavier liquid phase.
4. Heavy-liquid containing light phase
   - Interface radius too large. Decrease interface radius by adjusting ring dam discharge diameter or by adjusting back-pressure valve for discharge via pressurized centripetal pump.
   - Solids plugging up light-phase discharge. Clean up plugged area to allow continuous discharge of lighter liquid phase.
   - Fines in feed increased. Reduce rate or fix upstream fines.

Basket Centrifuge Process Troubleshooting

1. Filtrate solids too high
   - Cloth opening too large. Use tighter opening cloth.
   - Hole in filter cloth or worn cloth. Inspect and replace cloth.
   - Finer particle-size distribution in feed slurry. Take appropriate remedial action on upstream reactor, crystallizer, mill,..., or change to tighter cloth.
2. Very low filtration rate
   - Building up of cake heel, which is glazed by repeated contacting with plow. Need to wash heel; back blow cake heel; purge cake heel using air gun from cake plow.
3. Cake solids too wet
   - Solids rate too low. Reduce feed loading and cake height appropriately.
   - Inadequate dewatering time. Increase dewatering time.
   - Speed or G too low. Adjust bowl speed appropriately, look for higher torque and current.

- Finer particle-size distribution in feed slurry (for nonbiological solids). Reduce feed rate or make appropriate remedial action on upstream reactor, crystallizer, mill,....

4. Cake containing higher impurities
   - Impurity level in feed increased. Increase wash ratio, kg wash liquid/kg solids or kg wash liquid/kg mother liquor.
   - Inadequate wash. Increase wash ratio.
   - Cake too thick. Reduce cake height.

5. Basket not charging optimally
   - Cake depth too low. Reset load detector so as to fill solids to 70–80% of volume set by overflow weirs; may need this to be carried out in steps of feeding followed by filtration until target cake solids depth has been reached.
   - Cycle control mechanism malfunctions. Fix PLC or DCS controls.
   - Cake nonuniformly distributed. Reduce feed solids concentration to allow more liquid to smooth any uneven cake distribution during feeding; use improved feed distributor and accelerator.

6. Excessive vibration
   - Concentrated feed slurry eccentrically bringing out maldistribution of feed onto basket using J-pipe. Replace with multiple J-pipe manifold or use dilute feed if possible.
   - Plow interfering with cake prior to unloading. Bring plow's rest position to a smaller radius away from interfering with the cake.
   - Accumulated undischarged cake interfering with basket. Determine cause of cake accumulation.
   - Loose or defective isolators. Tighten or replace isolators.

7. Cake plow problem
   - Cake plow in the direction of basket rotation gets deflected by the cake during cake unloading.
   - Reverse plow direction to meet basket rotation.

While these guidelines are given only for pusher, solid-bowl, screen-bowl, disk, and basket centrifuges, most problems are also shared among other types of centrifuges and therefore may be applicable with some modifications.

It is of utmost importance to determine whether the process problems arise over a period of time or occur instantaneously. In the first case this may be related to wear and tear of the centrifuge, whereas in the latter, it is rather due to an abrupt change in feed condition.

Sometimes a slow gradual change in the upstream process units (such as crystallizer, reactor, mill, preliminary separation,...) can give rise to a corresponding gradual response from the centrifuge over the same period of time.

Appendix B

Symbols

Symbol	Definition	SI units (English units)
A	Cross-sectional area	m² (ft²)
A_{noz}	Nozzle cross-sectional area	m² (ft²)
$A_{Rec\,d}$	Equivalent settling area to obtain recovery of particle size d equal to Rec_d	m² (ft²)
a	Acceleration	m/s² (ft/s²)
a	Index constant, Eq. (17.10a)	*
B_o	Bond number, Eq. (7.19)	*
b	Axial width	m (ft)
b	Index constant, Eq. (17.11a)	*
C_f	Frictional coefficient	*
C_p	Particle shape factor	*
C_v	Viscous drag coefficient	*
D	Diameter of rotor or bowl	mm (in)
D_{fp}	Diameter of feed pipe nozzle	mm (in)
D_p	Polymer dosage	kg/t (lb/ton)
d	Particle size	mm, μm (mesh)
d_c	Cut-off particle size	mm, μm (mesh)
d_g	Grit size	mm, μm (mesh, in)
d_h	Hydraulic diameter	mm, μm (in)
d_m	Maximum size	mm, μm (in)
d_0	Reference particle size, $= 1$ μm	μm (in)
$d_{50\%}$	Median particle size	mm, μm (mesh, in)
d_{noz}	Nozzle diameter	mm (in)
Ek	Ekman number, Eq. (2.6)	*
f	Push frequency	cycle/s
F	Force	N (lb$_f$)

Appendix B

Symbol	Definition	SI units (English units)
F_e	Cumulative under size of centrate	%
F_f	Cumulative under size of feed slurry	%
F_G	Centrifugal force	N (lb$_f$)
F_g	Earth's gravitational force	N (lb$_f$)
F_p	Push force	N (lb$_f$)
F_s	Cumulative under size of cake solids	%
G	Centrifugal gravity	m/s^2 (ft/s^2)
g	Earth's gravity or acceleration	m/s^2 (ft/s^2)
gr	Gear ratio	*
H	Hindered settling factor	*
h	Cake height	mm (in)
h_{cap}	Capillary rise	mm (in)
h_p	Pool depth	mm (in)
h_{pe}	Penetration of rod	mm (in)
K	Cake permeability	m^2 (ft^2)
K	Coefficient, Eq. (16.8)	*
k	Exponent, Eq. (16.3)	*
k_0	Ratio of lateral compressive stress to radial compaction stress	*
L	Characteristic length or length of centrifuge	m (ft)
L_c	Clarifier length of decanter	mm (in)
L_i	Basket length of ith stage	mm (in)
L_{lead}	Lead length of screen conveyor	mm (in)
L_{screen}	Screen length	mm (in)
M_c	Moisture content, = wt. moisture/wt. dry solid cake	kg/kg (lb/lb)
M_{dry}	Mass of dry sample	g (lb)
M_{wet}	Mass of wet sample	g (lb)
m	Mass	kg (lb)
m_{se}	Centrate/effluent solids throughput (dry solid)	kg/h (lb/h)
m_e	Centrate/effluent throughput (solid + liquid)	kg/h (lb/h)
m_f	Feed throughput (solid + liquid)	kg/h (lb/h)
m_{sf}	Feed solids throughput (dry solid)	kg/h (lb/h)
m_{ss}	Cake solids throughput (dry solid)	kg/h (lb/h)
m_s	Cake throughput (solid + liquid)	kg/h (lb/h)
N_c	Capillary number, Eq. (7.17)	*
N_{noz}	Number of nozzles, Eq. (4.5), Eq. (7.23)	*
n	Exponent in Eq. (7.15a), Eq. (16.8)	*
n	Number of spaces between adjacent disks in disk stack, or number of leads	*
P	Power	kW (hp)
P_{acc}	Power for acceleration of feed to speed, Eq. (2.19)	kW (hp)

Symbols

Symbol	Definition	SI units (English units)
P_{con}	Power for cake conveyance, Eqs. (2.20), (2.21a)	kW (hp)
p	Pressure	Pa (lb$_f$/in^2)
p_{atm}	Atmospheric pressure	Pa (lb$_f$/in^2)
p_b	Back pressure or pressure at bowl/basket wall	Pa (lb$_f$/in^2)
p_d	Dimensionless pressure, Eq. (7.22)	*
p_s	Solids compaction stress	Pa (lb$_f$/in^2)
p_{sD}	Dimensionless compaction stress	*
Δp	Pressure superimposed in addition to centrifugation	Pa (lb$_f$/in^2)
Q	Volumetric liquid flow rate	L/min, cm^3/h (gal/min)
$\tilde{Q}$	Volumetric filtration rate/axial width of basket	L/min · m (gal/min · ft)
$\tilde{Q}_D$	Dimensionless flow rate	*
Q_f	Volumetric feed rate	L/min, m^3/h (gal/min)
Q_n	Total wash liquid rate	L/min, m^3/h (gal/min)
$\tilde{Q}_s$	Volumetric solids flow rate/axial length of basket (Chap. 17)	L/min · m (gal/min · ft)
q	Liquid flux	m^3/m^2/s (ft^3/ft^2/s)
q_n	Wash liquid rate per nozzle	L/min, m^3/h (gal/min)
q_s	Solids flux	m^3/m^2/s (ft^3/ft^2/s)
RF	Repulp factor, Eq. (2.18)	*
%Rec$_d$	Percent recovery of particle size under size d in centrate, Eq. (6.4)	%
%Rec$_e$	Percent recovery of solids in centrate	%
%Rec$_s$	Percent recovery of solids in cake	%
Re	Reynolds number	*
Ro	Rossby number, Eq. (2.7)	*
R	Radius from axis of rotation	mm (in)
R_b	Bowl or basket radius	mm (in)
R_c	Cake surface radius	mm (in)
R_D	Ratio of radius from axis to that of cake surface, $= R/R_c$	mm (in)
R_{intake}	Intake radius, Eq. (4.1)	mm (in)
R_o	Original radial location	*
R_p	Pool surface radius	mm (in)
r	Radius	mm (in)
r_m	Filter medium resistance	
r_0	Radius of capillary tube	mm (in)
S	Liquid saturation	*
S	Sedimentation coefficient, Eq. (3.1)	s

Appendix B

Symbol	Definition	SI units (English units)
S_c	Capillary saturation	*
S_F	Film saturation, Eq. (7.15a)	*
S_p	Pore liquid saturation or bound water saturation	*
S_T	Transient saturation	*
S_{total}	Total saturation	*
S_z	Pendular saturation at particle contacts	*
S_∞	Equilibrium saturation	*
s	Stand-off distance of feed nozzle from target	mm (in)
sg	Specific gravity	*
T	Temperature	°C (°F)
T_{in}	Input torque	$N \cdot m$ ($lb_f \cdot in$)
T_p	Torque at pinion of gear box, Eq. (2.21c)	$N \cdot m$ ($lb_f \cdot in$)
T_{sp}	Torque at spline, Eq. (2.21b)	$N \cdot m$ ($lb_f \cdot in$)
t	Time	s, min, h
t_b	Bowl wall thickness	mm (in)
t_d	Dimensionless dewatering time, Eq. (7.15b)	*
u	Relative velocity in rotating frame or percolation superficial velocity, Eq. (7.1)	m/s (ft/s)
u	Tangential velocity, axial or longitudinal velocity, and filtration velocity	m/s (ft/s)
u_0	Characteristic filtration velocity, Eq. (7.10)	m/s (ft/s)
V, v	Velocity	m/s (ft/s)
v	Volume	L (ft³)
V_c	Cake travel velocity	m/s (ft/s)
V_{dirt}	Volume of dirt holding space in intermittent-discharge disk	m³ (ft³)
V_θ	Tangential velocity	m/s (ft/s)
V_s	Settling velocity	m/s (ft/s)
V_{sg}	Stokes' settling velocity under 1g	m/s (ft/s)
V_{sG}	Modified Stokes' settling velocity under G	m/s (ft/s)
V_{g0}	Rate of decrease in settling flux with increase in solids concentration in feed slurry, $= -d(\phi_s V_s)/d\phi_s$, Eq. (3.4)	m/s (ft/s)
V_p	Push velocity	m/s (ft/s)
V_{rim}	Rim speed of rotor	m/s (ft/s)
WR	Wash ratio	*
W_f	Feed solids weight fraction	*
W_e	Centrate/effluent solids weight fraction	*
W_m	Cake moisture weight fraction	*
W_s	Cake solids weight fraction	*
$Y(d)$	Yield, same as size recovery, Rec_d	*

Special characters

Symbol	Definition	SI units (English units)
$\mathscr{L}_\varepsilon$	Leung number, Eq. (6.11)	*
l_c	Cake travel in each push stroke	mm (in)
l_s	Push stroke	mm (in)
v_c	Cake bulk volume	cm^3 (in^3)
Δ	Differential speed between conveyor and bowl	rpm, rad/s
Σ	Sigma factor, = equivalent sedimentation area	m^2 (ft^2)
Ω	Angular rotational speed	rpm
Ω_b	Bowl angular rotational speed	rpm
Ω_c	Conveyor rotational speed	rpm
Ω_p	Pinion rotational speed	rpm
Θ	Dimensionless flow resistance variable, Eq. (12.13)	*
α	Helix angle	deg, rad
α_s	Specific cake resistance	m/kg (ft/lb)
β	Beach angle	deg, rad
δ	Gap	mm (in)
δ	Ekman layer thickness Eq. (2.5)	
δ_{droop}	Gravitational droop	mm (in)
ε	Cake void fraction or porosity	*
ε_s	Cake solids volume fraction	*
ϕ	Cake angle of repose	deg, rad
ϕ_f	Feed solids volume fraction	*
$\phi_s(t)$	Feed solids volume fraction at time t	*
ϕ_{s0}	Initial feed solids volume fraction	*
ϕ_{smax}	Maximum solids volume fraction	*
γ	Climb angle	deg, rad
η_a	Accelerator efficiency, Eq. (11.1)	*
η_c	Conveyance efficiency for pusher, Eq. (9.3)	*
η_G	G efficiency, Eq. (11.3)	*
λ	Frictional coefficient	*
μ	Liquid viscosity	Pa · s (lbm/ft · s)
ν	Kinematic liquid viscosity	m^2/s (ft^2/s)
π	Constant, = 3.14159267	*
θ	Inclination angle of disk channel w.r.t. horizontal, Fig. 4.2	deg, rad
ρ_c	Bulk cake density	g/cm^3 (lb/ft^3)
ρ_{eff}	Effective cake density, = cake density − liquid density	g/cm^3 (lb/ft^3)
ρ_f	Feed slurry density	g/cm^3 (lb/ft^3)
ρ_m	Construction material density	g/cm^3 (lb/ft^3)
ρ_s	Solid density	g/cm^3 (lb/ft^3)

Symbol	Definition	SI units (English units)
ρ_{sl}	Slurry density	g/cm^3 (lb/ft^3)
σ	Interfacial tension	N/m (lb$_f$/ft)
σ_h	Hoop stress in rotor, Eq. (2.22)	Pa (lb$_f$/in^2)
τ	Shear stress	Pa (lb$_f$/in^2)
τ_y	Yield stress	Pa (lb$_f$/in^2)
ζ	Dimensionless time, Eq. (3.4)	*
ζ	Velocity factor, Eq. (4.5)	*
\$$_H$	Hauling costs	\$/ton wet cake
\$$_I$	Incineration costs	\$/ton water evaporated
\$$_P$	Polymer costs	\$/lb dry polymer
\$$_T$	Total costs	\$/ton dry feed solids

*Dimensionless.

Appendix C

Conversion Factors

To convert from	to	Multiply by
Area		
square centimeter (cm²)	square foot (ft²)	1.08×10^{-3}
square foot (ft²)	square meter (m²)	0.0929
square inch (in²)	square centimeter (cm²)	6.452
square inch (in²)	square meter (m²)	6.452×10^{-4}
Length		
centimeter (cm)	foot (ft)	0.0328
centimeter (cm)	inch (in)	0.3937
centimeter (cm)	meter (m)	10^{-2}
centimeter (cm)	micrometer (μm)	10^4
centimeter (cm)	millimeter (mm)	10
foot (ft)	centimeter (cm)	30.48
foot (ft)	inch (in)	12
foot (ft)	meter (m)	0.3048
foot (ft)	micrometer (μm)	3.048×10^5
foot (ft)	millimeter (mm)	3.048×10^2
inch (in)	centimeter (cm)	2.54
inch (in)	foot (ft)	0.083
inch (in)	meter (m)	0.0254
inch (in)	micrometer (μm)	2.54×10^4
inch (in)	millimeter (mm)	25.4
meter (m)	centimeter (cm)	10^2
meter (m)	foot (ft)	3.28
meter (m)	inch (in)	39.37
meter (m)	millimeter (mm)	10^3
meter (m)	micrometer (μm)	10^6
millimeter (mm)	centimeter (cm)	0.10
millimeter (mm)	foot (ft)	3.28×10^{-3}
millimeter (mm)	inch (in)	3.937×10^{-2}

Appendix C

To convert from	to	Multiply by
millimeter (mm)	meter (m)	10^{-3}
millimeter (mm)	micrometer (μm)	10^{3}
micrometer (μm)	centimeter (cm)	10^{-4}
micrometer (μm)	foot (ft)	3.28×10^{-6}
micrometer (μm)	inch (in)	3.937×10^{-5}
micrometer (μm)	meter (m)	10^{-6}
micrometer (μm)	millimeter (mm)	10^{-3}

Energy

British thermal unit (Btu)	joule	1.054×10^{3}
Calorie	joule	4.182
Erg	joule	10^{-7}
ft·lb_f	joule	1.356
kilowatt-hour (kWh)	joule	3.6×10^{6}

Flow rate

1. Mass rate

kilogram per hour (kg/h)	metric ton per hour (ton/h)	10^{-3}
kilogram per hour (kg/h)	pound per hour (lb/h)	2.2
kilogram per hour (kg/h)	(short) ton per hour (ton/h)	1.1×10^{-3}
metric ton per hour (t/h)	kilogram per hour (kg/h)	10^{3}
metric ton per hour (t/h)	pound per hour (lb/h)	2.2×10^{3}
metric ton per hour (t/h)	(short) ton per hour (t/h)	1.1×10^{3}
pound per hour (lb/h)	kilogram per hour (kg/h)	0.454
pound per hour (lb/h)	metric ton per hour (t/h)	0.454×10^{-3}
pound per hour (lb/h)	(short) ton per hour (ton/h)	5×10^{-4}
(short) ton per hour (ton/h)	kilogram per hour (kg/h)	9.08×10^{2}
(short) ton per hour (ton/h)	metric ton per hour (t/h)	0.908
(short) ton per hour (ton/h)	pound per hour (lb/h)	2×10^{3}

2. Volumetric rate

cubic meter per hour (m³/h)	gallon per minute (gal/min)	4.403
cubic meter per hour (m³/h)	liter per hour (L/h)	10^{3}
cubic meter per hour (m³/h)	liter per minute (L/min)	1.667×10^{1}
gallon per minute (gal/min)	cubic meter per hour (m³/h)	0.227
gallon per minute (gal/min)	liter per hour (L/h)	2.27×10^{2}
gallon per minute (gal/min)	liter per minute (L/min)	3.785
liter per hour (L/h)	cubic meter per hour (m³/h)	10^{-3}
liter per hour (L/h)	gallon per minute (gal/min)	4.403^{-3}
liter per hour (L/h)	liter per minute (L/min)	1.67×10^{-2}
liter per minute (L/min)	cubic meter per hour (m³/h)	6×10^{-2}
liter per minute (L/min)	gallon per minute (gal/min)	0.264
liter per minute (L/min)	liter per hour (L/h)	6×10^{1}

Force

dyne	gram (g)	1.020×10^{-3}
dyne	newton (N)	10^{-5}
dyne	pound (lb_f)	2.248×10^{-6}
kilogram force(kg_f)	newton (N)	9.807

Conversion Factors

To convert from	to	Multiply by
newton (N)	dyne	10^5
newton (N)	kilogram force (kg_f)	0.102
newton (N)	pound (lb_f)	0.225
pound (lb_f)	dyne	4.448×10^5
pound (lb)	gram (g)	454
pound (lb_f)	kilogram (kg)	0.454
pound (lb_f)	newton (N)	4.448

Mass

To convert from	to	Multiply by
kilogram (kg)	pound (lb)	2.203
metric ton (t)	kilogram (kg)	10^3
metric ton (t)	pound (lb)	2.2×10^3
pound (lb)	kilogram (kg)	0.454
short ton (ton)	kilogram (kg)	907.1
short ton (ton)	pound (lb)	2000

Power

To convert from	to	Multiply by
Horsepower (hp)	kilowatt (kW)	0.7457

Pressure

To convert from	to	Multiply by
atmosphere (atm)	dyne/cm^2	1.1033×10^6
atmosphere (atm)	foot of water at 39.1°F	33.9
atmosphere (atm)	inch of Hg at 0°C	29.921
atmosphere (atm)	lb_f/ft^2	2116.3
atmosphere (atm)	lb_f/in^2	14.696
bar	atmosphere (atm)	0.9869
bar	dyne/cm^2	1×10^6
lb_f/ft^2	atmosphere (atm)	4.725×10^{-4}
lb_f/ft^2	kg/m^2	4.882
lb_f/in^2	atmosphere (atm)	0.068
lb_f/in^2	kg/cm^2	0.07
pascal (Pa)	N/m^2	1

Torque

To convert from	to	Multiply by
in-lb_f	N-m	8.85
N-m	in-lb_f	0.1130

Viscosity

To convert from	to	Multiply by
centistoke (cSt)—kinematic	cm^2/s (= stoke)	0.01
centipoise (cP)—dynamic	N · s/m^2 (= Pa · s)	0.001
centipoise (cP)—dynamic	dyne · s/cm^2 (= Poise)	0.01

Volume

To convert from	to	Multiply by
cubic inch (in^3)	cubic centimeter (cm^3)	16.387
cubic meter (m^3)	cubic centimeter (cm^3)	10^6
gallon (US) (gal)	cubic meter (m^3)	0.003785
gallon (US) (gal)	liter (L)	3.785
gallon (US) (gal)	cubic inch (in^3)	231
liter (L)	cubic meter (m^3)	0.001
liter (L)	gallon (US) (gal)	0.2642

Appendix D

Sieve and Particle Sizes

Sieve size U.S. mesh	Sieve size Tyler	Opening/size micrometers	Opening/size inches
3.5	3.5	5600	0.223
4	4	4750	0.187
5	5	4000	0.157
6	6	3350	0.132
7	7	2800	0.111
8	8	2360	0.0937
10	9	2000	0.0787
12	10	1700	0.0661
14	12	1400	0.0555
16	14	1180	0.0469
18	16	1000	0.0394
20	20	850	0.0331
25	24	710	0.0278
30	28	600	0.0234
35	32	500	0.0197
40	35	425	0.0165
45	42	355	0.0139
50	48	300	0.0117
60	60	250	0.0098
70	65	212	0.0083
80	80	180	0.0070
100	100	150	0.0059
120	115	125	0.0049
140	150	106	0.0041
170	170	90	0.0035
200	200	75	0.0029
230	250	63	0.0025
270	270	53	0.0021
325	325	45	0.0017
400	400	38	0.0015
450	—	32	0.0012
500	—	25	0.0010
635	—	20	0.0008

Name Index

Berk et al., 387
Crosby, 49
Emmett et al., 49
Greenspan, 49, 70
Karolis and Krautlein, 121
Lamb, 49
Lee, 121
Lehmann, 121
Letki, 121
Leung, 121, 163, 191, 210, 272, 316, 387
Leung and Black, 227
Leung and Shapiro, 121, 272
Leung, Shapiro, and Yarnell, 121
Mayer, 210
Mayer and Stahl, 191
Miyano, Nishida, and Leung, 143, 387
Muller, Kompe, and Kluge, 121, 357
Retter, 357
Reynolds et al., 387
Richardson and Zaki, 49, 320, 337, 357
Sambuichi, Nakakura, Osasa, and Tiller, 70
Shapiro, 121, 143
Stahl et al., 227
Stadager and Stahl, 191
Suzuki, 121, 143
Svedberg and Pedersen, 70
Tiller, 399
Wilkie and Blackburn, 70, 210
Wilkie and Patnaik, 210
Wright, 210
Wuensch, Unkelbach, and Arhelger, 121, 143

Subject Index

Accelerator:
  cone, 250–251, 253, 256–257, 266
  double-disk, 251–252, 257–259, 265
  efficiency, 247–248, 255, 258, 263–266
  hub, 249–250, 254–256, 264–265
  vanes assembly, 252, 259–260
Amorphous, 273
Angle of attack, 260
Angular momentum, 7, 42, 248
Antibiotics, 137
Anti-Coriolis baffle, 254–255

Biotechnology, 2, 137
Bond number, 181–182, 184–186
Boundary layer, 7–9
Boundary layer model, 151–156
Buchner funnel, 274
Bucket tests, 286–289

Cake:
  bulk density, 297–298
  compactible, 29–32, 123, 280, 392–395, 397–399
  compaction:
    stress, 359–366, 389–399
    time, 365–366, 392
  conveyance, 43–44, 102–103, 105–106, 352–356, 367, 403–404
  desaturation, 29, 31–33, 172–173, 175–187, 285, 293–294
  discharge:
    (*see* conveyance)
  drainable, 29–32
  formation, 280–281
  height, 111–112, 118, 169–172, 180–185, 290–292, 324, 345, 360–367, 369
  impurities, 39, 187–188, 296, 298

Cake (*Cont.*):
  incompactible, 29–32, 395–397
  moisture, 28, 31, 32, 134, 173, 175, 290–293
  nondrainable, 29–32
  permeability, 166–171, 390–397
  pores, 29
  porosity, 30, 294
  saturation, 30–32, 173–175, 294
    capillary, 177–179, 181–187
    equilibrium, 173–174, 177, 179–180
    film, 180
    pendular, 180–181
    bound or pore liquid, 180
    test, 293, 298
    thickness (*see* height)
    transient, 179
    total, 179
  solid, 28, 30, 90, 105, 107, 117, 278
  void, 30, 292, 294
  washing, 39, 187–191, 298–299
  yield stress, 279
Calcium carbonate:
  ground, 131–132
  precipitated, 131–132
Capillary:
  number, 181–183
  rise, 181
CENSOR®, 119, 134–136
Centrate:
  solids, 33–34
  clarity, 61, 65, 367–369
Centridry®, 386–387
Centrifugal acceleration G and centrifugal force, 12, 46–48, 68, 71, 111, 114, 116, 177–182, 185–186, 188, 208, 222, 229–231, 247–248, 273–276, 287–292, 297, 306, 308–309, 311, 322–325, 328–336, 342, 344, 348–354, 357, 360,

Centrifugal acceleration G and centrifugal force (*Cont.*):
  364, 366, 371, 378–379, 395, 401, 403–404, 407–408
Centrifuge:
  batch filtering basket:
    acceleration, 193
    applications, 230
    base-bearing, 196
    bottom-driven, 201
    bucket, 52
    centri-Feed®, 203
    clean-in-place, 205
    cone-disk feeding, 201
    constant-head test, 282–284
    cycle, 193
    deceleration, 195
    dewatering, 195
    feeding, 193
    horizontal peeler, 206–207, 231–233
    j-pipe, 204
    link-suspended, 197, 200
    open-bottom, 199
    pendulum, 199–200
    pressurized inverting filter, 209–210, 231–233
    siphon horizontal peeler, 207
    solid-bottom, 196
    thermal drying, 210
    top-driven, 199–200
    troubleshooting, 408–409
    unloading knife:
      double-acting, 195
      single-acting, 195
    variable-head test, 284–285
    vertical peeler, 201–206, 231–233
    washing, 194–195
  batch sedimenting basket:
    clinical, 51
    high-G basket, 66–67
    multibowl, 62–63
    preparatory, 54
    solid-wall basket, 63–66, 281–282
    test-tube:
      angle-head, 51–54
      babcock bottle, 53
      horizontal-head, 51–54
    tubular, 56–58, 60–61
    ultracentrifuge, 54–56
    zonal, 56–59
  comparison, 309–310
  continuous, filtering

Centrifuge, continuous, filtering (*Cont.*):
  conical screen, 211
  shallow angle:
    oscillating, 214
    scroll:
      applications, 229–230, 237–240
      dewatering, 226
      operation, 215–216
    vibrating, 213–214, 229
      applications, 233–237
  steep angle, 212
  pusher:
    applications, 229–230, 240–242
    cake transport, 218–221
    dewatering model, 224–225
    double-acting, 222
    single-stage, 216–217, 221–222
    multistage, 217–218, 221–222
    optimization, 334–335
    push plate, 216, 219, 402
    retention time, 221
    troubleshooting, 401–402
  screen bowl, 222–224, 230
  screen-bowl dewatering, 226–227
  screen-bowl applications, 241–245
  screen-bowl troubleshooting, 406–407
  screen recycle, 224–225
continuous sedimenting:
  decanter, 99–121
    applications, 144
    baffles, 108
    back drive:
      torque, 369–371, 377
      torque-differential control, 371–373
    beach, 106–108
    bearings, 406
    bowl strips, 404
    cake conveyance, 106
    cake-solids control, 117
    climb angle, 107
    concurrent design, 110–111
    construction materials, 113
    conveyor drive, 102–103
    countercurrent design, 110–111
    differential speed, 111–112, 321–322, 345, 403
    dip weir, 108–109, 354
    disk-stack decanter, 114
    feed concentration, 320, 348
    feed leakage, 405
    feed rate, 343–344, 403

Centrifuge; continuous sedimenting, (*Cont.*):
  G-force, 111, 322–324, 403
  geometry, 99–101, 110
  helix angle, 105
  high-solids, 359–387
    characteristics, 359–369
    back drive torque, 369–371
    economics and optimum, 380–385
    examples, 373–380
    torque-differential control, 371–372
    with drying, 385–387
  hydraulic assist, 108–110
  lead, 105–106, 118
  lubrication, 406
  misalignment, 40
  nozzle decanter, 119, 356
  operation, 101–102, 110–113, 318
  optimization, 317–333, 336
  polymer dosage, 320
  pool, 112, 322–323, 346, 403
  ribbon, 104
  shear pin, 404
  sloping clarifier, 116
  solid-blade, 104
  solid-solid-liquid, 118
  temperature, 324, 403
  torque and power, 324–325
  troubleshooting, 402–406
  vibration, 405, 406
 disk:
  applications, 145
  dropping bottom, 93
  flow pattern, 75
  geometry, 72–75
  hermetic design, 84, 86–90
  intermittent, 92–95
  liquid seal, 76
  manual, 90–92
  nozzle, 95–99
  operation, 71–72
  sludge depth, 347
    solid bowl (*see* decanter)
  troubleshooting, 407–408
  selection based on:
    feed concentration, 307–308
    particle Size, 304, 306
    process function, 304–305
  sizing based on:
    filtration, desaturation and washing, 165–191, 224–227, 311

Centrifuge; continuous sedimenting; disk (*Cont.*):
  sedimentation, 147–163, 308, 311
  costs, 311–316
  batch filtering baskets, 313–314
  continuous filtering, 313–314
  decanters, 313
  disks and tubular, 312
  installation, 315
  maintenance, 315–316
Centripress®, 386–387
Centrisizer®, 118–120
  (*See also* nozzle decanter)
Centripetal:
  acceleration, 11–12
  force, 11
Chatter, 41
Clarification, 4, 153–157, 274–278, 303, 329, 342–349
Classification, 4, 157–161, 279–280, 303, 330
Clay (*see* Kaolin separation)
Clean-in-place, 3, 303
Coal, 2, 132, 233–238, 241–244, 267–268
Concentration, 4, 303, 332
Conservation of angular momentum, 17, 68, 248
Construction materials, 47–49
Coriolis, 16–18, 53, 254–256
Corrosive, 273
Cream separation, 128–130
Critical speed, 49
Crystalline, 273
Cut size, 152, 342–343

Darcy's law, 166
Darcy-Shirato equation, 390
Degritting, 4, 159–163, 281, 303, 331, 349–351
Deliquoring, 28
Desliming, 357
Dewatering, 4, 28, 333–334
Distributor Control Systems, 194
Distributor Target, 262
Drying, 385–386

Earth's gravity, 2
Ekman:
  layer, 19
  number, 19
Extraction, 123, 138

Feed:
  acceleration, 42, 247–272
  compartment, 249
  pipe standoff, 262
  ports, 249, 254–255
  solids concentration, 295
Fermentation broth separation, 137
Filtration:
  bulk, 165
  cake heel, 168
  cake permeability, 166
  characteristic velocity, 171
  Darcy's law, 166
  filter medium resistance, 168–169
  medium blinding, 399
  pressure distribution, 168
  rate, 168
  specific cake resistance, 166
  thin cake filtration, 170
Fine-particle separation, 3
Fish meal processing, 130
Flocculated solids, 275
Flocculated solids (shear sensitivity), 38–39
Flue gas desulfurization, 2, 231–232
Fluid:
  density 22–23
  inertia, 19
  viscosity, 19
Food processing, 136–137
Fractionation:
  (See also classification)
Friable, 273
Frictional force, 11, 16
Frictional coefficient, 44, 218

G Efficiency, 248, 264
Grade efficiency, 149, 263–264
Gravitational droop, 261–262
Gravitational sedimentation, 20–21

Hermetic design, 84, 86–88, 260
Hydraulic diameter, 181
Hydraulic drive, 3, 102
Hydrocyclone (cyclone separator), 12, 22, 15

Interface:
  cake-slurry, 68–69

Interface (Cont.):
  liquid-liquid, 82–85
  liquid-slurry, 68–69

Kaolin separation, 2, 267, 339–357

Lamella, 22, 53
Leung number, 308, 311, 326
  clarification, 152–157
  classification, 157–159, 280
  definition, 152–153
  degritting, 159–163
Liquid discharge:
  centripetal pump, 79
  pairing disks, 80
  skimming pipe, 80
  weir, 79
Liquid flux, 390–393
Liquid volume fraction, 390–391

Mass balance, 34, 277, 390
Mineral separation, 133
Molecular weight measurement, 54

Nitration separation, 136
Non-bleeding, 302
Non-rotating, 9

Oil recovery, 124–125
Oil clarification, 126
Overflow limit, 194
Overspeeding, 256, 260–261

Palm oil separation, 127
Particle size:
  distribution:
    bimodal, 159, 161
    monodispersed, 35
    polydispersed, 36, 157
  recovery, 36, 149, 157–158
Paste, 302
Perforate bowl, 6
Pharmaceuticals, 137
Pilot tests, 295
Plastics sorting, 134–136
Plug-flow, 147

Subject Index   429

Polyelectrolytes, 4
Polymer:
  activity, 37
  dosage, 37–39, 366–367, 402
  introduction, 373
  neat, 37
Polypropylene, 1
Polystyrene, 1, 270–271
Polyvinyl chloride, 1, 4, 41, 133–134
Posttreatment, 1, 4
Power, 42–44, 141–142
Pressure, 16, 82, 166–168, 185–186
Pretreatment, 1, 3
Processing:
  chemical, 1
  food, 2
  industrial, 2
  mining, 2
  mineral, 2
Production tests, 296–298
Prograde motion, 16
Programmable Logic Controller, 194, 409

Reaction forces, 15
Recovery:
  product in centrate, 341, 347–348, 350, 352
Reference frame:
  inertial, 12
  noinertial, 12
  rotating, 12, 166, 390
Relative centrifugal force, 12
Rendering, 127, 131
Repulp factor, 40
Reslurrying:
  (*see* separation and repulping)
Rotating flow, 7
Rotation:
  anticlockwise, 106
  clockwise, 106
  free vortex, 15
  meter, 262
  solid-body, 12
Retrograde motion, 16
Rim speed, 22, 46
Rossby number, 19

Sanitary-in-place, 3, 303
Scale-up:
  G x surface area, 308

Scale-up (*Cont.*):
  Leung number, 153–163
  pool volume, 308
  sigma factor, 150–151
  t_d, 180, 225–226, 291, 293–295, 311
  transient centrifugation time variable ζ, 68, 275
Screen openings, 401, 407
Screening tests, 273
Secondary flow, 7
Sediment, 52
Sedimentation:
  centrifugal, 21
  gravitational, 20
  time variable, 68
Sedimentation coefficient, 54
Separation:
  solid-liquid, 75, 303
  solid-liquid-liquid, 4, 76–77, 124–131
  solid-solid-liquid, 4, 118
Separation and repulping, 4, 39, 123, 303, 336
Settling area, 21
Settling velocity:
  centrifugal gravity, 21
  compaction, 26
  density difference, 20, 23
  discrete particle, 26
  Earth's gravity, 20
  flocculation, 26
  hindered, 27
  irregular particle, 24–25
  particle Size, 20–22
  Reynolds number, 20, 25
  Stokes' law, 20–22
  solids:
    concentration, 26–27
    temperature, 22
    viscosity, 22
Sewage, 265–267, 359
Sigma factor, 150–151, 308
Skin, 399
Smoothener, 256–260, 266
Soda ash, 239, 241–242, 269
Sodium chloride, 269
Sodium sulfate, 270–271
Solid Fuel, 2
Solids:
  dissolved, 39, 302
  flux, 390, 393
  recovery, 33, 38, 275–277
  suspended, 39, 302

Solids (*Cont.*):
  throughput, 34, 40, 47, 302
  volume fraction, 390–396
Spin-tube tests, 274
Specific power consumption, 141–142
Starch, 233
Stokes' law, 20–22
Stress:
  compression, 18
  hoop, 45
  tensile, 18
Superimposed pressure, 185–187
Supernatant, 52, 274, 277
Svedberg equation, 54
Swallowing capacity, 271–272

Tangential speed, 247–248, 253, 258
Tar sand, 2, 157–158
Taylor-Proudman column, 17
Tea cup, 8
Temperature, 273, 295–296, 324
Thickening, 4, 141–143, 145, 332
Torque, 41–44, 264, 324–325
Toxic, 273
Transient centrifugation, 67–69, 275

Ultracentrifugation:
  differential, 55
  density-gradient:
    settling-rate, 55
    isopycnic, 55

Variable-frequency drive, 3, 119–121
Velocity:
  circumferential (or tangential), 12, 248
  radial, 248

Vibration, 402
Viscous Diffusion, 9, 254, 256
Volumetric:
  throughput, 40

Wash:
  compartment, 189–191
  diffusion, 188
  displacement, 188
  dissolution, 188
  multispray, 187
  nozzle, 188–191
  rate, 188–189, 295, 298
  ratio, 187, 189
  stationary pipe, 188
Wastewater:
  dewatering, 139–140
  thickening, 140–143
Wastewater sludge types:
  aerobic digested, 380
  anaerobic digested, 378–380, 382, 384
  industrial, 139, 373–375
  heat treated, 139–140
  lime, 140
  mixed, 140, 375, 378, 380, 382, 384
  primary, 140, 380
  waste activated, 140, 380
Wastewater sludge handling, 381–384
Water:
  density, 23
  viscosity, 23
Wear, 268, 271–272
Wedge wire screen, 165, 217, 401

XL•PLUS®, 263–264, 267–271

Yield, 36, 280

ABOUT THE AUTHOR

Wallace Woon-Fong Leung, a leading international expert on centrifugation and fluid-mechanics on separation processes, is Director of Process Technology at Bird Machine Company (a Baker Hughes company) where he directs the process applications and R&D of centrifuge technology and solid-liquid separation, as well as consults and troubleshoots centrifuge installations worldwide. He is also a Course Director for the Center for Professional Advancement, New Brunswick, NJ, through which he developed and has taught in Europe and the United States numerous industrial short courses on centrifugation and solid-liquid separation. A contributor to Perry's and Green's *Chemical Engineers' Handbook, Seventh Edition* and Schweitzer's *Handbook of Separation Techniques for Chemical Engineers, Third Edition* (both titles published by McGraw-Hill), Dr Leung is the author of numerous technical papers and over 25 United States patents as well as many foreign counterparts on centrifugation. He serves on the board of Directors of the American Filtration and Separations Society where he has chaired annual meetings, technical sessions and presented papers on separation technologies. Previously, Dr. Leung was a scientist with Schlumberger and Gulf Oil. He was awarded the Cedric Ferguson Medal by the Society of Petroleum Engineers based on his research work on flow in porous media. He received the B.Sc. degree in aerospace and mechanical engineering from Cornell University, Ithaca, New York, and the M.Sc. and Sc.D. degrees in mechanical engineering from the Massachusetts Institute of Technology, Cambridge, Massachusetts.